AF522144

ENCYCLOPEDIA OF BIOPROCESS TECHNOLOGY

Vol. 2

PHARMACOLOGICAL BIOPROCESS TECHNOLOGY

By

Dr. Arvind N. Shukla

School of Studies of Zoology & Biotechnology

Vikram University

Ujjain

(India)

DISCOVERY PUBLISHING HOUSE PVT. LTD.

NEW DELHI-110 002

Published by:
Tilak Wasan
DISCOVERY PUBLISHING HOUSE PVT. LTD.
4383/4A, Ansari Road, Darya Ganj
New Delhi-110 002 (India)
Phone: +91-11-23279245, 43596064-65
Fax : +91-11-23253475
E-mail: parul.wasan@gmail.com
discoverypublishinghouse@gmail.com
web : www.discoverypublishinggroup.com

***First Edition:* 2012**

ISBN: 978-93-5056-026-6 (Set)
978-93-5056-027-3 (Vol. 1)
978-93-5056-028-0 (Vol. 2)
978-93-5056-029-7 (Vol. 3)
978-93-5056-030-3 (Vol. 4)
978-93-5056-031-0 (Vol. 5)

Encyclopedia of Bioprocess Technology

Printed at:
Shree Balaji Art Press
Delhi

Preface

The present title **"Encyclopedia of Bioprocess Technology"** has been written for undergraduate, post-graduate students and those engaged in pharmaceutical, environmental biotechnology, industrial and clinical research. Actually the explosion of new technologies with their vast potential has brought with it the need for emerging scientists to equip themselves and their laboratories with a whole array of new expertise. With the high discriminating power of the DNA systems has come high potential in evidentiary terms, high profile status for many investigations, and not least, a high degree of professional scruting of evidence produced by such technology.

This book is addressed to all who are curious and who want to understand fundamental biological processes that ensure our survival. The person who might be interested is probably someone just starting research who wants to learn about aspects of regulation, outside his chosen field. It could become a supplementary text for graduate students and post-graduate in biochemistry and molecular biology.

To make the work more comprehensive and informative, the author has consulted many authoritative books, research journals, abstracts, monographs etc., so there can be no claim to originality except in the manner of treatment.

The author express his thanks to his friends and colleagues whose continue inspirations have initiated him to bring out this book.

The author expresses her gratitude to Mr. Wasan and staff of M/s Discovery Publishing House Pvt. Ltd. for their whole hearted co-operation in the publication of this book.

The author will remain sincerely responsible for any shortcomings of the book and be grateful to the readers for their suggestions and constructive criticism for the continuous betterment of the book. He takes this opportunity to appeal to the readers to send their suggestions straightaway to this Publisher.

Arvind N. Shukla

CONTENTS

COMPUTERIZED DESIGNING OF CHROMATOGRAPHY

Currently, chromatographic purification processes are optimized empirically. Separation media, running conditions, and sequence of purification steps are selected according to the expertise of the responsible scientists. In biotechnology, computer-aided design is less far advanced than in certain areas of chemical engineering. The lack of expert systems is the reason for the traditional optimization, which is based on trial and error. Mathematical models for description of the chromatographic separation process have been around since the early 1940s. The common use of computers in the separation laboratory, the need of rapid process development, and the theoretical knowledge accelerated the development of expert systems and software tools. They are considered to be beneficial for design and optimization of purification steps. The area of computer-aided chromatography design includes the design of experiments, the appropriate interpretation of experimental results (by the aid of mathematical algorithms), the prediction of separation conditions, and the rescaling of unit operations.

Four levels of complexity have to be considered for optimization of separation steps. The first level is the general process economics. This level will be influenced by the second level: flow sheeting. The third level is the optimization of particular unit operations. The fourth level is the sectioning of unit operations into their thermodynamical and molecular considerations by the aid of molecular simulation. Simulated moving bed, countercurrent chromatography, and annular chromatography are not been considered in this chapter, because these modes of chromatography have not been used for bioprocessing to large extent.

NOMENCLATURE

The word *model* denotes a system of algebraic or differential equations, which are held to be a representation of a physical system. From this general point of view, any relationship between quantities is a model, and the principles of modeling are the principles of chemical engineering science. *Process models,* and here particularly chromatography, can be counted to a chemical process. They are established by application of the three fundamental conservation principles: conservation of mass, momentum, and energy. Because liquid chromatography is carried out under isothermal conditions, the conservation of energy can be neglected in general. The contribution of liberated or consumed thermal energy during adsorption and desorption is negligibly small. The conservation of mass and momentum and the virtual breakdown of a chromatography system to a small unit, where the adsorption process takes place, are the fundamentals for modeling of chromatography. In principle, there is no difference between *computer models* and other models.

Table 1.1 Important variables for optimization of preparative chromatography

Variables
Column size/shape
Height (Z)
Diameter
Height-to-diameter ratio
Height-to-particle diameter
Flow rate (u)
Flow rate during loading, during elution
During regeneration
Residence time (Z/u)
Physical conditions of packing
Packing quality
Packing density
Mobile phase
Modifier composition
Rate of change of mobile phase
Modifier composition (gradient shape)
Loading
Volume
Concentration
Flow rate
Stationary phase
Ligand type
Ligand density
Type of support
Porosity, tortuosity
Bead diameter
Beads size distribution (monodispers)
Temperature
Scarcely a process in biotechnology, which uses temperature gradients for elution

Many models are so complex that they can be solved only numerically. In these cases, computers can be used for reasons of convenience, independent of the complexity of the model applied. Further considerations about modeling are given by Vajda and Valko. We usually assume that the input, initial conditions, and parameters of a model determine the response (result). The *parameters,* which are the different coefficients in mathematical equations, must be evaluated. This process is called parameter estimation. Once a model is established, the performance of a system can be predicted by varying the input and initial conditions. The prediction will yield a correct answer (response) as long as the predictions are carried out within a certain range of validity. The words *prediction* and *simulation* describe, in this context of modeling, a similar process.

The forecast of process conditions, using process parameters acquired at a small scale, is called prediction. Frequently parameters cannot be estimated because of the lack of a determination method or accessibility. Thus, a hypothetical value is used to solve the set of mathematical equations. This process is called simulation. Depending on the task, some of the *operational variables* may be kept constant or may not be taken into consideration. The residence time of a product in a column is a critical factor, when the product undergoes a conformational change or a reaction in the column. Alteration of the residence time in a column can completely change the result of the purification. Reversed-phase chromatography (RPC) and hydrophobic interaction chromatography (HIC) are more susceptible to residence time changes than ion-exchange chromatography or affinity chromatography are. The *response values* depend on the strategy and the stage where the chromatographic purification is placed. Frequently used response values are:

1. *Productivity* (*P*) is defined as

$$P = \frac{Q_R \cdot C_0 \cdot V_F}{V_t \cdot t_C} \qquad \ldots(1)$$

where Q_R is the recovery ratio, C_0 is the sample concentration, V_F is the sample volume, V_t the column volume, and t_C the cycle time, which includes all time periods necessary to start the next cycle (process time). The purity has to be defined before optimization. The productivity in the aforementioned case does not consider the lifetime of a column.

2. *Yield* or *recovery* is simply the percentage of recovered material and material present in the feed. Yield is not easily accessable by mathematical models. In cases, product loss is caused by a very complex event. Unspecific (irreversible) adsorption to stationary phases, aggregation, and denaturation together with enzymatic degradation are responsible for incomplete recovery. These effects are also dependent on age and history of the column. These events are not covered by chromatography models.

3. *Dynamic capacity* is the capacity at a certain flow in contrast to the capacity determined by batch experiments. BTC experiments at 5% breakthrough are usually used to determine the dynamic capacity. Originally, the analytical solution of an idealized model, known as the Thomas solution, has been used for approximation and prediction of breakthrough curve (BTC). Under conditions applied in preparative chromatography, the BTC deviates significantly from the ideal model. Recently, models have been published that take effects of kinetics under overloaded conditions into consideration.

4. To minimize the number of purification steps, the highest *purity factor* should be achieved at each step. The regulations for production of pharmaceuticals by recombinant DNA technologies require very high purity standards. The standards can only be met when a purification power of each step is used to the optimal extent. Other factors, such as design of a plant, quality of the materials and buffers, operator training, etc., must also be obeyed. The optimization of the chromatographic process is only one piece of the puzzle.
5. A high *concentration factor* leads to a smaller volume for the next step, which has to be handled and the product becomes more stable. Also, an upper limit in the concentration can be of consideration, because the solubility of the product is exceeded. The concentration factor (CF) can be defined as:

$$CF = \frac{C_R \cdot V_R}{C_0 \cdot V_F} \qquad \text{...(2)}$$

   where C_0 is the initial (feed) concentration, V_F is the feed volume, C_R is the recovered concentration, and V_R is the recovered volume.
6. The *process time* may be limited by the stability of the product. Economical considerations also come into play, such as shift time and batches, which have to be produced by a certain time period. The process time includes the optimization of the gradient, flow rate, etc. Process time has to be optimized with regard to total process time. Cleaning, regeneration, and, if necessary, repacking have to be included.

Valid Mathematical Models

Today, broadening of scientific and technological knowledge is based on three cornerstones: theory, experiments, and computer simulation. Reliable computer simulation needs an appropriate mathematical model. Without a valid mathematical model that reflects the real physical situation, the most sophisticated computer algorithm will fail. A model will never be as good as the physical reality, because the complexity has to be reduced to a certain formalism. A fictional model of the complexity of the reality would be as slow and laborious as the reality. Vajada and Valko compiled a list of criteria for establishing a model in chemical engineering. These rules also apply for the chromatography of proteins.

1. First, scientific curiosity is a strong motivation for establishing a model. A priori curious-appearing models may be of considerable value, when they are integrated with the appropriate objective. This requires the innovative selection of parameters and design of experiments to determine those parameters.
2. The second reason for using models is the reduction information and indirect measurements. The most common information reduction in chromatography peak is the approximation of a peak by its first and second moment, instead of using the information from the entire peak curve. Using these information reduction processes, the appropriate model must be selected to describe the peak area, width, and the asymmetric shape.
3. The ability to design and optimize a separation scheme quickly and effectively is also a motivation for developing a model. To minimize process time and determine optimal column size and

optimal gradient shape, loading and sequence of the different process steps (equilibration, loading, washing, elution, and regeneration) are necessary.

4. The design of optimal particle properties is also an important goal for manufacturers of resins and for engineers selecting a resin. Modeling of packing provides a directive on the pressure drop and performance loss caused by irregularities.
5. Control and simulation for validation purposes will become an important issue in modeling of chromatography. The ability to calculate acceptance criteria for scale-up or scale-down will allow rapid, accurate validation of tasks, such as virus clearance. The correct column dimensions, gradient shapes, limits of operation parameters, and rapid diagnosis of separation deficiencies are also added benefits. Scale-up of the production of a biologic is often considered a major change of the process, requiring additional clinical trials that are time consuming and costly. However, if the appropriate scale-up process can be performed with the generation of the same product exhibiting identical qualities, no additional clinical trial would be required. Computer-assisted optimization evaluates the influence of the process change. It would save significant manufacturing costs if the altered scheme were not considered a major change.

The existing models for simulating chromatography are based on different core assumptions. Depending on these assumptions, different levels of complexity are handled. At each level of complexity, different simplifications are made in order to make the mathematical treatment accessible.

According to van Deemter et al. the following assumptions are made:

1. The equilibrium concentrations in the two phases are proportional (linear chromatography).
2. The exchange process is thermodynamically reversible, that is, the equilibrium between particle and fluid is immediate; the mass transfer coefficient is infinitely high; and longitudinal diffusion and other processes with similar effects can also be ignored. The latter processes are considered ideal chromatography.

From these assumptions one can categorize chromatography into four cases. Although the cases 1, 2, and 3 are of great value for understanding chromatography, only case 4 reflects the real situation in preparative chromatography.

Table 1.2 Modes of complexity in chromatography modeling

	Ideal chromatography	*Nonideal chromatography*
Linear isotherm	1	2
Nonlinear isotherm	3	4

Analytical and Preparative Chromatography

Besides the size and the shape of the separation column, the difference between analytical and preparative chromatography is the intended use. In analytical chromatography, the components are separated and identified according to their retention times. In some cases, a component-specific detector is used for the identification. Generally, a small volume (mass) is applied onto the column. Therefore, the separation is carried out under linear isothermal conditions. Frequently, the chromatography can be described as case. Preparative chromatography is intended for isolation of a compound. In contrast to analytical chromatography, the separation is carried out in the nonlinear regime. Peak-broadening effects resulting from the nonlinearity of the isotherm must be taken into consideration. The sample is applied as a high concentrated solution or in a large volume as a very dilute solution. The Dirac δ function can no longer be used for describing the sample volume and concentration. The Dirac δ function is a hypothetical approximation for an infinitesimal small volume containing one mass unit. The whole system is designed so that extra-column effects cannot be neglected. When the performance of a chromatographic separation should be predicted from small scale to large scale, those effects must not be neglected.

Table 1.3 Differences between analytical and preparative chromatography other than the intended use of the technique

Analytical	*Preparative*
Identification by retention or selective detection	Separation coupled with collection of separated proteins
High efficiency	High capacity
Small particles	Large particles
Small loading of columns	Overloading of columns
Linear adsorption isotherms	Nonlinear adsorption isotherms
Sample volume negligible	Sample volume not negligible, (Dirac δ function not applicable) Extracolumn space not negligible

Special emphasis is put on this problem, because the successful application of computer-assisted optimization is strongly dependent on the extrapolation from small scale to large scale. The extra-column effects can dominate a preparative separation in small scale. In the mathematical description of an analytical column, various mass–transfer problems can be handled by using fine adsorbent particles at a high flow rate. However in a large-scale column, fine particles introduce hydrodynamical problems such as high pressure drop, channeling, and fingering of flow. Consequently, in large chromatography, larger particles are used and mass–transfer resistencies are more pronounced.

General Objectives

The general objective for computer-assisted optimization of chromatography is the reduction of the number of experiments required to identify an optimum. The optimization is accomplished by an iterative process. Frequently, optimal running conditions must be identified. Optimal running conditions can be defined as best buffer for loading, washing, and elution while maintaining highest resolution. The same can be valid for the best throughput (productivity) under highest resolution. The latter require chemical parameters (i.e., buffer composition) physical and engineering parameters (i.e., flow rate, height equivalent to theoretical plate). In other words, selectivity, efficiency, and productivity (throughput) must be optimized. Depending on the task, one must rate selectivity, efficiency, and throughput in different ways.

Selectivity is mainly achieved by the chromatography sorbent and buffer composition and to some extent by engineering parameters. Efficiency depends on the physical characteristics of the sorbent, the overall setup of the chromatography system, and the chemistry of the mobile phases used. Throughput is influenced by the dynamic capacity of the sorbent, which depends on the composition of the sorbent and the buffers used.

All three important characteristics of preparative chromatography are interconnected. Changing one parameter influences the others. Traditionally, after selection of a particular sorbent, one can identify the appropriate separation conditions by seeking the appropriate buffer conditions. Then the throughput is maximized by overloading and increasing the flow rate. Another very important objective in preparative chromatography is the extrapolation from small-scale data to the performance of large scale (scale-up) and the extrapolation from a large scale to small scale (scale-down). When a process is translated from a laboratory scale to a pilot and industrial scale, very often the linear gradient elution must be changed to stepwise elution, because this elution mode is more simple.

After optimization, the single steps in the sequence of steps have to be optimized, and thus flow sheeting comes into play. In flow sheeting, the performance of the different steps is interconnected, and a strong reciprocal action is observed. However, multiple interconnected experiments must be carried out. The more steps and performance characteristics that are simulated, the fewer experiments needed and thus the faster an optimal separation process is identified. What are the prerequisites for computer-assisted optimization of preparative chromatography? Because experiments have to be performed in order to have access to parameters for simulation, the experiments should be carried out at the smallest applicable scale. This is not only a heuristic for computer-assisted optimization, but is also valid for all kind of optimization. Whenever possible, the experiments should be carried out with equipment that allows a high degree of automation.

Electronic data acquisition of a chromatogram is essential for mathematical treatment. The predicted data should be valid for a broad range of flow rates, column sizes, and loading conditions. The main parameters influencing the performance (including the extra-column parameters) should be known. Once the extra-column peak broadening can be discriminated from the parameters of the column, the simulation is fed with this information. During scale-up, the contribution of extra columns and column band broadening changes. Observing these facts improves the accuracy of the prediction of chromatography from small scale to large scale.

Utilities of Computer-aided Design

The objective in computer-aided design is the proper selection and order of chromatography steps in a purification process. Based on a primary knowledge on the properties (e.g., isoelectric point, molecular weight, surface hydrophobicity, biospecificity) of the protein of interest and its impurities, expert systems can assist in setting up a prospective purification design. In parallel, these systems have to consider some basic rules for the separation sequence consisting of capture, purification, and polishing:

1. Choose separation steps based on different properties.
2. Separate the largest impurity first.
3. Choose steps that use differences in the properties of the product and the impurities best.
4. Use a high-resolution step as soon as possible.
5. Set the most laborious step last.

After characterizing the chosen steps experimentally, economic considerations concerning productivity, capital, or production costs must be evaluated. Using computer- aided design routines, several characteristics of the product (amount required in a calculation period, purity required, recovery), resin (cost per liter, lifetime), labor constraints (labor costs, number of operators required), and production constraints (column dimensions available, costs of waste disposal, costs of overheads) can be evaluated to find the optimal scale of an individual purification step.

MECHANISM OF ISOLATION

Chromatographic theories have been applied for description of band broadening and retention of a solute while migrating through the column. In liquid chromatography, the column geometry can be described by the radius (r) and the length (z) of the column. Independently of the model applied, some basic assumptions are made in preparative chromatography of biomolecules:

1. The column is considered to be homogeneous in radial direction. The applied protein is distributed homogeneously over the cross-sectional area of the column. As a consequence, the r dimension can be disregarded, and the effective parameter for the column geometry is z.
2. The mobile phase is incompressible, and the mobile phase velocity is constant along the column.
3. There is no influence of thermal effects. The adsorption process is considered to be isothermal, although this is not consistent with regard to rigorous thermodynamic treatment. The temperature is constant during the experiment, independent of position and time.
4. Mass exchanges between mobile and stationary phase are infinitely fast, and the two phases are permanently in composition equilibrium. The need of this assumption is circumvented by kinetic models.
5. The ratio between the mobile and stationary phase volume is constant during the experiment, independent of position and time. It is also independent of the mobile phase concentrations of protein and salt.

Depending on the type of chromatography and the intended use, these basic assumptions can heavily deviate from the real conditions.

Theoretical Plates

Using these models, the column is divided into identical theoretical plates, and for each of these plates a mass balance is established. Originally, a linear isotherm was assumed, and in each plate mobile and stationary phase concentrations were in equilibrium. These models introduce the concept of the height equivalent of a theoretical plate (HETP). For a column of length (L) with N theoretical plates, HETP is

$$\text{HETP} = \frac{L}{N} \quad \ldots(3)$$

Comparison of HETP of different columns or different conditions is a well-accepted measure for the performance of a column. HETP is not a constant value for a sorbent species, it depends on the column length, packing quality, flow rate, solute used for determination, and calculation method.

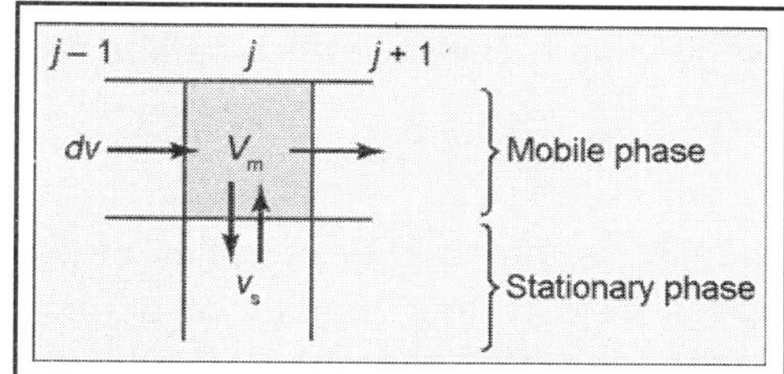

Figure 1.1 Schematic diagram of the distribution in a Martin and Synge plate model.

Martin and synge plate model

This model considers the column as a series of stages, and an infinitesimal fraction of the mobile phase volume passes from one stage to the next. Because an infinitesimal small fraction of volume is transferred from one stage to the next, this model is denominated a continuous plate model. The ratios between mobile and stationary phase volume in each vessel are identical and constant. The mass balance for the jth vessel can be written as

$$C_{m,j-1} \cdot dv = C_{m,j} \cdot dv + v_m \cdot dC_{m,j} + v_s \cdot dC_{s,j} \quad \ldots(4)$$

when dv denotes the movement of the mobile phase volume, v_m and v_s are the mobile and stationary phase volume of a stage and C is the concentration of the species under consideration. Martin and Synge introduced one mass unit into the first stage. Solving equation 4 with respect to a linear isotherm and sample only in the first stage gives a Poisson distribution.

$$C_N = C_1^0 \cdot \exp\left(-\frac{\nu}{\nu_m + K \cdot \nu_s}\right) \cdot \frac{\left(\dfrac{\nu}{\nu_m + K \cdot \nu_s}\right)^{N-1}}{(N-1)!} \quad \ldots(5)$$

which can be approximated by a Gaussian distribution for high plate numbers. In equation 5, C_N is the concentration of the Nth stage, C_1^0 is the initial concentration of the first stage, and v is the volume of mobile phase that has left stage N from the start.

Craig plate model

Analogously to the Martin and Synge model, this model divides the column into a series of discrete stages (plates), and each stage contains the same volume of stationary and mobile phase. But in contrast to the previous model, the flow is not continuous.

It is assumed that the whole amount of mobile phase moves to the next stage after equilibrium is attained. Then again, the distribution between stationary and mobile phase takes place immediately. Hence, the Craig model is a discontinuous plate model. The mass balance for the jth plate can be written as

$$v_m \cdot C_{m,j,t} + v_s \cdot C_{s,j,t} = v_m \cdot C_{m,j-1,t-1} + v_s \cdot C_{s,j-1} \quad \ldots(6)$$

The indexes t and $t-1$ denote for the actual and the previous time increment. For a linear isotherm and sample only in the first plate, this model gives a binomial distribution that also can be approximated by a Gaussian distribution for large plate numbers. In the Craig model, peak dispersion is smaller than in the Martin and Synge model. This difference can be related by

$$N_C = N_{M.S} \cdot \frac{k_0'}{1+k_0'} \qquad \text{...(7)}$$

when N_C is the number of plates of the Craig model, $N_{M,S}$ is the number of plates of the Martin and Synge model, and k_0' is the retention factor

$$k_0' = \frac{t_R - t_0}{t_0} \qquad \text{...(8)}$$

at infinite dilution. Here, t_R is the retention time at the conditions of interest and t_0 is the void time, that is, the retention time assuming nonbinding conditions.

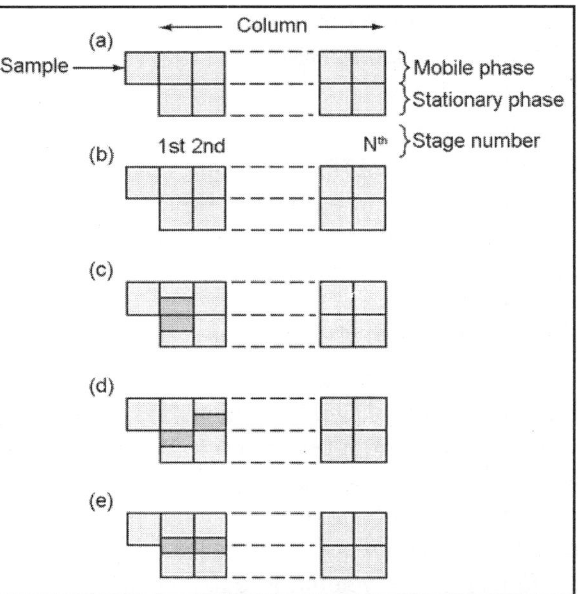

Figure 1.2 Schematic diagram of a Craig machine for N stages and a constant distribution coefficient. (a) A sample segment is arriving at the first stage. (b) The segment arrives in the mobile phase of the first stage. (c) The sample is distributed between mobile and stationary phases according to the distribution coefficient. (d) The mobile phase volumes are moving one stage further. (e) Mobile and stationary phases are distributed again.

Other plate models

The plate models described in this section are extended to kinetic models by introduction of mixing in the stages. Hence, the equilibrium in the stages is no longer instantaneous. The finite time for adsorption and transport kinetics between the phases are compensated by mixing kinetics within the stages. The sectional model is an extension of the Craig model. At the end of each time, increment the whole content of the liquid phase of each stage is transferred to the next stage. Therefore, the flow of the mobile phase is discontinuous. The tanks-in-series model is an extension of the Martin and Synge model. The flow of the mobile phase is continuous, and the column is considered as an array of well-mixed cells.

Plate Theories

Although the plate theory was originally developed for linear chromatography with the intention to describe the non- ideal effects, this theory was, in the early works, focused on ideal nonlinear chromatography.

Ideal model

The mass balance equation was derived independently by Wicke and Wilson. In their mass balance equations, axial dispersion was disregarded. DeVault introduced the influence of the pore volume, and Weiss discussed Wilson's approach for several types of adsorption isotherms. The mass balance equation of this ideal model is

$$\frac{\partial C_{m,i}}{\partial t} + H \cdot \frac{\partial C_{s,i}}{\partial t} + u \cdot \frac{\partial C_{m,i}}{\partial z} = 0 \qquad \ldots(9)$$

Here, C_m and C_s denote for the protein concentration of species *i* in the mobile and stationary phase, respectively. *H* is the phase ratio, *t* is time, *z* is column length and *u* is the mobile phase velocity. This model is of interest only for nonlinear isotherms. In the case of a linear isotherm and a column of infinite efficiency, the sample pulse travels through the column without any change.

Equilibrium-dispersive model

According to Van Deemter et al., Giddings, and Haarhoff and van der Linde, equation 9 can be replaced by

$$\frac{\partial C_{m,i}}{\partial t} + H \cdot \frac{\partial C_{s,i}}{\partial t} + u \cdot \frac{\partial C_{m,i}}{\partial z} = D_{a,i} \cdot \frac{\partial^2 C_{m,i}}{\partial z^2} \qquad \ldots(10)$$

when the mass transfer is not infinitely fast. Here, $D_{a,t}$ is the apparent dispersion coefficient that lumps together axial dispersion and nonequilibrium from adsorption and desorption.

$$D_{a,i} = \frac{HETP \cdot u}{2} \qquad \ldots(11)$$

It is assumed that the mobile and stationary phases are instantaneously in equilibrium. Assuming a column of infinite length, a Dirac $\delta(z)$ pulse as an injection profile, and a linear isotherm, the solution of this equation is a Gaussian distribution.

Lumped kinetic model

In this model, the dispersion coefficient D_L accounts only for the axial dispersion and is not lumped to the influence of nonequilibrium effects.

$$\frac{\partial C_{m,i}}{\partial t} + H \cdot \frac{\partial Cs_{,i}}{\partial t} + u \cdot \frac{\partial C_{m,i}}{\partial z} = D_{L,i} \cdot \frac{\partial^2 C_{m,i}}{\partial z^2} \qquad \ldots(12)$$

Depending on the assumptions, different models can be distinguished. If the influence of the mass transfer resistance is negligible and only the kinetics of adsorption and desorption are taken into account, the model is called reaction-dispersive. For a linear isotherm, the kinetic equation is

$$\frac{\partial C_s}{\partial t} = k_a \cdot C_m - k_d \cdot C_s \qquad \ldots(13)$$

when k_a and k_d are the rate constants of adsorption and desorption, respectively. When a Langmuir-type isotherm is assumed, the Langmuir kinetic model is obtained.

$$\frac{\partial C_s}{\partial t} = k_a \cdot (C_{s,max} - C_s)C_m - k_d \cdot C_s \qquad \ldots(14)$$

If we assume the adsorption-desorption kinetics as infinitely fast but not the mass transfer kinetics, the model is called transport-dispersive. Then, the kinetic equation for C_s is

$$\frac{\partial C_s}{\partial t} = k_m \cdot (C_m - C_s) \qquad ...(15)$$

when k_m is the apparent mass transfer coefficient.

Assuming a Dirac $\delta(z)$ pulse as an injection profile, a linear isotherm, and that mass transfer kinetics in the transport-dispersive model and adsorption-desorption kinetics in the reaction-dispersive model are fast, the solution of these equations is a Gaussian distribution. The general rate model consists of two partial differential equations for each component, one for the mobile phase in the interparticle area of the column and one for the fluid in the intraparticle area. It was used by Kucera for linear isotherms and by Lee et al. and Yu and Wang for nonlinear isotherms. For numerical solutions of these equations several different approaches can be used. The techniques generally applied are finite-difference and finite-element methods.

The direct calculation of numerical solutions of the mass balance equations by finite-difference models leads to stability problems of the calculations. A method applied extensively replaces the axial dispersion by introducing a numerical dispersion. In this case, a solution for the ideal model is looked for that gives a first-order error equivalent to the axial dispersion. The numerical dispersion can be adjusted by changing the space increment (z) or time increment (τ) of the iteration. The basis for all calculations is a modified form of equation 10

$$\frac{\partial C_m}{\partial z} + \frac{\partial\left[\frac{1}{u} \cdot (C_m + H \cdot C_s)\right]}{\partial t} = \frac{D_a}{u} \cdot \frac{\partial^2 C_m}{\partial z^2} \qquad ...(16)$$

which is identical with

$$\frac{\partial C_m}{\partial z} + \frac{\partial G(C_m)}{\partial t} = \frac{D_a}{u} \cdot \frac{\partial^2 C_m}{\partial z^2} \qquad ...(17)$$

For the ideal model this equation simplifies to

$$\frac{\partial C_m}{\partial z} + \frac{\partial G(C_m)}{\partial t} = 0 \qquad ...(18)$$

Replacing the first term by the backward finite difference and the second by the forward finite difference (backward-forward differences), this equation for the $j + 1$ at the position n is

$$\frac{Cj_{m,n} - C_m^{j-1}}{h} + \frac{G_n^{j+1} - G_n^{j}}{\tau} = 0 \qquad ...(19)$$

This calculation scheme is identical to the Craig model if the ratio of the calculation intervals h/τ equals u. This Craig model was widely used by several authors. In contrast to a forward-backward scheme, which is put to use for isocratic conditions, it can be applied to calculate gradient elution chromatography. More exact calculations with respect to the estimation of the axial dispersion effects are possible using finite-element methods. In chromatography, the method of orthogonal collocation on fixed elements is preferred to the method of moving elements.

Arve and Liapis call this approach the finite bath method instead of finite element. All these methods are limited to isocratic conditions by means of numerical stability of the calculations. Recently, a fast algorithm based on the assumption of elution as a collocation of point masses migrating with a velocity dependent on the local concentration and the flow rate was presented. This model gives comparable results to a Craig model, but the computation time is two orders of magnitude smaller.

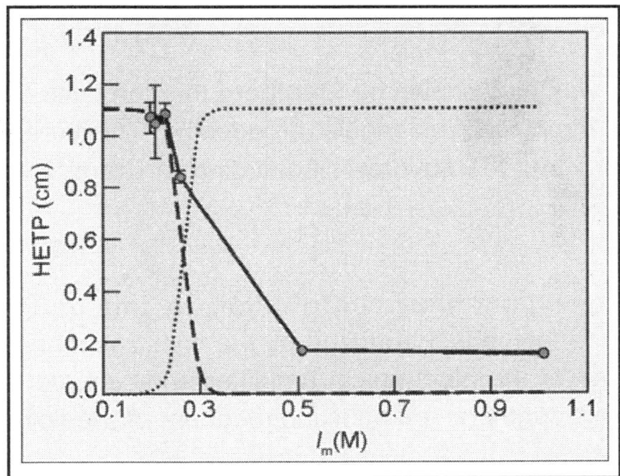

Figure 1.3 Dependence of HETP on I_m measured by BSA pulses applied to an anion exchanger.

Stochastic Theory

A statistical approach developed by Giddings and Eyring has been used for a stochastic model to describe migration and broadening of a sample in a chromatographic column. They considered the migration of a solute in a column as a Poisson distribution process and ignored the axial dispersion. They obtained following probability distribution:

$$P = \frac{k_a \cdot k_d \cdot t_0}{2 \cdot \sqrt{\pi} \cdot (t - t_0)^{0.75}} \cdot \exp\left[-(\sqrt{k_d \cdot (t - t_0)} - \sqrt{(k_a \cdot t_0)^2}\right] \quad \text{...(20)}$$

This model was extended to more complex situations accounting for axial dispersion and mass transfer resistances. For an increasing number of entries of a molecule into the stationary phase and a linear adsorption isotherm, this probability distribution tends to a Gaussian distribution.

Scales of Chromatography

To be able to describe the behavior of a protein sample in different scales of chromatographic systems, the influence of the extra-column space on the band broadening must be distinguished from the broadening occurring in the column. A comprehensive description of the extra-column effects on gas chromatography is given by Sternberg. Most of the explanations are also valid for liquid chromatography.

Broadening of extra-column effects

The extra-column effects can be subdivided into dispersion in tubes, contributions of dead volumes, finite detector volume, and dynamic behavior of transducers and electronics. According to Taylor, tubes introduce a symmetrical Gaussian-type broadening (σ_t^2) to the sample. The band spreads under the combined action of molecular diffusion and the parabolic axial flow velocity profile. This broadening can be described by

$$\sigma_t^2 = \frac{t_R \cdot r^2}{24 \cdot D_M} \quad \text{...(21)}$$

where t_R is the retention time, r is the radius of the tube, and D_M is the molecular diffusion coefficient. Dead volumes create a type of washout kinetic. Therefore, they introduce an exponential contribution to the band broadening. This exponential contribution (τ_{dead}) is a function of the dead volume (V_{dead}) and the flow rate (F)

$$\tau_{dead} = \frac{V_{dead}}{F} \quad ...(22)$$

It was also shown by Sternberg that the finite sensing volume of a detector introduces an additional symmetrical, rectangular broadening. This broadening (σ_d) depends on the sensitive detector volume (V_d) and the flow rate. According to Cram and Glenn it can be described by

$$\sigma_d^2 = \frac{V_d^2}{12 \cdot F^2} \quad ...(23)$$

An exponential contribution to the band broadening is introduced by the finite response rate of the electronical amplifier and recorder. This contribution (τ_{el}) is independent of the flow rate.

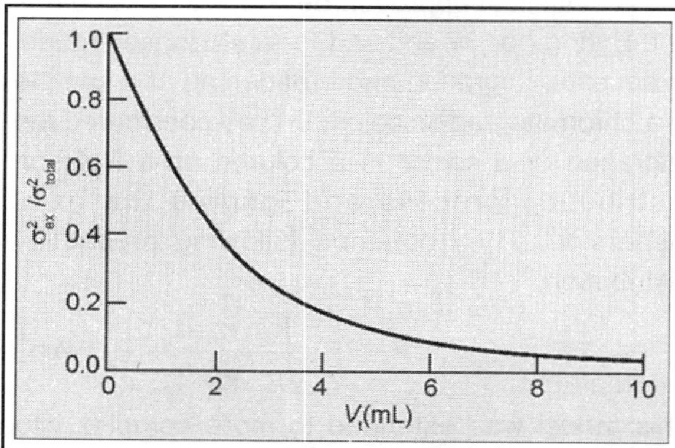

Figure 1.4 Contribution of extra-column band broadening to total broadening as a function of total column volume (V_t). Original data from experiments using a 0.5 cm i.d. column.

Mathematical treatment

Assuming additivity of the variances, the total extra-column broadening can be written as

$$\sigma_{ex}^2 = \sigma_s^2 + \sigma_t^2 + \sigma_d^2 + \tau_{dead}^2 + \tau_{el}^2 \quad ...(24)$$

where σ_{ex}^2 is the lumped extra column dispersion and σ_s^2 is the variance of the initial theoretical injection profile. Assuming that in liquid chromatography the variances introduced by the electronics and the detector volume are negligibly small compared to the other variances, equation 24 simplifies to

$$\sigma_{ex}^2 = \sigma_s^2 + \sigma_t^2 + \tau_{dead}^2 \quad ...(25)$$

Lumping together the *F*-independent variances (σ_A^2) and the *F*-dependent variances (σ_B^2), the extra-column broadening (σ_{ex}^2) is

$$\sigma_{ex}^2 = \sigma_A^2 + \frac{\sigma_B^2}{F^2} \quad ...(26)$$

An equation for the prediction of the extra-column contribution to the total broadening can be derived based on the definition of the plate number *N* of a system

$$N = \frac{V_p^2}{\sigma_{c.i.}^2} \quad ...(27)$$

If the total broadening (σ_{total}^2) is considered as the sum of the column internal broadening ($\sigma_{c.i.}^2$) and the extra-column broadening (σ_{ex}^2)

$$\sigma_{total}^2 = \sigma_{c.i.}^2 + \sigma_{ex}^2 \quad ...(28)$$

and *N* is replaced according to equation 3, the following relationship can be obtained

$$\sigma_{total}^2 = \frac{HEPT}{L} \cdot V_R^2 + \sigma_{ex}^2 \quad ...(29)$$

This equation was derived for analytical chromatography by Huber and Rizzi and has also been applied for preparative chromatography. Using this relationship it is possible to exhibit a quantitative picture of the contribution of extra-column broadening on the total broadening.

COMMERCIAL PROGRAMS

Computer-aided tools have been developed to different levels of complexity. Programs have been written for prediction of peak profiles, breakthrough curves, gradient shapes, and optimization of resolution. On the other hand, expert systems have been created for development of chromatographic separations, by using the prediction of peak profiles as tools for optimization.

Expert Systems

An expert system applies heuristic rules for the solution of a problem by using a knowledge base and an interference engine. The knowledge base is a collection of rules and facts gathered from practical experience. The interference engine applies domain knowledge to draw conclusions for the problem under consideration. Although the usefulness of expert systems was shown in the late 1960s for chemical applications with the expert system DENDRAL, expert systems for chromatography were not introduced until the early 1980s. The expert system ECAT (Expert Chromatographic Assistance Team) provides information for column packing and geometry, and known analyte properties determine the type of chromatographic mode (reversed phase, ion exchange, etc.). Optimizations based on the linear solvent strength model (LSS), assuming a linear relationship between the log *k*′ and the mobile phase modifier concentration, Φ, are offered. A linear relationship between the log *k*′ and the solvent strength of the mobile phase modifier is assumed

$$\log k' = \log k_0' - S\Phi \qquad ...(30)$$

where *k*′ is the capacity factor (normalized retention time in respect to the dead time) defined as

$$k' = \frac{t_R - t_0}{t_0} \qquad ...(31)$$

and φ is the volume fraction of the mobile phase modifier. *S* reflects the dependency of k' on φ, and k_0' is the capacity factor without the mobile phase modifier.

The expert system expert system for chromatography (ESC) was generated for method development in gas chromatography and HPLC. A set of about 500 chromatograms from the literature constitutes the basis for optimization. Peaks can be simulated, and mobile and stationary phases can be selected. For the peak identification, a curve-fitting tool based on the exponential modified Gaussian function is implemented. Fell et al. developed an expert system to optimize the mobile phase composition in RPC. Therefore, they applied simplex optimization and iterative regression techniques based on experimentally determined retention times.

The European project ESCA (Expert Systems for Chemical Analysis) provided three complex expert systems for method development in chromatography. The first deals with method ruggedness; the second, with method, and the third, with method development from initial method selection to optimization. The expert system HPLC Doctor and a similar system developed by Tsuji and Jenkins promise help when specific problems with HPLC arise. Numerous expert systems were developed for RPC. A comprehensive description of expert systems in chromatography is given by Hamoir and Massart. Presently, for preparative chromatography expert systems are not available. The starting material is very complex, and the purification is generally not completed in one cycle. Expert systems on downstream processing of fermentation products have been developed by Leser and Asenjo. These systems comprise optimal flow sheeting of the possible unit operation suited for recovery of bioproducts.

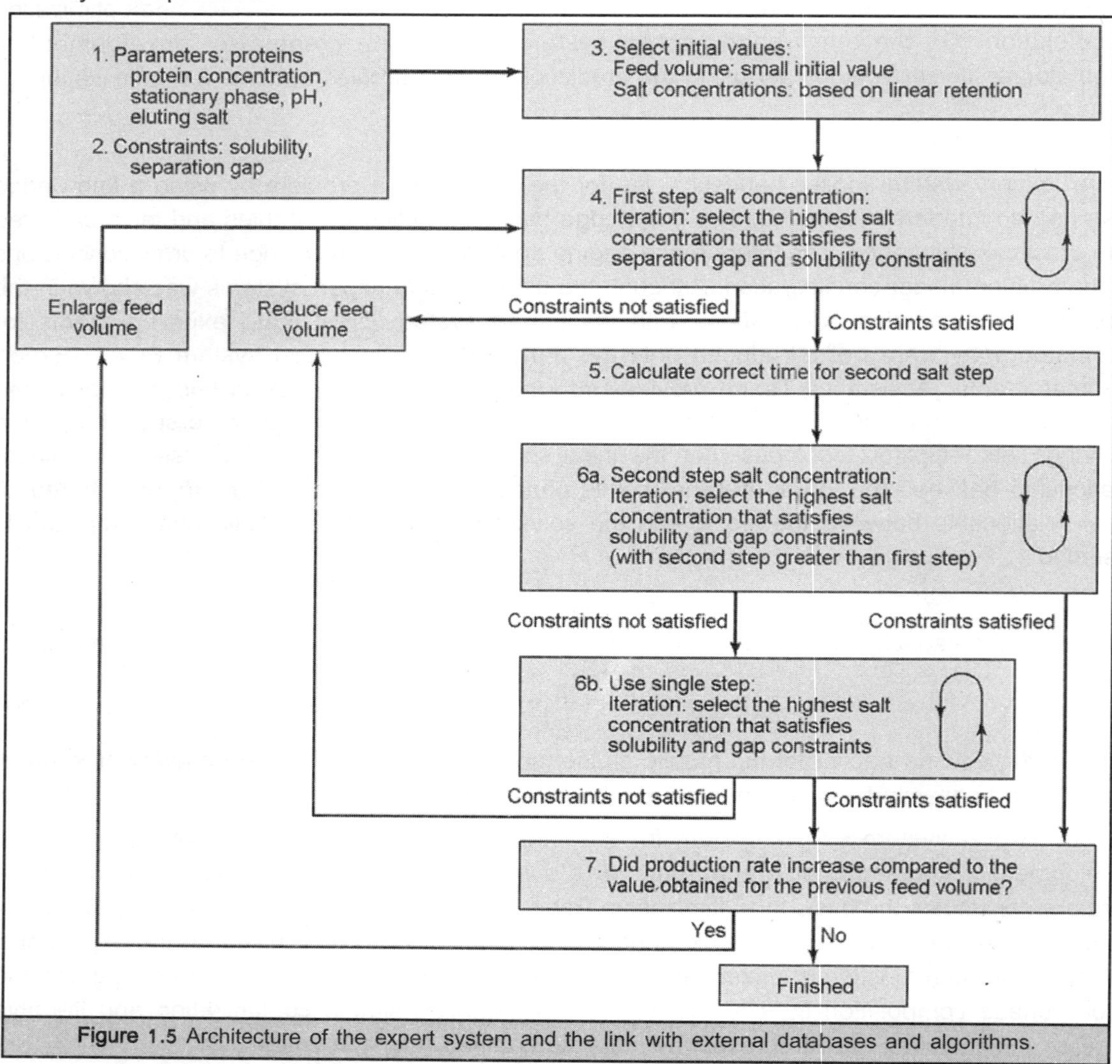

Figure 1.5 Architecture of the expert system and the link with external databases and algorithms.

Leser et al. built an expert system for selecting the sequence of processes for downstream processes, using a commercially available shell. The knowledge base contains around 600 logical rules, methods, and structures that emulate the expert's reasoning in selecting downstream processes. The creation of the rationale for the selection of a sequence of high-resolution purification steps must be based on heuristic rules because the existing models for chromatographic separations are not adequate for solving questions of selecting processes, they relate more to the mechanisms of interaction between proteins and matrixes. The heuristic rules are similar to those used to guide the selection of processes for multicomponent mixtures in chemical engineering. The rationale for selection of high-resolution purification operations uses the theoretical concept of separation coefficients.

Table 1.4 Classification of heuristics

Type	*Applications and examples*
Method heuristics	Specify choice between different unit operations.
	Describe conditions for preference of microfiltration over centrifugation.
Design heuristics	Specify order or sequence of process steps.
	Perform low-resolution step before high-resolution step.
	Use large particles at the beginning of a purification sequences.
Species heuristics	Specify based on the properties of the components.
	Choose separation method based on the greatest difference in properties of product and impurities.
	This rule tells us to use always affinity chromatography.
	There are certain other limitations.
Composition heuristics	Specify feed composition based on separation costs.
	This rule tells us that affinity chromatography may be too expensive.

Simulation Chromatographic Behavior

Since the early 1980s, several computer programs for simulating chromatographic behavior have been developed. Most of the programs are only of academic applicability because the experimental accessibility of model parameters is very cumbersome. Smit et al. describe a discrete mathematical model for nonlinear, nonideal chromatography. The nonlinearity can be represented by several different isotherms (linear, Langmuir, Freundlich, sigmoidal). The nonideality is represented by a lumped parameter reflecting axial molecular diffusion, Eddy diffusion, and different sample shapes. No dispersion occurs in the stationary phase. The conservation of mass is used to set up a mass balance. The resulting differential equations are solved by a fourth order Runge-Kutta method. Phillips et al. use a probabilistic approach for the simulation of a chromatographic separation process, called discrete event simulation.

All molecules are executed with the same adsorption–desorption mechanism. Each molecule undergoes a certain number of events (adsorption and desorption). The program can only simulate isocratic conditions and was originally developed for gas chromatography. Nonlinear isotherms are also taken into account by applying a molecular density map. The probability of adsorption is related to the accessible (free) binding sites. Whitley et al. established a very advanced computer program for simulation of peak profiles named VERSEC-LC. The calculation is based on an highly evolved rate model, where the column is virtually divided into different stages (vessels). This model treats the column as a series of *N* equal volume stages (equivalent to plates or series of vessels or transfer units). The mobile phase flows through these stages at a given rate. Each stage is divided into three fractions: the bulk solution, pore solution, and the solid phase. A gradient between the bulk solute concentration and pore solute concentration results from the combined effects of film and pore diffusion mass transfer resistances. Equilibrium assumed between the solute in the pore solution and the solute on the solid phase is described by nonlinear isotherms.

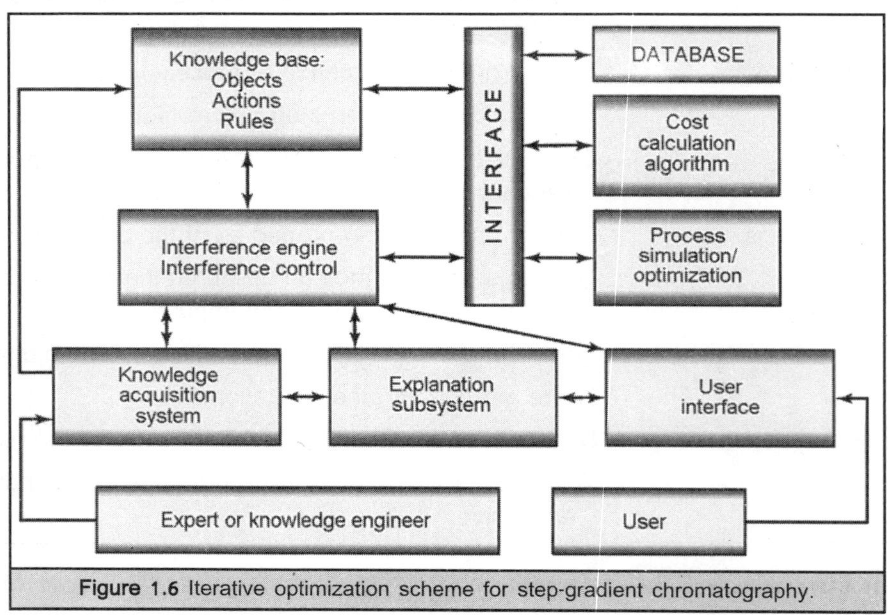

Figure 1.6 Iterative optimization scheme for step-gradient chromatography.

The program can simulate gradients, different loading concentrations, and volumes using different pore structures and diffusion coefficients. Several computer programs (e.g. Dry-Lab) use the linear solvent strength theory (LSS) for the calculation of peak positions. A column is calibrated by two runs, and the optimal gradient shape can be calculated by using a resolution map. The program is mainly suited for RPC, because ion-exchange chromatography (IEC) does not follow the LSS theory. The program is intended for optimization of analytical chromatography. Gallant et al. developed a computer model based on the stochastic mass action (SMA) formalism. SMA is a further improvement of the stoichiometric displacement model. The SMA formalism represents the chromatographic adsorption process as a stoichiometric exchange of mobile phase protein and bound counterions.

Furthermore, a fraction of counterions is shielded by the adsorbed protein. The accessible sites are corrected by a so-called shielding factor, which is proportional to the stationary phase concentration of protein. For acquisition of the parameters, describing the protein adsorption at different salt concentrations, the LSS formalism is applied, although this formalism is not completely valid for IEC. Prediction of protein zone migration and broadening is carried out by a single lumped dispersion model (which is described by equation 10). Gallant et al. propose an optimization routine using simulation of chromatography by the aforementioned models. The optimization runs through different stages. Initially, modeling parameters have to be acquired, and protein constraints (e.g., solubility, resolution) have to be defined. These constraints meet actual requirements of large-scale protein production.

The program Simulus uses a similar approach as described by Whitley et al., and diffusional resistances and a surface reaction resistance were included. The fundamentals for this program were published by Cowan et al. Furthermore, they have integrated moving grids for calculation of the peak profiles of Wiblin et al. The program is marketed by BIOSEP, AEA Technology. ProSys Simulation, distributed by BioSepra Inc. is a computer package using the simple Craig simulation. The computer package is connected to hardware that allows automated determination of the parameters required for optimization of preparative chromatography. Currently, it is the only commercially available integrated package of parameter acquisition and simulation software. For IEC, distribution coefficient values as functions of salt concentrations are predicted by linear gradient experiments as proposed by Yamamoto et al. The calculation of simulation parameters can be carried out semiautomatically, assisted by the software.

Validation of the simulated peak profiles can easily be accomplished because the software does not discriminate between experimental and simulated runs. So, comparison and superposition are simple. The software package implemented in the ProSys Workstation also contains a tool for economical evaluation of chromatographic unit operations. All the aforementioned computer programs do not take into account the extra-column band spreading. This mechanism may become very important when it comes to scale-down and parameter acquisition at very small scale. The authors of this chapter have developed a computer program that is an extended version of the simulation program used in the commercial system ProSys. This extended version allows the prediction of peak profiles using a big variety of different adsorption isotherms. Furthermore, the influence of extra-column inlet and outlet can be simulated. The modifications of gradient profiles by the mixing device is also part of the simulation. Rescaling and influence of the setup can be virtually studied. The simulation program can also be used for validation of chromatographic conditions.

Manmade Neural Networks

Artificial neural networks have been recently applied for chromatographic peak classification (92). Neural networks are specialized computer systems having processing capabilities similar to the human brain. They are distinct from conventional computer systems, because they process entire patterns, rather than successive individual items of data. They are trained by example to carry out their tasks, instead of being explicitly programmed. The processing patterns and learning capabilities displayed by human brains are difficult to mirror using conventional computer programming. Nevertheless, a neural network is, ultimately, a particular kind of computer program, far removed in its theory or implementation from any biological system. Modern digital computers built with traditional designs have a fundamental limitation, the so-called von Neumann bottleneck. Traditional computation requires a problem to be broken down to a set of operations that are performed in serial fashion, that is, one instruction at a time. Typically, each instruction must be completed before the next instruction is executed.

Artificial neural networks represent a fundamentally different approach to computation. They are explicitly designed to mimic the basic organizational features of biological nervous systems: parallel, distributed processing. Artificial neural networks consist of a large number of simple interconnected processing elements, where the processing elements are simplified models of neurons and the interconnections between the processing elements are simplified models of the synapses. The processing of information in such networks therefore occurs in parallel and is distributed throughout each unit composing the network. Neural networks are particularly appropriate for a defined class of tasks, which maybe characterized as follows:

1. Pattern recognition, rather than sequential data processing
2. Tasks demanding nonlinear capabilities in producing their classification, which can be a problem for other techniques
3. Technically ill-defined tasks, even though the decision may be obvious
4. Tasks with sufficient examples of the target data available to allow a learning system to be considered

Peak-shape classification is a pattern recognition task that is pertinent to chromatography. The automated determination of the peak profiles in a chromatogram gives an objective answer when two chromatograms are identical or different. It may be a valuable contribution to validation for a purification sequence or validation of scale-up.

IF AND BUTS OF MODELING

For a conventional software, exact definition of the conditions leading to the optimum is necessary. Mathematical algorithms (explicit equations or differential equations) must be available to calculate the optimum. The major obstacle in preparative chromatography is the complexity of the variables, the problem to estimate the parameters, and the response value. It may be also possible that an optimum is not found within the investigated range of variables. Independently from the starting point (initial pair of variables), an optimum will be found. Systems, where the variable and the response value are related by a peak function or where the variable consists of several variables lumped together, produce a surface with a unique maximum of the response value.

The lumped variables or reciprocal action with other variables should be carefully investigated. Minor changes in the experimental conditions may cause a big variation in the optimal response value. The optimum obtained in a system depends on the starting point. Several optima and even local minima are present. Such a system can be characterized by carrying out an optimization strategy with starting points over the whole area in small increments of the considered variables and comparing the reached minima and maxima. This situation is often observed, when the

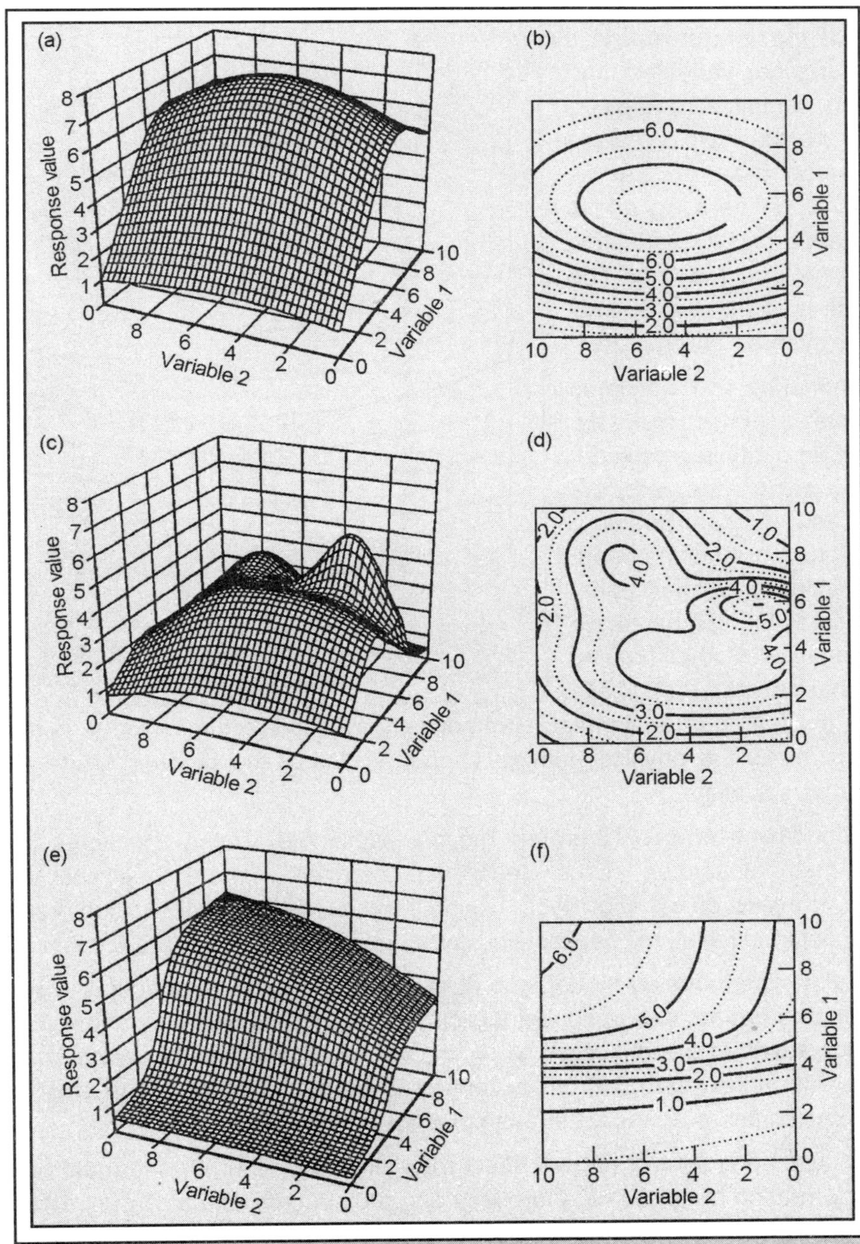

Figure 1.7 (a to d) Evaluation of optimum with two variables and one response value. (e and f) An optimal response value is not found in the considered range. In chromatography, the variable can be, for example, flow rate or loading, and the response value can be, for example resolution, productivity, or purity.

productivity of a chromatographic separation is optimized in respect to flow rate and loading. When the separation is not sensitive to resolution, the system is hypothetically only constrained by the range of the variables. The performance of chromatography is influenced by far more variables than two, but it is very complicated to plot a function with more than two variables in a nonconfusing manner. Let us assume that a chromatographic separation depends on two variables, such as flow rate and loading. The hypothetical task is to find the optimal response value, such as productivity.

Using experimental and simulated results, the optimal response values are searched. A certain purity is presumed. Then the productivity and the purity are plotted versus flow rate and loading. Both contour plots are superimposed. The area, where the presumed purity is reached, indicates the optimal range of operational variables. Sometimes the conditions are constrained by pragmatic issues, such as availability of equipment (commercially available columns with a certain diameter, costs of a buffer) or by the physicochemical characteristics of the product. Many proteins have limited stability and solubility.

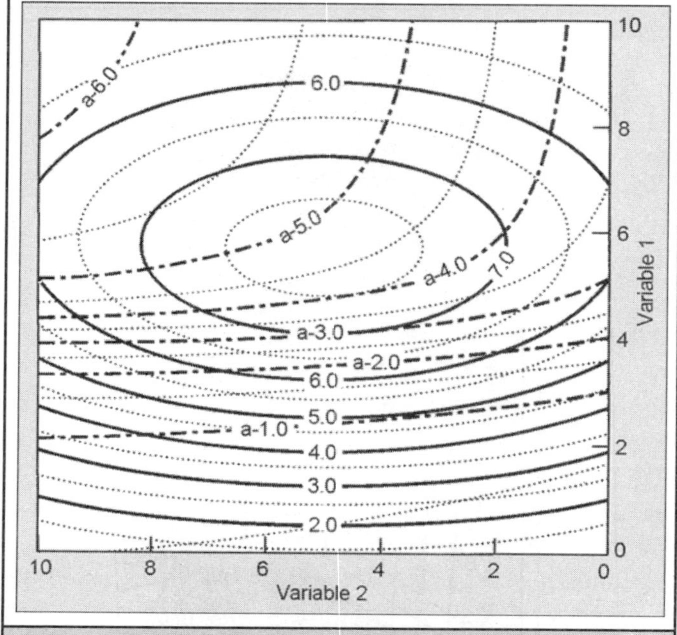

Figure 1.8 Evaluation of an optimal operational range, when two different response values are related to the two variables.

There are conditions where the protein maybe inactivated. These conditions are not easily transferred into mathematical equations. Sadana summarized possible causes for inactivation of biological macromolecules during separation. During the adsorption and desorption process, proteins may undergo a conformational change. Proteins can associate or they can precipitate on the surface.

Self-association was modeled by Lemque et al. They could derive quantitative information on cooperativity effects from the adsorption isotherms. From their model, they showed that protein–protein interaction exists in the adsorbed state and led to an increase in protein self-adsorption. Considering a crude protein sample from a clarified fermentation broth, many proteins and other biological macromolecules are present in this solution.

Changing the concentration or pH will affect the self-association. On the other hand, one can also expect the formation of hetero di- and oligomeric protein complexes. Protein-DNA complexes can also occur. The capacity and the lifetime of a column are affected by the aforementioned aggregation phenomenon. This short discussion of protein interaction shows a very important aspect in optimization of chromatography, and complications arise when they have to be translated into a mathematical formulation. One possible way to circumvent these obstacles is the standardization of the sample as far as possible.

Computer Assisted Optimization

Conventionally, resolution has been optimized, because the major activities in computer-assisted optimizations have been carried out in analytical applications. Though some rules may also be applied for optimization of preparative chromatography, in analytical chromatography resolution (R_S) is defined as:

$$R_S = \frac{2 \cdot (t_1 - t_2)}{w_1 + w_2} \quad \ldots(32)$$

where t_2 and t_1 are the retention time of the separated substance, and w_1 and w_2 are the peak width at the base. A resolution map is the response value of the optimization. A resolution higher than 1 should be achieved throughout the whole chromatogram. A high resolution is not desired, because it increases the analysis time. The optimization of the running conditions (mobile-phase modifier composition, gradient shape, flow rate, etc.) can be carried out by factorial design, simplex optimization, and others.

Simplex optimization

Although there are many variants of the simplex procedure, they are all based on the basic procedure of Spendley et al. A simplex is defined as a geometric figure having one more point (vertex) than the number of variables being optimized. Thus, for two variables, a simplex is a triangle and for three variables the simplex is a tetrahedron. Although it is difficult to visualize a simplex for more than three variables, the mathematics does not become significantly more complex and the procedure is easily handled by manual or digital computation.

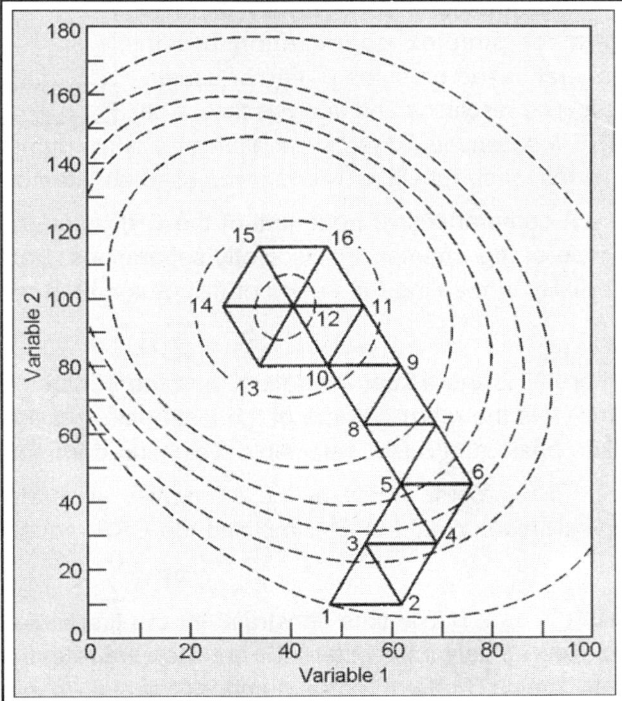

Figure 1.9 Fixed-size simple optimization of two variables. Initial simplex is 123 and the optimum region is at vertix 10, 11, 12.

The optimization proceeds by rejection of the vertex, which has the worst experimental response, and by reflecting its coordinates through the midpoint of the hyperplane. Unfortunately, this original, fixed-step size procedure suffers from several severe limitations. In particular, progress across the response surface is made at a constant rate because the step size is fixed; this fixed size also means that the optimum may not be precisely located because the simplex will be forced (in two dimensions atleast) to circle around it. Additionally, the fixed-step size simplex is prone to failure on a surface ridge, in that it will become stranded and not make progress toward the optimum. These limitations have been largely overcome in the modified procedure of Nelder and Mead. They introduced two new operations to the fixed-step size procedure, expansion and contraction.

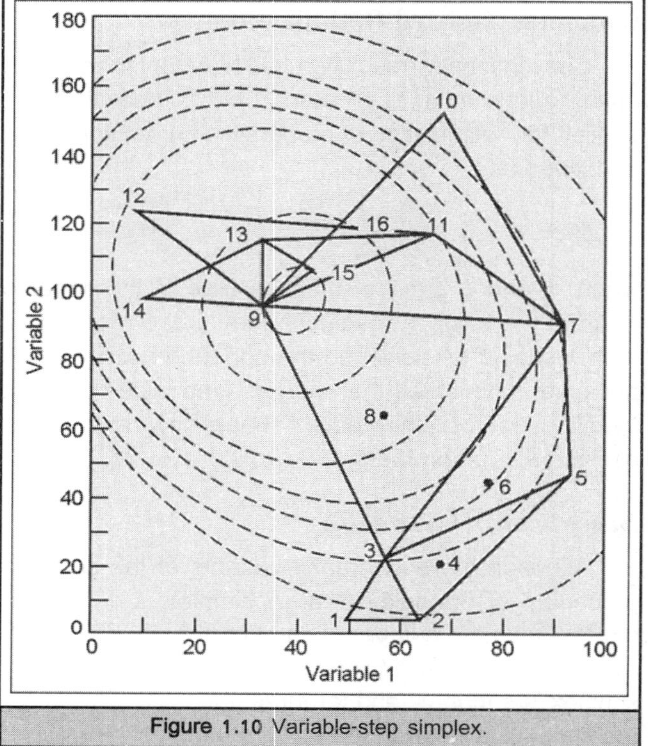

Figure 1.10 Variable-step simplex.

The latter operations allow the simplexes to expand and thus accelerate toward the optimum region. When they have approximately located the optimum region, the simplexes contract and reduce the search region until the optimum is precisely located. Berridge reviewed the chromatography response functions (CRF) used in analytical chromatography. These functions can be also applied to preparative chromatography.

The functions successfully used to date for simplex optimization are in general based on sums of terms, usually reflecting resolution and analysis time. The CRF are derived for the separation of a small number of components from a complex sample. A few response functions will be discussed as examples.

A comprehensive treatment of the CRF used in analytical chromatography would exceed the scope of this chapter. For isocratic separations, Berridge developed a function to provide greater flexibility in meeting the needs of fully automated optimization:

$$\mathrm{CRF} = \sum R_i + n^a - |t_m - t_n| - c(t_0 - t_1) \qquad \ldots(33)$$

where R_i is the actual resolution, n is the number of detected peaks, t_m is the desired retention time, t_n is the retention time of the last peak, t_0 is the dead time, t_1 is the retention time of the first elute peak, and a is a selectable weighting parameter.

To overcome some of the anomalies generated by using a multicriterion CRF in simplex optimization, Wright et al. simplified the CRF (equation 33) to

$$\mathrm{CRF} = \sum R_i + n - (t_m - t_n) \qquad \ldots(34)$$

with $t_m - t_n > 1$. The time constraint for the first eluted peak is deleted and that for the last peak is considered only if the difference between the retention time and the target retention time is greater than 1 min. The term for the number of peaks (n) is not raised to a power to ensure that when the term is included, it has a significant influence on the CRF value.

Isocronal analysis

Atamna and Grushka describe the concept of optimization by isochronal analysis, formerly known as the time normalization for HPLC. Briefly, in this mode of optimization, two experimental conditions are changed simultaneously. One is altered to improve resolution, whereas the other is changed to maintain the analysis time constant. The constant analysis time means:

$$t_{R1} = t_{R2} \quad ...(35)$$

where the subscripts 1 and 2 indicate two different experimental conditions. From equation 36, the following conditions can be derived:

$$\frac{Z_1(1+k_1')}{u_1} = \frac{Z_2(1+k_2')}{u_2} \quad ...(36)$$

where Z is the column length, u is the linear velocity, and k' is the capacity factor. The need to change two experimental conditions simultaneously is clearly seen from equation 36. The most practical pairs of experimental conditions that can be exploited in isochronal analysis are mobile phase velocity and composition, mobile phase velocity and column temperature, or mobile phase composition and column temperature. When the residence time of a product is of concern, this type of optimization may have some value also for preparative chromatography.

Purification of Protein

Flow sheeting in bioprocessing is a very complex task. A process for purification of a protein consists of several steps. Usually there is a link between the performance of each step and the performances of the previous step and the subsequent steps. The complexity of the problem is obvious when considering a tree structure, assuming a three-step purification with only two possibilities at each step. Even in this simplified case we end up with eight different purification schemes. Additionally, an economical evaluation of the individual arrangement is hard because the steps of one sequence are influencing each other. The use of expert systems is suggested for the rational selection of purification processes starting from different microbial sources, such as harvesting, preconditioning, concentrating, purification, and polishing. Artificial neural networks are suited to improve the quality of expert systems for flow sheeting. The numerical certainty for defined steps can be fine tuned by feedback with existing results. A successful application has not yet been shown today. However, the large number of software tools for artificial neural networks will promote this application. The Biopro Designer is a computer program allowing the economical evaluation of each separation step. As in real life, chromatography is limited to other separation steps, such as ultrafiltration, precipitation, etc.

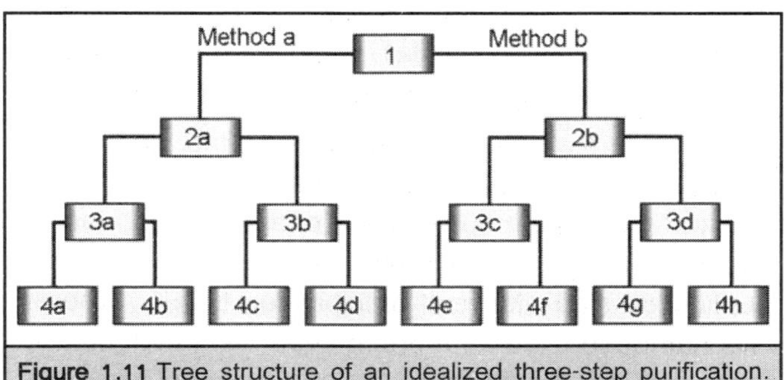

Figure 1.11 Tree structure of an idealized three-step purification. Each step offers only two possibilities.

2

SEPARATION OF BIOMOLECULES

Hydrophobic interaction chromatography (HIC) is an important chromatography technique used for the separation of biomolecules. The interaction is based on physicochemical characteristics of the molecules other than the charge, as in *ion-exchange chromatography* (IEC), or the size, as in gel filtration. Although its importance is increasing in industrial applications, the start has been slow. The technique is certainly very powerful when the optimal conditions for a certain separation are determined, but to achieve this has often been far from easy. One reason is that the mechanism of the interaction is complex, depending on several properties of the separation medium as well as of the separated substances, and not yet fully understood.

There are a number of hypotheses that approach the phenomenon of hydrophobic interaction from different points of view, but the practitioner will still find that there is no simple theory that will always facilitate the optimization of a separation, as is the case when using other chromatography techniques. Hydrophobic interaction is favored by high salt concentration and is thus a logical purification step after salt precipitation or IEC, where the environment has a high ionic strength. HIC has been used as a group separation method, rapidly separating what retards on the column from what is not interacting, and as a powerful high resolution technique with gradients of descending salt concentration.

In addition to what is usually designated as HIC, there are several related chromatography techniques based entirely or partly on hydrophobic interaction. Reversed-phase chromatography (RPC) is the most important, working with strong hydrophobic interaction that is often achieved by high substitution with long aliphatic chains. The principal difference between HIC and RPC is methodological in nature. In HIC, the interaction is increased by salt in order to retard the sample on the column and, basically, a return to native conditions will elute the compounds, whereas in RPC the sample is retarded on the column and eluted by addition of organic solvents. The methods based on what are described as salt-mediated and thiophilic interactions gain in importance, whether they should be considered as HIC techniques is perhaps a matter of semantics. Moreover, the binding in affinity chromatography and to dye-ligand matrixes often have significant elements of hydrophobic interaction.

NATURE OF HYDROPHOBIC INTERACTION

There are several approaches to theoretically elucidate the nature of hydrophobic interaction, ranging from thermodynamics over surface tension to van der Waals forces. From different angles they all try to explain the observed influence on hydrophobic interaction by ions with salting-out effect, chaotropic ions, type of salt, temperature, pH, and additives. No single theory, however, manages to fully cover the mechanisms of interaction. In HIC the added complexity, contributed by matrixes, spacers, ligands, and the effects of buffer environment on different species of biomolecules, implicates that, apart from very general recommendations, the design of a HIC procedure still has to be founded on careful practical and empirical studies.

Salting Out Process

There is definitely a relationship between the salting out process used for the precipitation of proteins and hydrophobic interaction as used in chromatography. The Hofmeister series list anions and cations according to decreasing ability to precipitate and increasing ability to dissolve proteins, and the combinations (salts) from the beginning of the list are well known to promote hydrophobic interaction. Probably precipitation by salting-out and HIC are based on the same phenomenon, the principal difference being that in precipitation the hydrophobic interaction has to take place between molecules of the same kind or others that happen to be present in the solution, whereas in chromatography a well-defined counterpart is hydrophobic enough to bind some of the passing compounds, preferably at lower salt concentration than is needed for precipitation.

(a)

Anions: Increasing "salting-out" effect ←

PO_4^{3-}; SO_4^{2-}; CH_3 $COOH^-$; Cl^-; Br^-; NO_3^-; ClO_4^-; I^-; SCN^-

(b)

Cations: Increasing chaotropic effect →

NH_4; Rb^+; Na^+; Cs^+; Li^+; Mg^{2+}; Ca^{2+}; Ba^{2+}

(c)

Salts promoting the model surface tension of water:

$Na_2SO_4 > K_2SO_4 > (NH_4)_2SO_4 > Na_2HPO_4 > LiCl > KSCN$

Figure 2.1 The Hofmeister series of anion (a) and cations (b). The arrows indicate increasing salting-out effect and chaotropic effect, respectively. Salts with high ability to promote hydrophobic interaction are combinations of ions found to the left in the series. The same salts are found early in the series of salts listed in order to decreasing ability to promote the surface tension of water (c).

Theory of Thermodynamics

The main and very elegant theory of hydrophobic interaction refers directly to the Gibbs free energy equation (equation 1), which describes the change in energy when a chemical system is transferred from one state to another:

$$\Delta G = \Delta H - T\Delta S \quad ...(1)$$

where ΔG is the change in Gibbs free energy, ΔH is the change in enthalpy or heat content, T is the temperature in Kelvin, and ΔS is the change in entropy or disorder of the system. IfAG is negative, the second state has a lower energy and the system will adapt to it, provided that there are no energy barriers. In hydrophobic interaction, the entropy change is of greatest importance. If hydrophobic molecules such as aliphatic carbon chains are immersed in water, the monolayer of water molecules in contact with a carbon chain will be in higher order than those in the bulk of the water. If two or more hydrophobic structures come together, the surface that has to be covered by ordered water molecules decreases, some water molecules join the less-ordered bulk water, and the entire system gains in entropy and thus, according to equation 1, decreases its free energy. This state will be favorable for energetic reasons and thus promoted. If some of the hydrophobic molecules are attached to a chromatographic matrix as ligands, the setup for HIC is obvious.

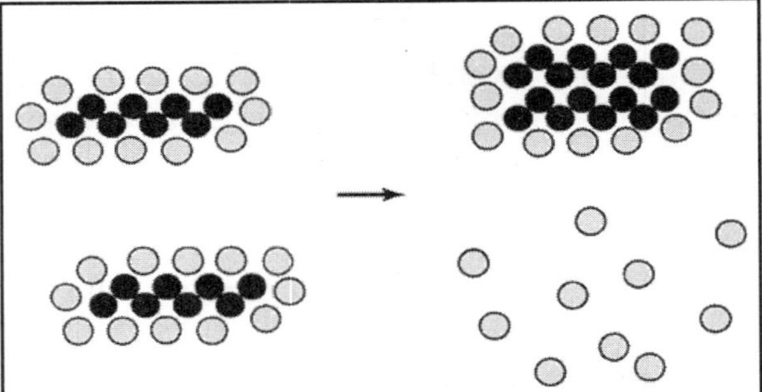

Figure 2.2 Hydrophobic interaction. Two aliphatic carbon chains (black circles) are immersed in water. Only the water molecules (gray circles) with high order in the monolayer adjacent to the nonpolar molecules are indicated.

The phenomenon of increased interaction in the presence of salting out ions is explained by a higher gain in entropy when water is transferred from the surface of a nonpolar molecule to the bulk of water. Whether this is achieved because the surface water in the presence of salt is initially more ordered or the bulk water less ordered or both is not quite clear. Equation 1 reveals that at a higher temperature an increase in entropy will give a higher energy gain because the entropy factor of the equation will be multiplied by a higher number. Consequently, if the binding to a medium has been shown to increase with the temperature, this has been taken as an indication that the interaction is really hydrophobic.

van der Waals Forces

Another theory suggests that the surface tension of the water surrounding a nonpolar structure and the tendency to minimize it are responsible for the interaction. If salt is added to the water, it has been shown that the surface tension increases. As a consequence, the energy gain in minimizing the surface exposed to the hydrophobic molecules is larger and thus the interaction is stronger. Although the salts that increase the surface tension most are among those that should promote hydrophobic interaction according to the Hofmeister series, the order is not quite the same. A third theory suggests that van der Waals forces are responsible for the hydrophobic interaction in HIC. This is supported by the fact that these forces should be increased as the order of water increases in the presence of salt. Probably the theories mentioned here all describe one part of the complicated interaction, and a combination of them is more relevant to describe the phenomenon of hydrophobic interaction.

CHROMATOGRAPHICAL TECHNIQUE

As early as 1948 Tiselius presented what he called Adsorption separation by salting out, but not until the beginning of the 1970s were reports on the synthesis of media published. Some researchers were working with a mixed mode because of their medias' content of charges, but charge-free hydrophobic media were also synthesized, and the first commercially available products were presented some years later. At present, a wide range of products, including HIC-media for HPLC, are available but phenyl, octyl, butyl, and alkyl ligands are still by far the most common.

Factors Affecting Hydrophobic Interaction Chromatography

In addition to the environmental parameters, including ionic strength, temperature, and pH that influence the hydrophobic interaction as such, there are several factors that have to be controlled to use the hydrophobic interaction for chromatographic purposes. The chromatographic medium is constructed by coupling hydrophobic ligands to a chromatographic matrix, often by use of a short spacer to enhance the sterical possibilities of interaction with large molecules. The type of ligand and the ligand density affect the characteristics of the medium. The choice of buffer ions has been found to change the interaction pattern more profoundly than generally anticipated. The optional use of organic solvents and detergents have to be taken into consideration. The molecules themselves have certain characteristics and will often change chromatographic interaction behavior with the buffer salts in an individual manner.

Finally, the choice of dimensions of the chromatographic column and the mode of elution will depend on the size range of biomolecules that are subject to processing and the desired throughput of the separation step. One approach, to describe the complex interaction of HIC from a practical point of view, has been to characterize a HIC medium by the interaction pattern of a significant set of protein molecules and to evaluate the result by the use of principal component analysis. As described in later sections, the media can be classified by this approach, and the importance of different characteristics of the proteins can be correlated to the interaction of different media.

Matrix

Agarose-based matrixes with varying degrees of cross- linking are most commonly used for HIC, but organic resins and silica-based media are also used, especially in HPLC applications. All common chromatographic matrixes, such as cross-linked agarose, organic polymers, and silica, can be derivatized with hydrophobic groups to give HIC media. Normally, the influence of the by-design hydrophilic chromatographic matrix will just add slightly, if at all, to the performance of the medium, but it is a well-known fact that unsubstituted agaroses intended for gel filtration will expose what can only be characterized as hydrophobic interaction at extremely high salt concentrations.

This allows for using HIC even for rather hydrophobic proteins that would stick irreversibly to conventional hydrophobic gels. It has to be noted, though, that under normal salt concentrations (e.g., those used in gel filtration) no hydrophobic interaction will be taking place because the gels were successfully designed to be as hydrophilic as possible.

Ligands

Alkyl, butyl, phenyl, and octyl groups are the most common ligands in commercially available HIC media. They are listed in what is commonly considered as in order of increasing hydrophobicity, but as mentioned earlier there is no linear or single dimension relationship between their interaction with different species or sets of molecules. It is obvious that the aromatic part of the phenyl group can bind with π-to-π binding as well thus adding to the complexity of the interaction. Other ligands of interest include mercapto derivates, which have yet another spectrum of interaction. They have been claimed to work partly with a special thiophilic interaction, but there is some doubt whether their content of sulfur is really responsible for their interaction pattern or whether they are just another type of hydrophobic medium among the others and working by a combination of the same mechanisms as other hydrophobic media.

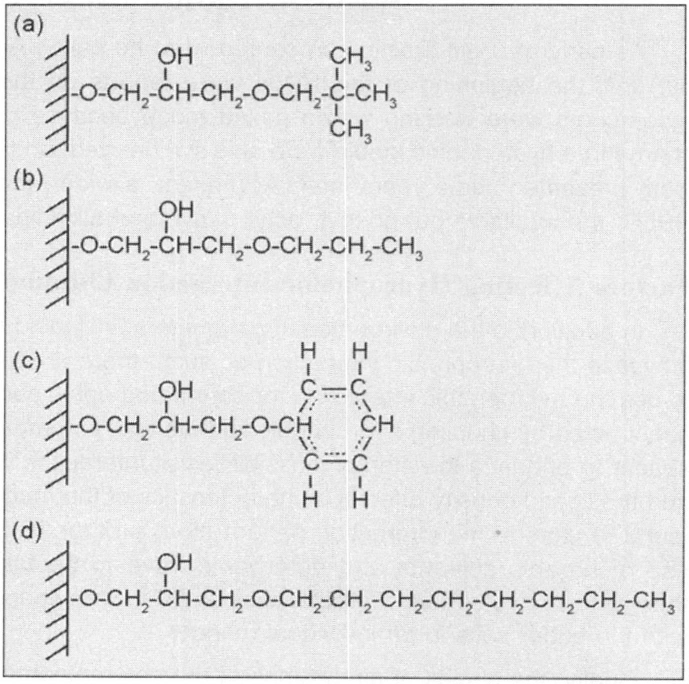

(a) $-O-CH_2-CH(OH)-CH_2-O-CH_2-C(CH_3)_2-CH_3$

(b) $-O-CH_2-CH(OH)-CH_2-O-CH_2-CH_2-CH_3$

(c) $-O-CH_2-CH(OH)-CH_2-O-C_6H_5$

(d) $-O-CH_2-CH(OH)-CH_2-O-CH_2-CH_2-CH_2-CH_2-CH_2-CH_2-CH_2-CH_3$

Figure 2.3 Common hydrophobic ligands. Alkyl (a), butyl (b), phenyl (c), and octyl (d) ligands coupled to a chromatographic matrix by glycidyl ether.

Coupling Reactions

Glycidyl ether is the coupling agent used most often to immobilize hydrophobic ligands, but in principle all the common coupling procedures can be used to attach the ligands to the chromatographic matrix. An extensive review of coupling methods is presented in the work by Carlsson et al. The coupling procedure will almost inevitably result in the ligands being attached via an often very short spacer, which is an advantage because conformational restrictions when large protein molecules bind are decreased. On the other hand, the choice of coupling chemistry will always influence the performance of the chromatographic medium by introducing hydrophobic or charged moieties, and the selectivity and the capacity may be very different even if the coupled ligand happens to be the same.

Binding Capacity

The binding capacity of an HIC medium depends partly on the ligand density, but it is important to remember that any figures given will always be related to the specific protein that was used to determine the capacity and to the conditions under which the determination was performed. The capacity can be determined as total capacity, with the protein exposed to the medium without flow, or as dynamic capacity, where the linear flow and the residence time of the molecules in the column will influence the result. There may well be a difference of magnitude of 10 to 20 between the capacities for different proteins, a fact that has to be considered when choosing a medium from a catalog where the dynamic capacity is often determined using serum proteins like albumin.

The capacity that can be used is dependent on the contamination picture as well and whether the main part of the impurities are more or less hydrophobic than the desired substance, so in practice it is necessary to determine the dynamic capacity in every specific application; this should be seen as a normal part of an optimization process. Much work is being done to optimize the dynamic capacity of media, both by suppliers and by users at research sites, but the question is whether the most convenient and uncostly way of increasing the capacity of a column is to increase the amount of gel!

Hydrophobicity and RPC

HIC and RPC are both based on hydrophobic interaction, and the borderline between the techniques is more defined by difference in methodology and the type of sample than by the interaction as such. In RPC the hydrophobicity of the media is higher because the ligands are more hydrophobic and the ligand density is higher. When the sample is run through an RPC column at conditions in the range of physiological pH and salt concentrations, most compounds are retarded on the column. Elution has to be achieved by using increasing gradients of organic solvents. Commonly used solvents are, in order of increasing elution ability, methanol, isopropyl alcohol, and acetonitrile. In contrast, in HIC very little of the sample is retarded at physiological conditions. The ionic strength of the sample has to be increased to promote attraction, and the elution is normally done by decreasing salt gradients.

If some components are still attached to the matrix in weak buffer, a mild organic solvent such as ethylene glycol will be used to wash them out. As a consequence, HIC is more often used for protein separation where the task often is to retain the tertiary conformation and the biological activity of the separated molecules, whereas RPC is often used for smaller molecules, such as peptides, which lack tertiary structure and subsequently will withstand the elution conditions without denaturation. Historically, RPC was developed as an HPLC technique to separate small molecules using mainly silica and organic polymer-based matrixes, and HIC was developed as a technique for the separation of proteins, as an alternative to IEC and gel filtration and using the common agarose matrixes as carriers of the hydrophobic ligands. The difference in molecular size range of the molecules to be separated leads to several important differences in the methodologies of the two techniques.

The differences in behavior between small and large molecules are important in all attraction techniques, such as IEC, HIC, and RPC. But because of the restrictions to expose large molecules to organic solvents, the borderline between separation of small and large molecules happens to coincide more or less with the borderline between the two hydrophobic techniques. But there are other technical reasons behind the difference of methodology based on molecular size. Small molecules interact with the matrix by single point attachment. A molecule is bound to the gel for a certain fraction of the time and is free to move with the buffer the rest of the time. Another way to express the same fact is that at a certain time one fraction of the molecules attach to the column while the rest are moving with the flow. Even if the equilibrium is shifted far toward binding, the macroscopic effect will be that small substances move down the column at some rate in a broad range of salt concentrations, although the speed may be rather low. In a mixture of substances, the rates of movement down the column will be different, as described by the individual equilibrium constants, and a separation will take place.

To allow for the molecules to separate, a long path length is needed, and thus long columns and keeping the separating zones sharp are essential for high effectivity. This fact has led to a hunt for high plate numbers, a feature that can be achieved by using small matrix particles. This in turn will cause a high back-pressure and the need to work under HPLC conditions. The full utilization of the selectivity of the column will be achieved by elution under isocratic conditions or by shallow gradients. The resulting extremely resolving techniques have been very successful and impressive and have been used for peptide fingerprinting and other types of qualified analysis of small organic molecules. Proteins, on the other hand, interact with the matrix by multipoint attachment. Even if at some salt concentration, some of the interacting parts of the large molecules may be unbound, there are often several points of interaction left that keep the protein attached to the column. Not until a certain elution condition is reached, such as higher ionic strength in IEC or lower salt concentration in HIC, where all binding sites of the protein are simultaneously unbound, will the protein move down the column.

The range between when the protein is not moving at all and moving at full speed is very narrow compared to when small molecules are separated. Isocratic elution is not possible because at constant buffer concentration some proteins will move without retardation and some will hardly move at all. A gradient is needed to elute the proteins in order. Because the resolution is not very dependent on the effectivity of the column, an HIC step where proteins are separated should be performed in short columns, and the linear flow should be kept low by increasing the diameter of the column. In large scale, the use of short and wide columns in protein separation based on IEC and HIC opens up the possibility to run with a high total flow and throughput and still keep the linear flow low enough to allow for the adsorption and desorption to take place without being too far from equilibrium.

INTERACTION OF MOLECULES OF HIC MEDIA

The performance of HIC media can be monitored by multivariate analysis, using a set or probe of protein molecules with different hydrophobic properties. The interactions of the molecule set with the HIC media could be correlated to molecular weight, isoelectric point, fraction of hydrophobic amino acids, surface charge, and several hydrophobicity indexes without the need for detailed knowledge about the parameters or the mechanisms involved. Instead, the different combinations of media and chromatography conditions can be fingerprinted by their performance in practice and easily compared. As well as artificial mixtures of proteins, serums with 15 to 20 of the proteins identified have been used as probes to determine adsorption patterns. Inversely, biomolecules can be classified by their hydrophobic characteristics as measured by their interactions with a standard set of different hydrophobic media. Molecules with similar adsorption patterns over the media should have similarities on a molecular level. The results hint that in the tested buffer system, there are at least two main mechanisms that have an impact on HIC. One mode of interaction was typical on highly hydrophobic media, for example, with an octyl chain as ligand, where the binding seems to be entirely correlated to the fraction of hydrophobic amino acids and to the hydrophobicity index of the protein molecule.

An explanation may be that a higher content of hydrophobic amino acids automatically leads to more hydrophobic amino acids being exposed on the surface of the protein. But because the fraction of exposed hydrophobic areas of proteins is always in the vicinity of 40 to 50%, another possible explanation takes into consideration that proteins are not rigid bodies with rigid surfaces but that the conformation should be considered as a statistical mean around which the molecule must have the property to breath or change the microconformation, thus allowing a long hydrophobic chain to come in contact with hydrophobic structures that statistically are not always exposed at the surface. An interesting finding is that the binding to the octyl medium is highly correlated with the molecular weight as well. This phenomenon is logical, though, taking into consideration that larger protein molecules generally have to have a higher fraction of hydrophobic amino acids to keep the tertiary structure.

The other mode of action does not correlate with the content of hydrophobic amino acids of the proteins at all, but instead correlates inversely with the surface charge of the molecules as found in biochemical handbooks. This interaction was displayed on weakly hydrophobic matrixes with short alkyl chains but also with some thiophilic media. An explanation is that charged groups will mask the type of hydrophobic interaction taking place on these gels under the tested conditions. Probably this interaction will be influenced significantly by altering pH. Very commonly used HIC media, based on phenyl groups and butyl groups as ligands, were shown to work with a mixed mode, that is, combination of the two pure mechanisms.

This fact adds to the explanation of the difficulty, especially using these HIC media, to find simple rules after which the separation should be optimized. It should also be noted that the performance of a HIC medium will always be the result of the combination of ligand, spacer, and matrix. Phenyl gels from different manufacturers have quite different patterns of interaction if tested by a set of proteins as described earlier. This has to be remembered when, after the optimization in laboratory scale, alternative media are tested before scaling-up of a HIC procedure. This optimization is only valid for the medium used. There is thus a risk that competing media, which might be advantageous, will be tested under non- optimal conditions and rejected under false premises.

FACTORS AFFECTING CHROMATOGRAPHY

Ionic Strength

A high ionic strength is needed by definition to attach biomolecules to the column. Very often, 1.5 to 1.7 M ammonium sulphate is used as initial buffer concentration without further trials to optimize. Another common approach is to equilibrate the column in weak tris-buffer to preserve the pH and increase the ionic strength by the addition of sodium chloride to a concentration of up to 3 M. It has to be noted, though, that ammonium sulphate, at a certain concentration, is between three and four times as effective in promoting the hydrophobic interaction compared to sodium chloride, so if the substances in question need a stronger interaction to be retarded on the column than can be achieved by almost saturated sodium chloride, ammonium sulphate is a natural choice. Using ammonium sulphate extends the range of HIC to be used for the separation of less hydrophobic molecules. In production-scale operations, the use of ammonium sulphate has been questioned because of the effect the nitrogen of used buffers may have on the environment by increasing the overnutrition. The choice between overnutrition and possible toxic effects of substitute buffers such as sodium sulphate may not be easy, and all the factors, from the conveniency of the separation procedure to the cost of purifying the sewage, will have to be considered.

Buffer Ions

Recently, it has been shown that the choice of buffer ions greatly influences the function of an HIC medium. The overall behavior of a buffer-ligand-spacer-matrix system could be switched in a wide range, such as from typical phenyl-like to typical butyl-like or vice versa, by just changing the buffer ions. Optimization work by testing different buffers might then be more beneficial than usually anticipated but, as with pH, the task still includes much work based on trial and error.

pH

It has been shown that hydrophobic interaction between proteins and an octyl gel generally increases with lowered pH and most often decreases with high pH, but the pH dependence in this case is not very profound from pH 6 to pH 8.5. For some proteins, though, the interaction increases with an increase in pH. One reason for the different results might be that sodium sulphate was used above pH 8.5 and ammonium sulphate buffer was used below pH 8.5. According to Oscarsson and Karsnas, the adsorption behavior of serum proteins on octyl gel is quite different in the two environments. The reason for this complexity again has certainly to do with the multi- mechanism nature of HIC, with irregularities implied by different kinds of interactions between buffer ions and protein surfaces (18) and thus, secondarily, interaction with the HIC matrix. In practice, changes in pH to get a desired separation can be very powerful because it may alter the entire adsorption spectrum, but, because of the trial-and-error nature of the task, pH is not the primary parameter to use for an optimization.

Temperature

High temperature promotes the binding, and HIC is preferably run at room temperature if the compounds to be separated can withstand this condition without losing activity. In room temperature, slightly lower buffer concentrations can be used than if the separation is made in a cold room. Elution of hardly bound substances may also be facilitated by performing the elution step at cold room temperature. In practice, however, the increase of temperature at the reequilibration of the column in room temperature inevitably produces air bubbles that will destroy the packing of the column, so if a two-temperature scheme is chosen, repacking of the column will most certainly have to be included in the separation cycle.

Flow, Residence Time

Hydrophobic interaction shows signs of being a slower process compared to other interactive chromatography techniques. The desorption of compounds from the column is often run under nonequilibrium conditions, resulting in long tailing peaks that seriously damage the resolution. It is also generally found that the adsorption of molecules sometimes seems to be a two-step reaction, where the initial binding is weaker and more readily broken under the elution conditions. After some time the bound molecules seem to adapt themselves to the column, possibly involving conformational changes. More harsh conditions may have to be introduced for the elution, and there is a risk that the recovery will decrease. On the other hand, there are reports showing that an elution pattern can be quite reproducible even when the bound substances have been immobilized on the column for a long time, and HIC is generally considered a high-recovery technique.

Elution

Generally, elution of substances bound to an HIC column can be achieved by altering one of the parameters that promote binding. It is possible to elute by increasing pH, if the adsorption was performed at low pH, or by lowering the temperature, although the effect of this may be too small. But most commonly, elution is performed by a descending salt gradient or, in large scale, by stepwise lowering of the salt concentration. If substances remain on the column at very low buffer concentration, an organic additive such as ethylene glycol or a detergent can be used for the elution. The ethylene glycol will alter the water structure and weaken the hydrophobic interaction, and the detergent will act by competing with and displacing the protein on the chromatographic medium.

PURITY AND ACCURACY

For process economy, the regeneration of the HIC column is essential. Fouling of the column and irreversible binding of contaminants to the column are common obstacles in HIC. There are three degrees of regeneration: removal of strongly bound or precipitated proteins, removal of hydrophobic substances like lipids and lipoproteins, and sanitation, which means removal or deactivation of microorganisms. Preferably, the regeneration is performed as a cleaning-in-place (CIP) procedure without costly and cumbersome repacking of the column. All washing steps should be performed with reversed flow to give the contaminants the shortest distance out of the column and avoid secondary interaction with the rest of the column. First, all conditions that will decrease possible hydrophobic interaction can be used for the regeneration.

Detergents that will compete with the attached substances and ethanol and isopropyl alcohol, buffers normally used in RPC, are obvious choices NaOH, 0.5 to 1 M, has been shown to almost universally sanitize the column and has a good cleaning ability as well. It is important to note that washing with narrow zones of the washing solution is often more effective in removing the column contaminants than exposing the medium for a constant concentration of the washing agent for an extended time. It is not unusual when cleaning HIC media that material will be washed out both in the beginning and at the end of a washing zone where the concentration is increasing and decreasing, respectively. Probably, a lot of equilibria are present, including hydrophobic interaction and interactions involving charges, and if one type of binding is minimized another may be enhanced and keep the contaminants on the column. By wobbling the concentration, the differences of reaction velocities are probably utilized, with elution as the result.

METHODS OF OPTIMIZATION

Although all the parameters mentioned above can be used to optimize a hydrophobic interaction chromatography step, for many of them the effect of a change is not easy to predict because it will also be dependent on the individual properties of the proteins subject to separation. Thus, a pH change may alter the entire elution pattern or just influence the binding of a single component. Other parameters that may change the entire interaction picture are the buffer ions and the chosen chromatography medium.

Choice of Medium

Ideally, it is preferable that the desired components attach to the medium under moderate salt concentrations (e.g., up to 1 M ammonium sulphate), and a screening of the most common commercially available media is always advisable. This screening is easily performed in small scale and should include a test of whether elution with a high recovery is possible with water or weak buffer. If a lower salt concentration is sufficient on a certain gel, this is a preferable choice for economical and environmental reasons provided that the resolution obtained is still acceptable.

Buffer Concentration

An HIC run is normally performed by equilibrating the column in a starting buffer, applying the sample, and eluting with a stepwise descending ionic strength or a descending continuous gradient. Optimizing the starting concentration, or the concentration at which the proteins are adsorbed to the column, is important to really utilize the HIC step. In laboratory scale and analytical experiments, where the resolution can be achieved by gradient elution, the ionic strength should be high enough to retard the desired component and to let it elute somewhere in the middle of the descending gradient. In this way, the combination of the capacity for the component of interest and the resolution will be favorable. In large-scale runs, well-defined and reproducible gradients are more difficult to achieve, so the elution is preferably done by stepwise decrease of the ionic strength. In some cases, it will work beautifully to chose an ionic strength that is just enough to retard the substance of interest, but in other cases, the peak containing the desired substance will broaden unacceptably at this condition, and the salt concentration will have to be increased.

Gradient Volume

A gradient volume between 5- and 10-column volumes will give enough resolution in most cases. By increasing the gradient volume, the resolution will increase, but the components will be eluted in a larger volume.

Rapid and Slow Reactions

It is often found that the reaction of binding to the column is rather rapid and selective, whereas the elution of a compound from the column is slower. As a result of this, tailing of zones at elution

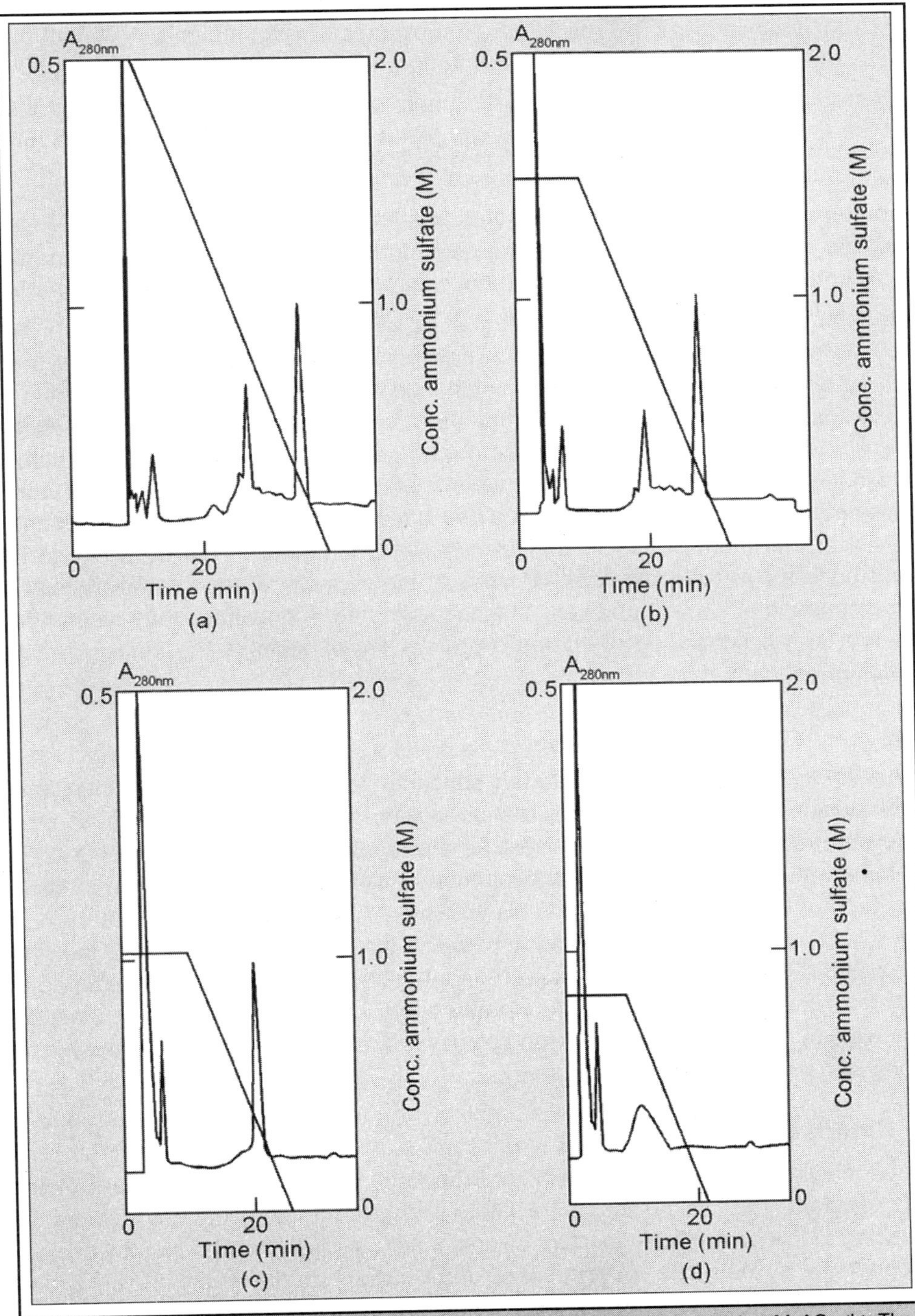

Figure 2.4 The effect of starting concentration. (a) The sample is applied in 2M AS. (b) The sample is applied in 1.5 M AS. (c) In 1 M AS only IgG is binding, and (d) 0.8 M AS the IgG is still retarded but eluted in a broad peak.

is very often experienced and the resolution of substances is jeopardized. A way of optimizing a method fully, utilizing the selectivity of the adsorption, is to use the following scheme:

1. Equilibrate the column with a starting buffer that has a lower concentration than is needed to retard the desired substance, adjust the sample accordingly, and run it through the column.
2. Elute everything that was retarded on the column with water or low salt. Check for activity.
3. Add some salt to the buffer and the sample and repeat steps 1 to 3 the desired substance is eluted from the column. To speed up the procedure, a prerun with a gradient starting at high salt concentration will hint at a proper initial concentration when step 1 is run the first time.

An optimized separation scheme includes as its first step running the column in a salt concentration just below the one where the desired substance was found to be retarded. All impurities that bind harder than the desired substance attach to the column and should be eluted. The next step is to bind the desired substance using a slightly higher salt concentration. If this is done carefully, very few contaminants will bind, and subsequently the desired substance can be eluted in high yield by a steep descending gradient, if any contaminating substances were retarded as well, or by a step of very diluted buffer. The described method is advantageous not only because the selectivity of the interaction is fully utilized but also because of the simplicity to recondition sample and buffers by just adding salt. Moreover, the capacity of the column is utilized to bind mostly the compound of interest and very little contaminants. A drawback may be that the capacity of the column for the substance of interest might be lower because the binding is taking place near the elution concentration.

Sampling

There is always a risk that the high salt concentration needed to attach the sample to the column will cause precipitation in tubings, pumps, columns, and detectors. This may be prevented by the following scheme: The column should be equilibrated with the proper buffer concentration but the sample with slightly lower salt concentration. This is easily accomplished by dilution of the sample. Instead of injecting the whole sample volume at one time, the injection should be done in aliquots divided by zones of the equilibration buffer. If the volume of the aliquots is kept less than 10% of the column volume and the volume of the intermediate zones with full buffer concentration exceeds 10% of the column volume, the sample condition will adapt to that of the equilibrated column. In this way, the precipitation in wrong places is avoided and the interaction with the column is still taking place in the correct environment.

Constant Parameters

To get a similar result in large scale as in laboratory scale, it is important to keep some parameters constant. The column length, the linear flow, and the ratio between gradient and column volumes belong to this category. A constant column length leads to the fact that the desired capacity will be proportional to the cross-sectional area of the column and actually decide it, because the same proportion of the column length should be occupied by the sample in both scales. Because of the relationship between the volumetric flow, the linear flow, and the cross-sectional area of the column on one side and the gradient-to-column-volume ratio on the other side, the gradient volume will be decided when the diameter of the column is set. In principle, a gradient elution in process scale will then take the same time as in laboratory scale. In practice, a decrease in resolution caused by a more ineffective solvent distribution system at the inlet of a process scale column will sometimes have to be compensated for by a lower flow.

CONCLUDING REMARKS

HIC has been used for the purification of serum proteins, nuclear proteins, membrane-bound proteins, and viruses and cells, and recombinant proteins and insulin receptors have been successfully prepared. Interesting applications include the use of HIC media as biochemical reactors by coupling enzymes by hydrophobic interaction and a method for the exchange of detergents bound to membrane proteins.

3

ION EXCHANGE CHROMATOGRAPHY

Preparative ion exchange chromatography is a technique for separating biomolecules based primarily on their electrostatic interactions with charged stationary phase materials. This technique is the most widely employed chromatographic separation technique in the biotechnology industry today. It is used for both capture and high- resolution separations. The last decade has seen dramatic advances in the state of the art of chromatographic bioprocessing with ion exchange materials. This article presents recent advances in resin materials, modes of operation, and modeling. Ion exchange chromatography (IEC) is the most widely employed chromatographic separation technique in the biotechnology industry today. It is based on the principle of electrostatic interactions between charged surfaces. In IEC, one employs a charged stationary phase material to separate oppositely charged solutes. Cation exchange employs a negatively charged surface to separate positively charged solutes, and anion exchange employs a positively charged surface to separate negatively charged solutes.

The binding of solutes to ion exchangers maybe viewed as an exchange process in which the solutes compete for binding with salt counterions and other charged components in the mobile phase. This article focus on preparative protein separations in ion exchange systems, beginning with a brief discussion of various factors that affect protein retention (e.g., pH, ionic strength, mobile phase modifiers). A detailed treatment of the state of the art of stationary phase materials is then presented. A description of different modes of preparative IEC is given, along with separation examples from the literature. Finally, the modeling of preparative ion exchange is addressed.

FACTORS AFFECTING THE ION EXCHANGE

Surface Chargers

Proteins are ampholytes, possessing both negative and positive surface charges. In fact, the zwitterionic nature of proteins is what makes ion exchange such a powerful tool for these separations.

The charges on the proteins arise because of the presence of acidic and basic amino acid residues on the protein surface as well as the amino and carboxyl termini of the protein. Thus, the net charge on a protein is a strong function of the pH of the solution. Under acidic conditions, the basic side chains are charged, under basic conditions, the acidic residues will be charged. At the isoelectric point (pI), the net charge on the protein will be zero. In general, when the pH of the mobile phase is greater than the pI of the protein, the protein will have a net negative charge and will be retained on an anion exchanger. On the other hand, when the pH of the mobile phase is less than the pI of the protein, the protein will have a net positive charge and will be retained on a cation exchanger. It turns out that the net charge of the protein is not the sole determinant of its retention.

The distribution of charges on the surface of the protein is also important. The protein may have patches of positively or negatively charged residues that may interact with the stationary phase even when its net charge is zero. Furthermore, the microenvironment pH often can differ slightly in the resin and bulk solutions. Thus, a protein may be retained even at its pI. In addition, it is important to consider protein solubility constraints in the development of an ion exchange separation, since proteins are sparingly soluble near their isoelectric point. In practice, to prevent protein denaturation, the operating pH in IEC is limited to the range 2–9. The choice of buffer to establish the operating pH is of paramount interest in ion-exchange chromatography. It is desirable to employ a buffering agent that has the same charge as the stationary phase material or is neutral. This precludes the binding of these ions onto the stationary phase and thus prevents pH shifts within the column. Hence, it is desirable to use cationic buffers (e.g., alkyl- amines, Tris, aminoethyl alcohol) in anion exchange chromatography and anionic buffers (e.g., phosphate, acetate, citrate, barbiturate) in cation exchange chromatography.

Strength of Ions

The chromatographic retention of proteins is modulated by the ionic strength of the mobile phase. In cation exchange systems, the relative strength of cations is as follows (2): $Ba^{2+} > Pb^{2+} > Sr^{2+} > Ca^{2+} > Ni^{2+} > Cd^{2+} > Cu^{2+} > Co^{2+} > Zn^{2+} > Mg^{2+} > Ti^{2+} > Ag^{+} > Cs^{+} > Rb^{+} > K^{+} > NH_4^{+} > Na^{+} > H^{+} > Li^{+}$. In anion exchange, the relative strength of anions is citrate > sulfate > oxalate > I^{-} > nitrate > $CrO_4^{2-} > Br^{-} >$ $SCN^{-} > Cl^{-} >$ formate > acetate > OH^{-}. Thus, Br^{-} will bind stronger than Cl^{-} in an anion exchange column. In cation exchange, an equivalent solution of KCl would elute a protein faster than NaCl. Continuous and step gradients of ionic strength are often employed to effect protein separations. While protein retention is generally reduced as the ionic strength is increased, at very high ionic strength (>1 M), hydrophobic interactions can become pronounced.

Role of Modifiers

Addition of modifiers to the mobile phase can often influence the selectivity of a given separation. Recently, Gagnon *et al.* reported an enhancement of the selectivity of proteins on the addition of poly(ethylene glycol) to the mobile phase. Other additives include chelating agents such as EDTA (to protect enzymes against metal impurities), reducing agents (e.g., dithiothreitol) to retard oxidation, and surfactants and chaotropic agents (e.g., urea) to solubilize membrane proteins.

Temperature

There are conflicting reports in the literature on the effect of temperature on resolution and band broadening. Increasing the temperature may reduce the viscosity of the mobile phase and increase the kinetics of the adsorption process. This may result in more efficient separations. However, changes in the temperature may also induce conformational changes in the protein. This may denature the protein or expose more hydrophobic patches to the stationary phase, enhancing nonspecific interactions and causing dramatic increases in protein retention. In general, the practice in industry is to employ either room temperature or low temperatures (4°C) to minimize protease activity and to ensure protein stability.

PARAMOUNT IMPORTANCE

The stationary phase is of paramount importance in IEC. Ideally, the stationary phase should be rigid enough to withstand high flow rates, provide sufficient selectivity and capacity for the problem at hand, and not denature the protein of interest. Furthermore, it should be stable over a wide range of operating conditions (pH, solvents, etc.) and amenable to rigorous cleaning-in-place with acid or alkali. The choice of the manufacturer is also important. The manufacturer should be able to ensure a regular supply with little lot-to-lot variation. Finally, the resin should not leach materials into the process stream. Selectivity in ion exchange materials is affected by the ligand density, spacer arm chemistry, and the backbone resin material. Unfortunately, state-of-the-art chromatographic modeling cannot a priori predict the selectivity of different resin materials for various classes of separation problem. Thus, selection of an ion exchange resin is problem specific and requires the screening of a large variety of materials for a given separation. In bioprocessing applications, IEC is used for both capture and high-resolution separations. Capture applications demand that the resin possess sufficient dynamic capacity and be rigid enough to allow operations at high flow rates. On the other hand, selectivity is paramount for high-resolution separations.

Size of Particles

While reversed-phase chromatography often employs materials of relatively small particle diameter for preparative applications, stationary-phase materials of 30- to 200-µm diameter are typically employed for preparative IEC. Higher resolutions can be achieved with smaller particles at the cost of increased back-pressure. Finally, the size of the particle can also influence the mode of packing.

Functional Groups

The stationary phase of an ion exchanger consists of a base matrix that is functionalized with basic (anion exchange) or acidic (cation exchange) groups. The charge on the stationary phase is pH dependent. This pH dependency allows one to categorize ion exchange stationary phases into strong and weak exchangers. Strong ion exchangers retain their charge over a very wide range of pH, whereas weak ones do so over a narrower range of pH. Thus, the capacity of weak ion exchangers is a function of the pH. Typically, strong exchangers have functional groups with very low pK_a (sulfopropyl in cation exchange) or very high pK_a (quaternary amino–type anion exchangers).

Table 3.1 Functional groups used in IEC

Name	*Abbreviation*	PK_a
	Anion exchangers	
Diethyl aminoethyl	DEAE	9–9.5
Quaternary aminoethyl	Q	9.5
Dimethyl aminoethyl	DMAE	9
Trimethyl aminoethyl	TMAE	>13
	Cation exchangers	
Carboxymethyl	CM	3.5–4
Phosphate	Phospho	<2 & 6
Sulfonate	S	2
Sulfoethyl	SE	2
Sulfopropyl	SP, SO_3^-	<1

Capacity and Porosity

The *capacity* of the ion exchanger can be measured as both the total number of charges per unit stationary-phase volume or the mass of protein adsorbed per unit volume. When one is determining protein-binding capacity, it is important to ensure that the activity of the adsorbed protein is not compromised. The capacity expressed in terms of adsorbed protein is the effective capacity of the column. It is dependent on both the specific protein and the operating conditions (pH, ionic strength). It is also dependent on the pore size and the *porosity* of the particles. If the particle is pellicular (i.e., not porous), the surface area is relatively low, resulting in reduced capacity. On the other hand, if the particle is highly porous, the capacity increases, but the resistance to mass transfer also increases. Pores that are too small may be inaccessible to the proteins. The pore size distribution is usually unavailable, and a nominal pore size is supplied by the manufacturer. Conventional ion exchange media have nominal pore sizes between 100 and 1,500 . At times, the pore size is also specified as the molecular weight of the largest globular protein that can be accommodated in the material.

Nature of Resins

The resins can be roughly divided into five categories. Polysaccharide gels have been in existence since the 1950s. These matrices are inert toward proteins, can be easily derivatized, have very high capacities, and are stable over a wide range of pH values. Polysaccharide gels based on weakly cross-linked agarose, dextran, or cellulose have weak mechanical properties and are limited to low-pressure regimes. In contrast, Sepharose Fast Flow ion exchangers (Pharmacia) are composed of a higher degree of cross-linked agarose, which increases their mechanical stability. High-performance polysaccharide materials (Pharmacia) are also available in relatively small particle diameter (34 μm) and have been shown to be particularly useful for difficult high resolution separations. These polysaccharide-based resins are currently the most widely employed chromatographic materials in the biopharmaceutical industry. Although silica-based ion exchangers have good mechanical properties, they are not as widely used for bioprocessing because their instability under alkaline conditions precludes alkaline cleaning.

Furthermore, surface silanol and siloxane groups can have adverse effects on proteins. To overcome this problem, silica supports have been coated with a polymeric layer. Resins based on porous particles of zirconium oxide have been synthesized and characterized for biochromatography. Zirconia possesses all the mechanical advantages offered by silica. In addition, it has good thermal and chemical stability. Unfortunately, this backbone material tends to bind proteins, which necessitates coating with a biocompatible layer. Clausen and Carr recently employed a phosphonate analog of EDTA for the modification of zirconia supports to provide a biocompatible stationary phase. As described shortly, a potential advantage of zirconium is its relatively high density, which has been shown to have potential for expanded-bed applications. The polymer-based matrices have good mechanical properties and are stable over a wide pH range. Furthermore, they have a relatively high loading capacity because of their macroporous structure.

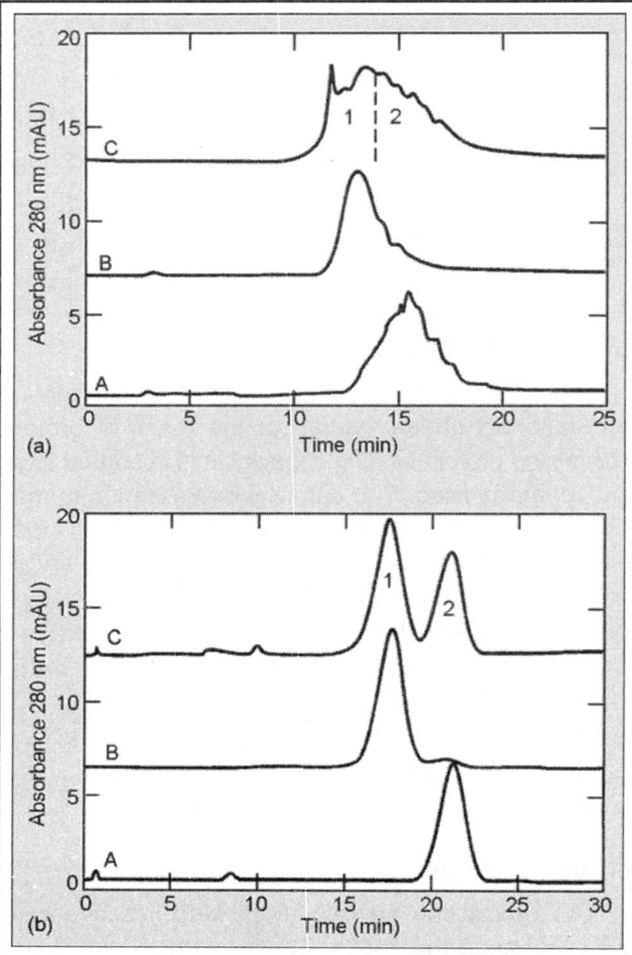

Figure 3.1 (a) Strong cation exchange chromatography on a solfopropyl column of three samples: A, rhDNAse; B, its deamidated variant; and C, a mixture of the two variants. (b) Cation exchange chromatography of three samples: A, rhDNAse; B, its deamidated variant; and C, a mixture of two variants on a tentacle ion exchanger.

The hydrophobic polymer matrices are typically coated with a hydrophilic polymer to reduce nonspecific interactions. Polymeric supports are being used increasingly in the preparation of ion exchangers for bioprocessing. In addition, many of the novel materials described below are based on polymeric supports. Each of the stationary phases in this category possesses unique features that distinguish it from more traditional materials. The POROS supports are synthesized by a suspension polymerization process. These chromatographic materials consist of a rigid polystyrene–divinylbenzene base matrix covered with a polyhydroxyl layer that can be further derivatized to produce various functional groups. This type of resin belongs to the class of particles termed "perfusive". Such particles have a bimodal pore distribution containing macropores (6,000–8,000) that transect the particle and interconnecting "diffusive" micropores (500–1,500). The macropores facilitate mass transport, while the diffusive pores increase the available surface area. Their permeability has been shown to be independent of the particle size. The solute breakthrough curves for the POROS media have been shown to be relatively unaffected by the flow rate in contrast to conventional ion exchangers.

Gerstner et al. have demonstrated that displacement chromatography of proteins can be carried out on these particles at very high linear velocities. The SOURCE materials have also been shown to exhibit chromatographic behavior that is independent of flow rate. In addition, the SOURCE resins have relatively high efficiency owing to their monodisperse particle size distribution. Recently, the development of agarose-based perfusive materials has also been reported. Hyper-D materials are composite stationary phases composed of a rigid macroporous support filled with a weakly cross-linked ion exchange gel. The rigid particle provides the necessary mechanical strength for high flow rate operations, while the soft gel provides high binding capacity. The equilibrium uptake of model proteins and the mass transfer rates in Q-Hyper-D has been studied by Fernandez and Carta. These materials have a very high binding capacity even at elevated flow rates. In fact, it has recently been shown that large-particle-diameter Hyper-D materials (200 μm) also exhibit high binding capacities even at elevated linear velocities.

Table 3.2 Representative commercial ion exchange media

Trade name	*Manufacturer*	*Function groups*	*Particle sizes (μm)*	*Matrix*
		Polysaccharide gels		
Bio-Gel A	Bio-Rad	DEAE, CM	80–150	Cross-linked agarose
Sepharose CL-6B	Pharmacia LKB	DEAE, CM	45–165	Agarose
Sephacel	Pharmacia LKB	DEAE	40–160	Cross-linked cellulose
Express-Ion	Whatman	DEAE, Q, CM, SE	60–130	Microgranular cellulose
Sephadex	Pharmacia LKB	DEAE, Q CM, SP	40–125	Cross-linked dextran
Sepharose Fast Flow	Pharmacia LKB	Q, DEAE, CM, SP	45–165	Cross-linked agarose
Sepharose High Performance	Pharmacia LKB	Q, SP	24–44	Cross-linked agarose
Ultrogel	BioSepra	Hydroxy-apatite	60–180	Cross-linked agarose
		Inorganic matrix		
Accell	Waters	CM, QMA	40	Polymer-coated silica
TSKgel 3SW	TosoHaas	CM, DEAE	5, 10	Silica
Spherodex	BioSepra	DEAE, SP, CM	40–300	Silica–dextran
Spherosil	BioSepra	DEAE, QMA	40–300	Polymer-grafted silica
Zephyr	Sepracore	DEAE, S	20	Silica–dextran
		Syntheticpolymers		
TSKgel 5PW	TosoHaas	SP, DEAE	30	Methacrylate copolymer

Trade name	Manufacturer	Function groups	Particle sizes (μm)	Matrix
Mono Q/S	Pharmacia LKB	Q, S	10	Cross-linked polystyrene–divinylbenzene
Macro-Prep	Bio-Rad	S, Q, CM, DEAE	50	Methacrylate copolymer
Protein-Pak	Waters	DEAE, Q, SP, CM	8,15,40	Rigid polymeric gel
Trisacryl	BioSepra	Q, DEAE, CM, SP	40–160	Macroporous synthetic polymer
Trisacryl Plus	BioSepra	DEAE, SP	40–160	Base-resistant polymer
Toyopearl	TosoHaas	DEAE, CM, SP	35,65,100	Methacrylate
Super Q	TosoHaas	Q	35,65,100	Synthetic polymer
		Novel media		
POROS	Perseptive	Q, S	10, 20	Coated polystyerene
SOURCE	Pharmacia LKB	Q, S	15, 30	Cross-linked polystyrene–divinylbenzene
Hyper D	BioSepra	Q, S	10,20,35,60	Polystyrene shell and hydrogel
Fractogel EMD (tentacle resins)	Merck	TMAE, DEAE, DMAE, SP, CM	20–40	Vinylic grafted polymer
STREAMLINE	Pharmacia	SP, DEAE	200	Macroporous cross-linked agarose with a quartz core
		Membrane ion exchangers		
MemSep	Millipore	DEAE, Q, CM		Cellulose
QuickDisk	Knauer	DEAE, Q		Poly(glycidylmethacrylate)
Sartobind	Sartorius	Q, S		Synthetic copolymer

It has been proposed that the mass transport mechanism in these materials is surface diffusion due to the high concentrations obtained in the gel. The "tentacular" stationary phases comprise another novel sorbent. These chromatographic materials have polyelectrolyte chains (typically 5–50 monomers long) grafted onto porous matrices. In contradistinction to the supports mentioned thus far, the charge moieties in tentacular phases are attached to the matrix via relatively long, flexible spacer arms. This results in reduced nonspecific interactions between the proteins and the base matrix. Furthermore, since the ligands are able to move relatively freely in space, they can explore the steric distribution of charges on a protein in a more efficient manner. This can potentially result in higher-resolution separations. Tentacular ion exchangers have been employed to differentiate between native and deamidated recombinant human DNAse which differ in a single surface charge.

Effective Capacity

The dynamic capacity of the column is the effective capacity determined from the break-through curve (i.e., when the effluent concentration reaches 10–50% of the inlet protein concentration). Table 3.3 lists the dynamic capacities of several anion exchange media obtained at low linear velocities; these capacities are seen to range from 30 to 165 mg/ mL. The cellulosic anion exchangers such as Sepharose Fast Flow have high dynamic capacities under these conditions.

Importantly, dynamic capacities can be strongly affected by the flow rate. In Figure 3.2, which compares the flow rate dependence of the dynamic capacities of five Q exchangers, the dynamic capacities of the tentacular (Fractogel EMD) and perfusive (POROS) resins are relatively low at low flow rates, but their dependence on the flow rate is rather weak. On the other hand, the dynamic capacity of the Sepharose FF resin exhibits a stronger dependence on the flow rate. The Hyper D and the Super materials exhibit a moderate flow rate dependence. However, the Hyper D material has a capacity substantially higher than those of the other materials at elevated flow rates.

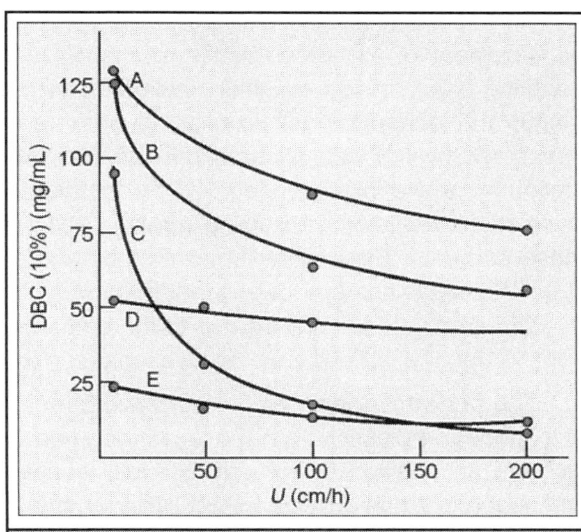

Figure 3.2 Variation of dynamic binding capacity (DBC) as a function of the flow (U) for Q anion exchangers.

An important caveat to this discussion on dynamic capacity is that it is often desirable to use low operating pressures in a bioprocessing environment. This practice may affect the linear velocities that can be employed under actual operating conditions and may make performance at elevated velocities less important.

Table 3.3 Ionic capacities of commonly used anion exchangers

Ion exchanger	*BSA sorption capacity (mg/mL)*
QMA Accell	111
DEAE Bio Gel A	38
DEAE Fractogel 650	32
DEAE Sephacel	131
DEAE Sepharose CL 6B	164
DEAE Sepharose Fast Flow	88
DEAE Spherodex	98
DEAE Trisacryl	86
DEAE Fractogel EMD	100
Q Fractogel EMD	55
QPOROS II	40
Q Sepharose Fast Flow	115
Q Super	127
QHyperD	125
Express-Ion Q	55

An important recent advance in the state of the art of chromatographic bioprocessing is the development of expanded-bed systems, in which the bed expands during the loading phase, enabling it to directly process unclarified feedstock. This can potentially obviate the need for membrane or centrifugal processing prior to the ion exchange capture step. The Streamline adsorbent, which has a mean particle size of approximately 200 μm, is a specially designed stationary phase for expanded-bed chromatography. It is stable over a wide pH range but unstable in solutions containing oxidizing agents. Accordingly, in situ sterilization may be a problem with these sorbents. Streamline forms a stable expanded bed without the need for external magnetic fields. Its capacities are lower than those of typical commercial adsorbents such as Sepharose FF and Spherodex. This may be due to the presence of the quartz core, which provides the necessary particle density but reduces the available surface area.

The performance of this sorbent has been shown to be similar in packed-bed and expanded-bed modes, indicating that the expanded-bed mode behaves like a packed bed with increased voidage. In addition to the commercially available Streamline material, several other expanded-bed supports are in development. Flickinger and Carr are exploring the potential of zirconia stationary phases for expanded-bed applications. These materials have the desirable property of having sufficient density without further modification. Gilchrist et al. synthesized a polysaccharide–ceramic composite sorbent for use in expanded-bed chromatography. Thommes et al. derivatized Bioran (controlled pore glass) with sulfonic acid residues. These particles compared well with Streamline media in initial evaluations.

Charged Membrane Systems

In contrast to all the chromatographic supports discussed thus far, an alternative approach to ion exchange is the use of charged membrane systems. Mass transfer in membrane exchangers is typically dominated by convection, which enables these systems to be operated at relatively fast linear velocities resulting in high throughputs. The typical configuration of membrane ion exchangers is the "pancake" configuration, which has a large width-to-length aspect ratio that provides higher surface areas and lower backpressures. The principal disadvantage of these configurations is the difficulty of achieving a uniform inlet flow distribution and collection of the eluate with minimal back mixing. Monoclonal antibodies and human recombinant antithrombin III were subjected to preparative purification on Sepharose FF columns and Sartobind units. The membrane exchanger yielded higher throughputs (>13-fold in the case of monoclonal antibodies) and comparable or higher recoveries. However, higher concentration factors were achieved on the Sepharose columns.

Cleaning and Sanitization

A critical issue in ion exchange bioprocessing is the ability to clean the column. This includes column regeneration (between each separation cycle), cleaning-in-place operations (after every 1–10 cycles), and sanitization (between batches). Any commercial ion exchange material must be stable enough to withstand high pH as well as oxidizing conditions. Column regeneration and cleaning-in-place operations are important for establishing long-term use of ion exchange columns. The conditions are usually determined empirically and are strongly affected by the feed composition. Typical conditions include high salt (1 M NaCl) and/or 1 M NaOH. The monograph by Sofer and

Nystrom offers a more detailed discussion of column maintenance procedures. In situ sanitization of the column is necessary to prevent microbial contamination between batches. Formalin, peracetic acid, and sodium hypochlorite are effective chemical disinfectants. Alkaline solutions of ethanol are effective for spore destruction. Clearly, the stability of the resin must be kept in mind when the sanitization conditions are chosen.

Polishing Steps

In process chromatography, it is important to distinguish between initial capture (recovery) and high-resolution or polishing steps. For product recovery, the capacity and production rate are paramount. In addition, it is often necessary to process large volumes of relatively dilute feedstock, which often contain particulates. It is precisely this challenge that has catalyzed the development of many of the "novel" materials described above. The advantage of perfusive-type materials (e.g., POROS) and membrane chromatography is that they can be operated at high linear velocities. However, these materials often have relatively low capacities. While the problem can be overcome by carrying out many column cycles for a given batch, this approach has not been widely adopted by the industry. Composite stationary phases (e.g., Hyper-D) possess high binding capacities and can be operated at reasonably high flow rates without a drastic loss of dynamic capacity or increased back-pressure. This makes their use for capture operations potentially very attractive. Expanded-bed sorbents are particularly attractive for capture operations because they combine clarification, concentration, and capture into a single step. It turns out that the stability of expanded beds is quite sensitive to such factors as the solids content, pH, ionic strength, and viscosity of the feed. Since the effect of these factors has not been fully elucidated, the choice of the mobile phase conditions in expanded-bed applications will be highly problem specific.

Nevertheless, expanded-bed chromatography has generated tremendous interest from the industry. As mentioned in the beginning of this section, the effect of stationary phase chemistry on selectivity is still unpredictable, and the selection of an ion exchange resin is problem specific. Recently, Levison et al. evaluated the functional and physical performance of a range of ion exchange media based on polysaccharide and polymeric or composite materials. Although the data suggest significant differences in the performance of various media for a common set of criteria, the selection of an ion exchanger remains problem specific. Thus, until the state of the art of chromatographic modeling is further improved, the biochromatographic engineer is left with the daunting task of screening a large variety of materials for a given separation. Fortunately, there are several automated chromatographic workstations that can readily screen a large number of resins in a relatively short period of time. Once the resin with the optimal selectivity has been identified, the task remains to develop an efficient, high-throughput chromatographic process. This requires the selection of the appropriate mode of operation.

OPERATING MODELS

Preparative ion exchange can be carried out by means of a variety of resin materials, column configurations, and modes of operation. While the selection and development of resin materials often receives significant attention, the choice of the proper mode of operation is often given less attention. In fact, the selection of the optimal mode of operation for a given bioseparation problem can have a profound effect on the production rate, yield, and purity for that separation. To date, linear gradient or step gradient modes have been used to carry out most industrial chromatographic bioprocessing. This section briefly describes these gradient modes and presents examples from the recent literature. In addition, the displacement and chromatofocusing modes of operation are discussed.

Recovery Operations

Step gradient is often employed for initial capture or recovery operations where the primary objective is to recover and to concentrate the bioproduct of interest. This operation is usually carried out in short, large-diameter columns packed with relatively large-particle diameter stationary-phase materials (>90 μm). This mode of chromatography consists of first loading the column with a large volume of low ionic strength feed solution, followed by a wash step, and finally an elution step based on an elevated salt and/ or modified pH eluent. This type of chromatography can be characterized as essentially an on/off operation. The most important stationary-phase characteristic in preparative step gradient chromatography is the dynamic capacity of the resin. Thus, these separations do not require very efficient columns and can be readily carried out in short "pancake"-type columns. These capture steps can be carried out by means of conventional packed-bed columns or with "expanded-bed"-type configurations.

As described earlier, it is important to screen a wide variety of ion exchange resins to identify a material that has sufficient selectivity and capacity for a given separation. Several examples of step gradient ion exchange chromatography are now presented. The use of an expanded bed in the pilot-scale recovery of the anticoagulant recombinant Annexin V from unclarified *E. coli* homogenate has been reported. The optimal binding and elution conditions for Annexin V were determined in a 1.6 cm x 20 cm packed bed of Streamline DEAE adsorbent using clarified homogenate. Then 1.7 L of unclarified homogenate was applied to the laboratory-scale Streamline 50 column (5 cm x 100 cm, Streamline DEAE adsorbent) under the mobile phase conditions established with the small packed bed. The protein was eluted using a 250-mM step gradient of NaCl. The elution was carried out in the sedimented mode at a mobile phase velocity of 100 cm/h. The results were very similar to those obtained with the small packed bed. The presence of cell debris did not seem to affect the process.

The biomass dry weight and the viscosity of the homogenate were 3.6% and 10 mPa/s, respectively. Finally, the process was scaled up to a pilot-scale expanded bed: Streamline 200 (20 cm x 95 cm, Streamline DEAE adsorbent). The adsorbent volume, feed volume, and flow rate were scaled up by a factor of 16 to account for the increased diameter of the column, resulting in a feed load of 26.5 L. The rest of the conditions were kept the same. Figure 3.3 shows the chromatogram from the pilot-scale recovery of the protein. Yields greater than 95% were obtained with both the Streamline 50 and Streamline 200 columns. The expanded bed was cleaned after each run using 0.5 M NaOH with 1 M NaCl followed by distilled water, 30% isopropanol, and 25% acetic acid. This example illustrates the potential of expanded-bed chromatography for ion exchange step gradient chromatography. The use of step gradient in the production and purification of

Pseudomonas aeruginosa exotoxin A (PE) from *E. coli* has been reported. The initial recovery of this protein from the fermentation broth involved cell lysis and centrifugation. Purification was achieved in three steps: weak anion exchange followed by hydrophobic interaction and a strong anion exchange step.

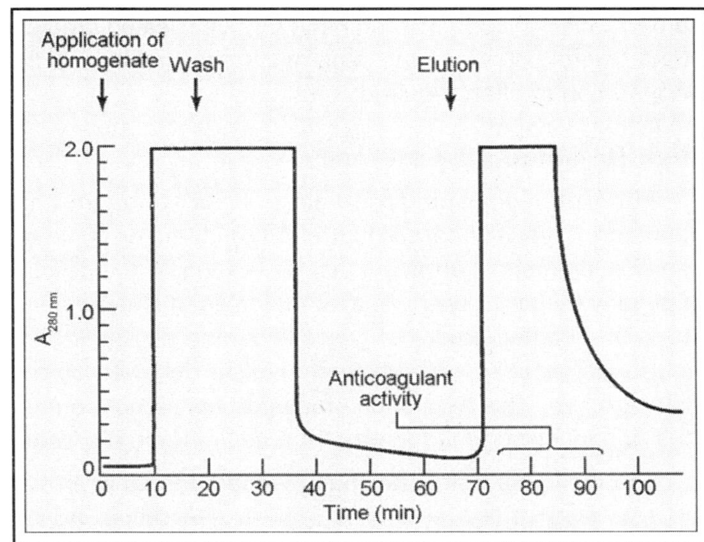

Figure 3.3 Chromatography from pilot-scale, expanded-bed recovery of recombinant Annexin V from unclarified *E. coli* homogenate.

The weak anion exchange step was carried out by means of a DEAE Sepharose Fast Flow column in the flow through mode to remove DNA and pyrogens. In flow through column operations the impurities are retained on the column, but the product is not. Thus, this type of operation is quite simple to operate and can result in a rapid reduction of impurities from the feedstock. This flow through step resulted in a 10-fold reduction in DNA and a protein purity of approximately 25%. In fact, it is quite common to add a flow through anion exchange step in a process for DNA clearance. The final anion exchange operation was carried out on a Q Sepharose high-performance (10 cm x 10 cm) column using multiple step gradients. The protein of interest was eluted with the second step, resulting in the purification of 0.3 g of protein at a purity of greater than 99%. Since the column loading was relatively low, multiple cycles were necessary to produce sufficient amount of material.

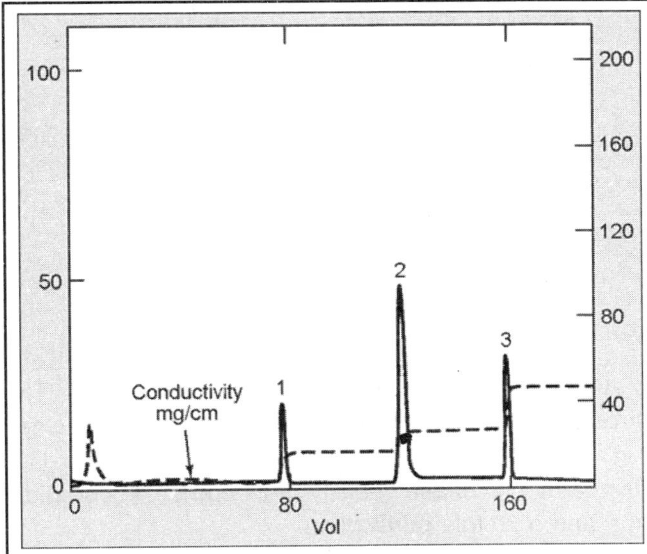

Figure 3.4 Purification of *Pseudomonas aeruginosa* exotoxin A and Q Sepharose HP: buffer, 20 mM Tris, 1 mM EDTA (pH 7.4)

The ion exchange columns were cleaned and regenerated between runs with 1 M NaOH and 3 M NaCl. This example illustrates how step gradient chromatography can sometimes be employed for high-resolution separations for systems exhibiting sufficient separation factors. However, as shown in this example, these operations are often limited by relatively low column loading. Step gradient has many advantages, namely, ease of operation, no need for complicated gradient pumping apparatus, and concentration of the bioproduct. When the separation problem is more difficult, however, a higher-resolution mode of operation is often required.

High Resolution Separations

For high-resolution separations, linear gradient chromatography is often employed. This operation is usually carried out in large-diameter columns using a range of column lengths and particle diameters. It turns out that depending on the difficulty of the separation, it may be necessary to employ longer columns and/or materials of smaller particle diameter for these separations. This mode of chromatography is typically carried out by first loading the column with a relatively large volume of low-ionic-strength feed solution, followed by a continuous linear gradient of increasing ionic strength. This type of chromatography cannot be characterized as an on/off operation and is thus dependent on the length of the column and the column efficiency.

The most important stationary-phase attribute in preparative linear gradient chromatography is the selectivity of the resin for the given separation. Accordingly, it is again essential that a wide variety of resin materials be screened to identify the material with the highest selectivity. In addition to the selectivity, the dynamic capacity of the stationary phase is also important for preparative linear gradient separations and must be considered when selecting a resin. The major advantage of linear gradient is its ability to resolve difficult separation problems. Several examples of linear gradient ion exchange chromatography are now presented. The use of linear gradients in the purification of L-asparaginase from *Vibrio (Wolinella) succinogenes* has been reported. The purification of this enzyme was achieved through a cation exchange step followed by an anion exchange polishing step. Approximately 13 g of protein (184,000 units) was loaded onto a 11.3 cm x 24 cm CM Sepharose cation exchanger at a mobile phase velocity of 24 cm/h. This gradient resulted in the recovery of 91% of the enzyme and a 28-fold purification.

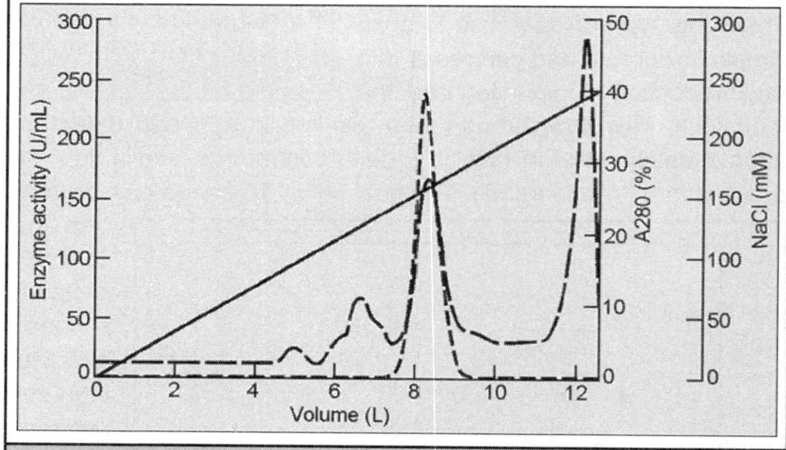

Figure 3.5 Cation exchange of L-asparaginase on a CM-Sepharose column (24 cm x 11.3 cm i.d.): buffer, 5 mM succinic acid-NaOH (pH 6); feed, dilute solution of cell extract with a conductivity of less than 2 mS; flow rate, 24 cm/h; gradient, 0-300 nM NaCl in 16 L of buffer.

An anion exchange polishing step followed. Approximately 0.4 g of protein (167,000 units) was loaded onto a 5 cm x 10.2 cm Q Sepharose FF column at a mobile phase velocity of 15 cm/h. A linear gradient of 0–200 mM NaCl was used to elute the enzyme. This separation scheme resulted in an 87% recovery and a 47-fold purification of the enzyme. These separations required quite low mobile-phase velocities. In addition, the high-resolution step was limited to a column loading of 2g/L, which is typical of high-resolution linear gradient separations. Often, the purification scheme for a protein involves both step and linear gradient operations. The purification of factor VIII (FVIII) and von Willebrand factor (vWF) from human plasma illustrates the use of both step and linear gradients in the purification protocol.

The FVIII–vWF complex is a complicated system in which the two proteins are highly glycosylated, have low pIs, and exhibit strong binding even under conditions of relatively high ionic strength. Furthermore, FVIII is prone to proteolytic degradation without its carrier protein vWF. The naturally adhesive vWF exhibits strong nonspecific interactions with organic, hydrophobic supports. A wide variety of stationary-phase materials were evaluated for this separation. Silica-based supports could not be used in the purification scheme because they tended to activate FVIII. Supports such as cellulose and poly(styrene–divinylbenzene) were unable to effect the separation owing to the strong nonspecific interactions of vWF. Columns packed with traditional media such as Toyopearl DEAE, MonoQ, MonoP, and TSK DEAE 5PW were subject to fouling. On the other hand, tentacular supports gave good results for this separation, possibly because their extended tentacular arms prevent the protein from interacting with the base matrix. The purification scheme for this complex involved three anion exchange steps and two virus inactivation steps. All the anion exchange steps involved the use of tentacular supports.

Weak Fractogel EMD-DEAE supports were used in the step gradient mode for the first two steps, and strong Fractogel TMAE in the linear gradient mode was employed for the final purification. The FVIII–vWF complex was strongly retained on the DEAE support, enabling the use of step gradient chromatography to remove a variety of contaminating proteins, solvents, detergents, and stabilizers (employed during viral inactivation steps). The FVIII–vWF complex was dissociated and separated in the final step by means of a linear gradient. Calcium ions were added to the buffer to cause the dissociation of the complex. Figure 3.6, which illustrates the separation achieved with the strong tentacular exchanger, indicates that this difficult separation was complicated by the heterogeneity of the protein glycosylation and incomplete dissociation. This example illustrates how linear gradient ion exchange with highly selective resins can be employed for very difficult separations. This process resulted in a purification factor of 8,000 with an overall yield of approximately 20%. According to the conventional wisdom, the more difficult the separation, the lower the column loading that can generally be processed by linear gradient. Thus, linear gradient is often limited to lower column loadings. In addition, linear gradient requires more complicated gradient pumping apparatus. In contrast, displacement

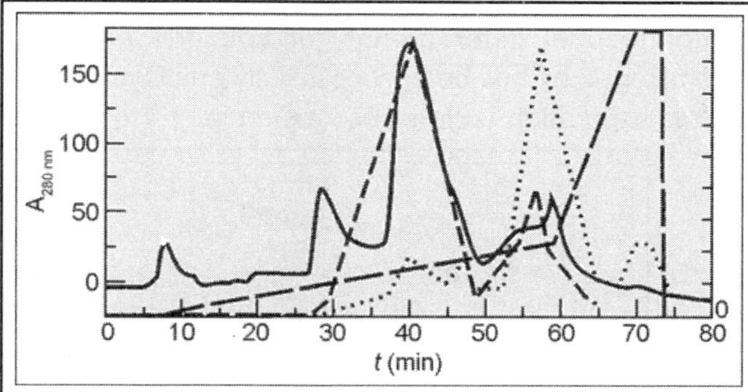

Figure 3.6 Dissociation and separation of FVIII and vWF from their complex in the presence of 50 mM calcium; buffer, 20 mM Tris-HCl (pH 7.4): feed, 300 units of FVIII in buffer; column, Fractoget EMD TMAE (20-40 µm, 120 nm x 16 nm i.d.); flow rate, 3 mL/min; gradient, 0-200 nM NaCl.

chromatography is able to carry out high-resolution separations at very high column loadings, using the simple pumping systems employed for step gradient chromatography.

Displacement Chromatography

The displacement mode of chromatography, first identified in the 1940s by Tiselius, combines the high throughput and concentrating power of step gradient chromatography with the high resolving power of linear gradient chromatography. Although displacement chromatography has been shown to hold significant promise for high-resolution preparative separations, a major impediment to the implementation of displacement chromatography has been the lack of suitable displacer compounds. Recent advances in this field have generated significant interest in displacement technology. Operationally, displacement chromatography is performed in a manner similar to step gradient chromatography, where the column is subjected to sequential step changes in the inlet conditions. The column is initially equilibrated with a carrier buffer of relatively low ionic strength. The feed mixture is then introduced into the column, followed by a constant infusion of the displacer solution. The displacer is selected such that it has a higher affinity for the stationary phase than any of the feed components. Under appropriate conditions, the displacer induces the feed components to develop into adjacent "square-wave" zones of highly concentrated pure material.

After the breakthrough of the displacer from the column effluent, the column is regenerated and reequilibrated with the carrier buffer. The separation of oligonucleotides from their failure sequences presents a significant challenge to chromatographic engineers. The failure sequences are typically very closely related to the main product. Furthermore, there is considerable conformational heterogeneity of both the product and the failure sequences. Displacement chromatography has been successfully employed for resolving oligonucleotide mixtures characterized by low separation factors. For example, 1.2 g of a 24-mer phosphorothioate oligonucleotide was

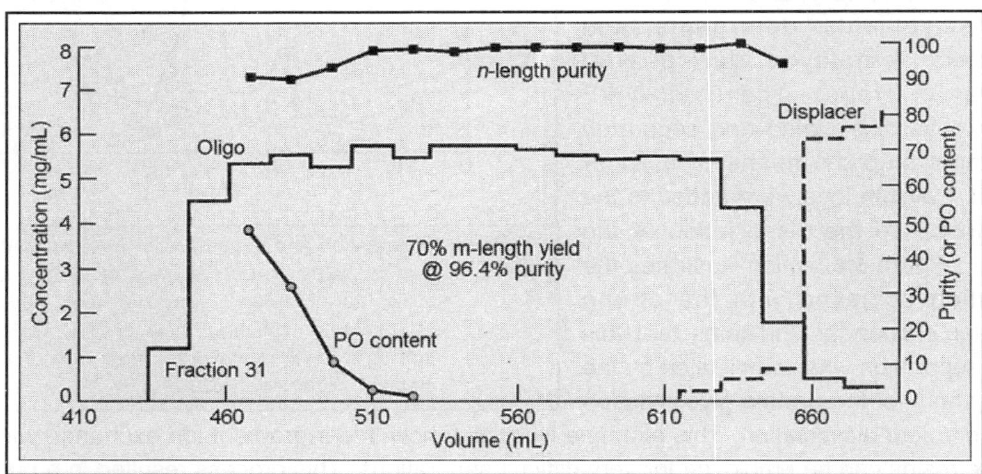

Figure 3.7 Preparative-scale purification of a 24-mer phosphorothioate oligonucleotide: feed, 1.2 g in 100 mM NaCl, 10 mM NaOH, 5% methanol; displacer, 800 mL of 7 mg/mL dextran sulfate; column, POROS HQ/M (20 μm, 100 mm x 50 mm i.d.); flow rate, 50 mL/min; fraction size, 14 mL.

purified on a POROS HQ/M column using dextran sulfate as the displacer. The separation yielded 70% of the product at a purity of about 96%. Methanol was added to the mobile phase to eliminate hydrophobic interactions between the oligonucleotide and the stationary phase and to improve the recovery. The column was regenerated with 3 M NaCl in 0.5 M NaOH. On a semipreparative scale, the entire separation (equilibration + loading + displacement + regeneration) could be accomplished in approximately 32 mins. An important recent advance in the state of the art of displacement chromatography has been the discovery that low molecular weight displacers can be successfully employed for protein purification in ion exchange systems.

Pentaerythritol-based dendritic polyelectrolytes ranging in molecular weight from 480 to 5,100 Da have been shown to be effective displacers for protein purification. In addition, protected amino acids, antibiotics, and a variety of anionic displacers have been developed. Low molecular weight displacers have been successfully employed for very difficult protein separations. These novel displacers have significant operational advantages over large polyelectrolyte displacers. First and foremost, if there is any overlap between the displacer and the protein of interest, the low molecular weight materials can be readily separated from the purified protein during post-displacement downstream processing involving size-based purification methods (e.g., size exclusion chromatography, ultrafiltration). The relatively low cost of synthesizing low molecular weight displacers can be expected to significantly improve the economics of displacement chromatography. Furthermore, the salt-dependent adsorption behavior of these low molecular weight displacers greatly facilitates column regeneration.

In addition, these displacers can be employed in "selective" displacement chromatography, where the product(s) of interest can be selectively displaced while the low-affinity impurities can be desorbed in the induced salt gradient ahead of the displacement train and the high- affinity impurities either retained or desorbed in the displacer zone. Figure 3.8 shows an example of a

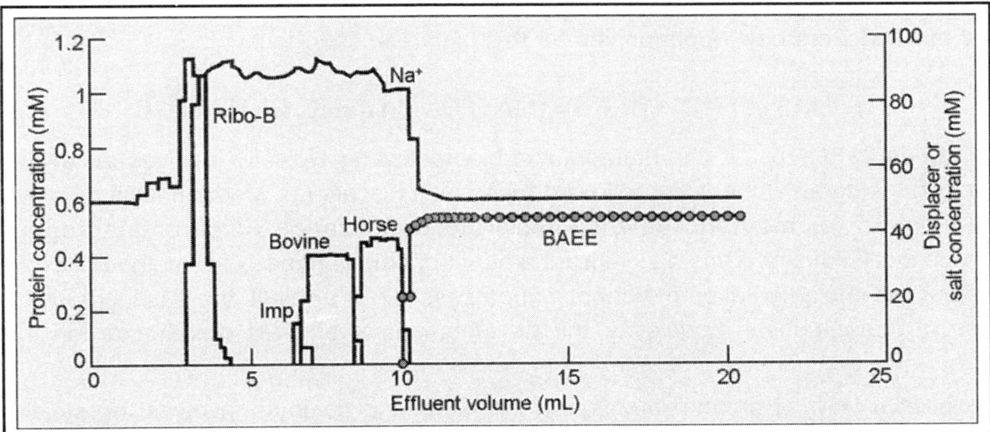

Figure 3.8 Displacement separation of a four-component protein mixture using BAEE as a displacer-column, 100 mm x 5 mm i.d. strong cation exchanger (8 μm); buffer, 50 mM sodium phosphate (pH 6.0); feed, 1.6 mL of 0.38 mM ribonuclease B, 0.45 mM horse heart cytochrome c, 0.37 mM bovine heart cytochrome c, and 0.34 mM lysozyme in buffer; displacer concentration, 45 mM; flow rate, 0.2 mL/min; fraction size, 200 μL.

selective displacement separation: a displacement experiment was carried out using a feed mixture of ribonuclease B, bovine heart cytochrome *c*, horse heart cytochrome *c*, and lysozyme with *N*-α-benzoylarginine ethyl ester (BAEE) as the displacer. As seen in the figure, the low-affinity protein ribonuclease B eluted ahead of the displacement train in the induced salt gradient, whereas the two cytochrome *c*'s were well resolved in the displacement train. Thus, in spite of the extreme closeness in their adsorption behavior, displacement chromatography resulted in excellent separation of the two cytochrome-*c* proteins with both high yields and high purity.

The highest-affinity protein, lysozyme, remained on the column under the conditions of this experiment. This experiment demonstrates that selective displacement chromatography can result in base-line separation of the product of interest from the low- and the high-affinity impurities, while resolving the target bioproduct from very similar affinity impurities in an isotachic displacement train. Recent results have indicated that displacement chromatography may have significant utility for extremely difficult bioseparations such as protein variants and glycoforms.

Separation of Protein Mixtures

In chromatofocusing retained, internal pH gradients are employed to effect the separation of protein mixtures. These pH gradients are produced internally by running a buffer of a particular pH through an ion exchanger equilibrated at a different pH. The focusing effects produced by these pH gradients were originally studied by Sluyterm and coworkers. Typically, polyampholyte buffers are employed in this mode of chromatography. These buffers are not only expensive but also contain components that can associate with proteins. However, Hearn and Lyttle and Frey et al. demonstrated the efficacy of employing multicomponent mixtures of common buffer systems to produce internal pH gradients in ion exchange columns. Recently, Strong and Frey demonstrated the chromatofocusing of model proteins under trace- and volume-overloaded feed conditions using mixtures of common buffers. The ability to carry out chromatofocusing with inexpensive, commonly employed buffers offers new opportunities for this high-resolution technology.

EVALUATION OF CHROMATOGRAPHIC BEHAVIOR

Methods development for chromatographic bioprocessing typically involves an evaluation of the linear chromatographic behavior followed by empirical scale-up. Mathematical models are not widely employed for methods development and optimization of chromatography in the biopharmaceutical industry. Thus a significant amount of time is needed for methods development, as well as for the implementation of nonoptimal processes. As a result, there is increasing interest in the use of mathematical models for the development of efficient preparative ion exchange processes.

The classical view of protein retention in ion exchange columns involves the stoichiometric exchange of proteins and mobile-phase counterions. This is referred to as the stoichiometric displacement model (SDM). In this model, the retention of proteins on ion exchange stationary phases is described as follows (for a 1:1 salt, e.g., NaCl)

$$C_{pb} + vQ_m \Leftrightarrow Q_p + vC_m + vC_b \qquad \ldots(1)$$

where p is the protein molecule, v is the ratio of the characteristic charge of the protein to the valence of the counterion, b is the co-ion (the ion with the same charge as the stationary phase) associated with the protein, m is the mobile-phase counterion (ion with the opposite charge of the stationary phase), C denotes the concentrations in the mobile phase, and Q denotes the concentrations on the stationary phase. The characteristic charge of the protein is a statistical average of the electrostatic interactions of the protein with the stationary phase as it travels down the column.

Some of the counterions remain bound to the protein on adsorption, while the rest make way for the interaction of the protein with the stationary phase. Drager and Regnier argued that the equilibrium between the protein and the co-ion can be neglected. This results in the simplified equilibrium reaction

$$C_p + mQ_m \Leftrightarrow vC_m + Q_b \quad \ldots(2)$$

The SDM predicts that the retention of the protein under isocratic, linear conditions (low solute concentrations) is related to the counterion concentration as follows:

$$\log(k) = -v\log(C_m) + \text{const} \quad \ldots(3)$$

where k is the retention factor under isocratic linear conditions, and C_m is the concentration of the counterion in the mobile phase. This relationship has been experimentally demonstrated under conditions usually employed in the IEC of proteins. Equation 3 forms the basis for the quantification of the effect of the ionic strength on the selectivity of linear separations.

Linear Chromatography

In linear chromatography, the various solutes migrate down the column independent of each other. The retention is characterized by a linear isotherm and an absence of intersolute competition for the adsorption sites. The plate model is used extensively to characterize the efficiencies and resolutions obtained in these systems. In the plate model, all the nonidealities of the column are lumped into one parameter: N, the number of plates [or, alternatively, in height equivalent to one theoretical plate, HETP = (length of column)/N]. The HETP is widely used for the comparison of different sorbents and the validation of packing procedures. It is usually determined from the effluent profiles of isocratic pulse injections of a tracer. For a gaussian effluent peak, the HETP can be determined from the following equation

$$\text{HEPT} = \left(\frac{w_h}{t_r}\right)^2 \frac{L}{5.545} \quad \ldots(4)$$

where w_h is the peak width at half-height, t_r is the retention time of the peak, and L is the length of the column. In ion exchange systems, the HETP measurement is affected by the salt in the background. It is recommended that pulse injections be made in a relatively high salt background to reduce Donnan exclusion. The HETP is also a function of the flow rate. For conventional, macroporous sorbents, the plot of HETP versus flow rate (termed a Van Deemter or Knox plot) exhibits a minimum, indicating the presence of an optimum flow rate for operation. The HETP can be viewed as a sum of all the mass transfer, kinetic, and diffusional resistances in the column.

Yamamoto and coworkers have employed a continuous-flow plate model to characterize and scale up linear gradient and step gradient operations. Graphical techniques have been developed for scaling up linear gradient elution in linear chromatography. In this approach, the ionic strength at which the peak will elute in a large column is obtained directly from a plot of *GH* (gradient slope times the stationary phase volume) versus I_r (ionic strength at the elution time of the peak), which is established with the use of a relatively small column. The dimensionless retention time and the peak width in the large column can then be determined from analytical expressions. These methods have been applied to the scale-up of β-galactosidase purification in a 30-L column based on linear gradient data obtained with a 23-mL column. Although the foregoing approach appears to work well for linear gradient scale-up of single solutes and multicomponent separations with base-line resolution, an alternative method is required for general multicomponent linear separations. The resolution R_s is related to the operating conditions and the column dimensions as follows:

$$R_s \propto \left(\frac{L}{G\ \text{HETP}}\right)^{1/2} = \left(\frac{D_m L_a L}{GHud_p^2}\right)^{1/2} \quad \ldots(5)$$

where D_m is the molecular diffusivity, *u* is the mobile-phase velocity, d_p is the particle size, *L* is the length of the column, *G* is the gradient slope times the column dead volume, and I_a is a constant, introduced to dedimensionalize the right-hand side.

The HETP in equation 5 is the HETP in linear gradient elution. An algorithm to estimate the HETP from a set of linear gradient experiments at various flow rates has been developed. Thus, for scale-up,

$$\left(\frac{L}{G\ \text{HETP}}\right)_{\text{small column}} = \left(\frac{L}{G\ \text{HETP}}\right)_{\text{large column}} \quad \ldots(6)$$

While these approaches have proved useful for the scale-up of linear chromatography, the production rate in these ion exchange systems can be dramatically enhanced by operating these columns under nonlinear conditions. To model various modes of operation under nonlinear (high loading) conditions, an alternative approach must be employed.

Nonlinear Chromatography

To study the effects of nonlinear chromatography, one needs a suitable isotherm that accurately describes nonlinear protein adsorption. Guiochon et al. have employed the Langmuir isotherm extensively to study the nonlinear chromatographic behavior of both small molecules and macromolecules. Berninger et al. have developed the VERSE-LC model, which solves the complete mass transfer equations in conjunction with the Langmuir or modulator-modified Langmuir isotherms to describe nonlinear chromatography. While the Langmuir isotherm has many advantages, it is inadequate for describing protein adsorption in ion exchange systems. Protein adsorption is multipointed in nature and is affected by the presence of mobile-phase modifiers. The Langmuir isotherm is unable to accurately describe these features. The multivalent ion exchange model incorporates both the multipointedness of the adsorption process and the effect of the mobile phase modifier on the process. It is derived from the SDM in conjunction with the law of mass action. Thus the multivalent ion exchange formalism predicts an isotherm of the form

$$Q_p = K_{eq} C_p \left(\frac{\bar{Q}_1}{C_1} \right)^{v_p} \quad \ldots(7)$$

where $\bar{Q}_1$ is the number of sites available for the adsorption of proteins, C_1 is the mobile-phase concentration of the counterions, K_{eq} is the equilibrium constant for the exchange process, and the subscript p denotes the protein.

The effect of the size of the protein molecules and lateral interactions on the adsorption process can be considerable. Furthermore, the adsorption of a protein molecule will result in the steric shielding of a number of stationary phase sites underneath the molecule. Brooks and Cramer have defined a steric factor, σ, to account for such steric hindrance and have used the postulate that the steric hindrance will be proportional to the concentration of proteins on the surface. A steric mass action (SMA) model was developed that essentially employs equation 7 in concert with the following electroneutrality condition for single component equilibrium:

$$\Lambda = \bar{Q}_1 + (v_p + \sigma_p) Q_p \quad \ldots(8)$$

The parameters for the SMA model (K_{eq}, v, and σ) can be readily obtained using standard chromatographic techniques. The equilibrium constant K_{eq} and the characteristic charge (for a monovalent counterion) v are obtained from either isocratic or linear gradient elution chromatography. The steric factor can be obtained from nonlinear frontal chromatography. Knowledge of the bed capacity and the protein parameters allows the generation of nonlinear isotherms at any salt concentration. The SMA formalism can be exfended to multicomponent mixtures. In this case, the electroneutrality condition becomes

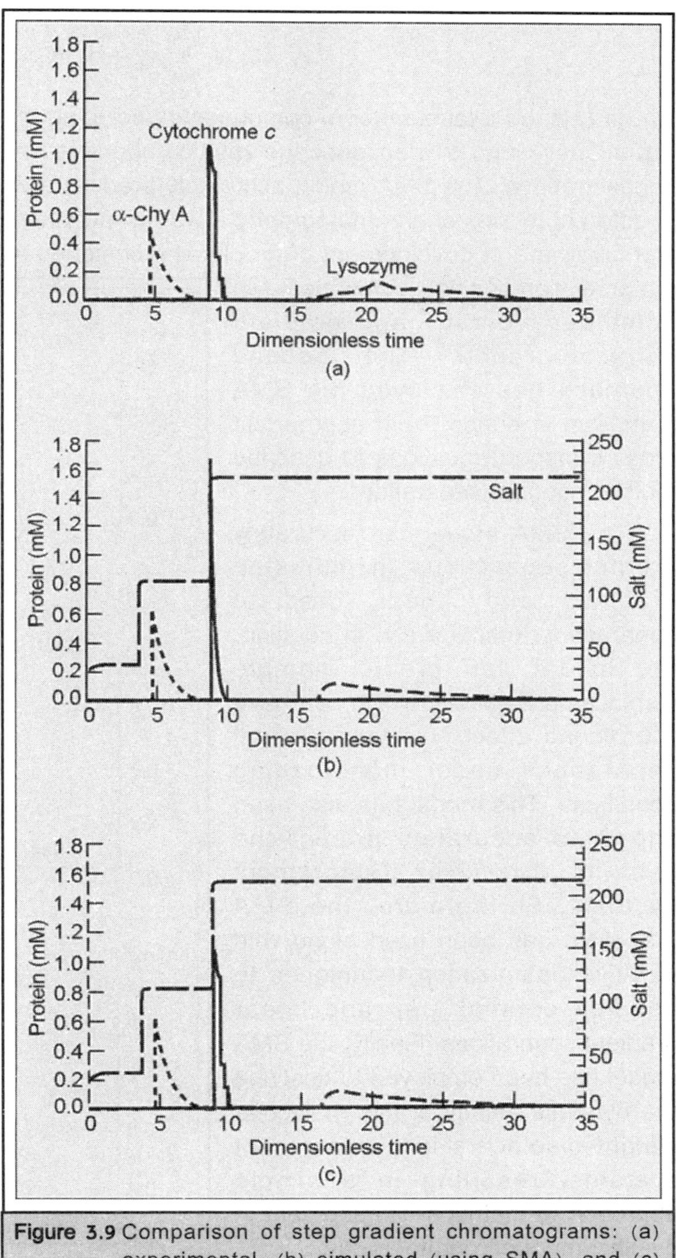

Figure 3.9 Comparison of step gradient chromatograms: (a) experimental, (b) simulated (using SMA), and (c) simulated (using SMA) with fraction collection.

$$\Lambda = \bar{Q}_1 + \sum_{i=2}^{N} (\nu_i + \sigma_i) K_{eq,i} \left(\frac{\bar{Q}_1}{C_1} \right)^{\nu_i} C_i \qquad \ldots(9)$$

where *N* is the total number of components (including the mobile phase counterion) in the system. Equations 7 and 9 then describe multicomponent nonlinear protein equilibria at various salt concentrations. The SMA model accurately predicts a variety of nonlinear ion exchange behavior in different modes of chromatography. Two distinct approaches have been employed to date. The first involves the development of graphical techniques based on the method of characteristics for the prediction of single solute peaks in nonlinear isocratic and gradient chromatography. The second approach has employed the SMA formalism in concert with appropriate mass transport equations to describe multicomponent separations.

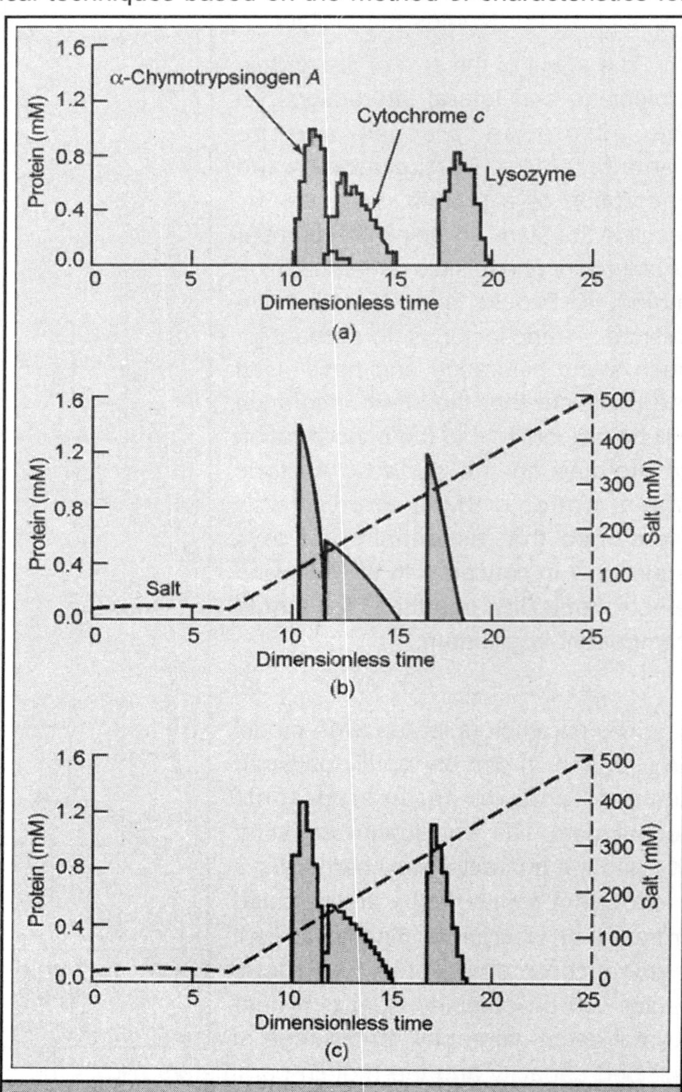

Figure 3.10 Comparison of step gradient chromatograms: (a) experimental, (b) simulated (using SMA), and (c) simulated (using SMA) with fraction collection.

The SMA model can accurately predict separations in multistep gradient and linear gradient separations, respectively. In addition, the model can predict sample displacement effects, which can have a profound effect on linear gradient separations under high loading conditions. This model has also been shown to accurately predict the behavior of complex displacement systems. Furthermore, the SMA formalism has been used along with iterative optimization techniques to develop optimal step and linear gradient separations. Finally, the SMA model has been employed to describe the dynamic affinity of low molecular weight displacers in displacement systems, resulting in a simple approach to methods development in displacement chromatography.

Nonstoichiometric Models

Though the SMA model is capable of accurately describing the behavior of nonlinear ion exchange chromatographic systems, it assumes that the individual charges on the protein molecule interact with the discrete charges on the ion exchange surface. In reality, the retention in ion exchange is more complex and is primarily due to the interaction of the *electrostatic fields* of the protein molecules and the chromatographic surface. Haggerty and Lenhoff demonstrated a correlation between the retention of various proteins with their surface potential.

Stahlberg and coworkers have developed the electrostatic interaction model (EIM), which treats the retention process as a coulombic interaction between two charged flat plates. Roth and Lenhoff employ van der Waals forces in concert with electrostatic interactions to quantify the adsorption of proteins onto charged stationary phases. Recently, a mechanistic model has been developed that incorporates stationary-phase properties such as the surface charge density and short-range interaction energies to predict retention on various stationary phases. The drawback with the nonstoichiometric approaches reported thus far is that they are limited to describing retention in the linear mode of chromatography.

CONCLUDING REMARKS

The last decade has seen dramatic advances in the state of the art of chromatographic bioprocessing with ion exchange materials. This article has presented recent advances in resin materials, modes of operation, and mathematical modeling. The need to process large volumes of relatively dilute feedstock has catalyzed the development of an array of "novel" chromatographic materials. While many novel materials have been developed, the effect of stationary-phase chemistry on selectivity is still essentially unknown, and thus the selection of an ion exchange resin requires significant column screening. Once a resin with sufficient selectivity has been identified, the task remains to develop an efficient, high-throughput chromatographic process. This requires the selection of an appropriate mode of operation. In fact, the selection of the optimal mode of operation for a given bioseparation problem can have a profound effect on the production rate, yield, and purity for that ion exchange step. Finally, the development of mathematical models capable of predicting scale-up and nonlinear protein chromatography now make it possible to employ modeling for bioprocess methods development.

4

CULTIVATION OF PHOTOSYNTHETIC ALGAE

Photosynthetic microorganisms play an important role in the conversion of solar energy into chemical energy through their photosynthetic activities. Photosynthetic conversion can be expected to be an efficient and alternative process in several industrial fields; for example, microalgae have been studied because they produce a wide variety of metabolites. Attempts to develop large-scale methodologies for the cultivation of microalgae have been summarized in a document produced by the Carnegie Institute of Washington, in which algae were predicted to provide an alternative to fermentation and agriculture products.

Table 4.1 Commercial products from microalgae

Products	*Uses*	*Value (dollars/kg)*	*Market (millions of dollars)*
Phycobiliproteins	Research	10,000	1–10
	Food coloring	100	
β-Carotene	Provitamin A	500	1–10
	Food coloring	300	
Xanthophylls	Feed	200–500	10–100
Vitamin C	Vitamin	10	10–100
Vitamin E	Vitamin	50	10–100
Polysaccharides	Viscosifiers	5–10	10–1,000
Amino acids			
Proline		5–50	1–10
Arginine		50–100	1–10
aspartic acid	Gums	2–5	100–1,000
Single cell protein	Animal feeds	10–20	100–2,000

Algal biomass has historically served as fertilizer, and as a food source for both humans and animals. They are also used for secondary wastewater treatment. This use of algal biomass is an important industrial application of microalgal cultures. With advances in processing technology, algal biomass has come to be seen as a possible source of fuel, as well as of fine chemicals and pharmaceuticals.

Several microalgae that produce useful chemicals, such as amino acids, vitamins, carotenoids (β-carotene), fatty acids (DHA, EPA, γ-LA), polysaccharides, and antibiotics have been reported. Several microalgal products have already been commercialized. It is estimated that there are roughly over 10,000 species of microalgae in the natural environment. Unique algal strains may be found in highly diversified marine environments.

We have discovered several marine microalgal strains that produce useful substances, including novel plant-growth regulators that promote redifferentiation, germination, and plantlet formation; tyrosinase inhibitors; a sulfated polysaccharide showing anti-HIV activity; novel antibiotics with light-regulated activity; and a UV-A absorbing compound.

The development of an efficient culture system is necessary for algal mass production and the industrial application of microalgae. Hence, we describe several technologies for microalgal mass cultivation under open and closed systems. Furthermore, recent studies of applied microalgal culture investigating the possible use of biological CO_2 fixation are presented. Also discussed are examples of the technical designs for on-site CO_2 removal systems combined with algal cultures, which promise to play a role in developing technologies for environmental protection.

Table 4.2 Production of useful materials by marine microalgae

Product	*Strain*	*Amount (mg or unit/g dry wt)*
Coccolith	*Emiliania huxleyi*	740
	Pleurochrysis carterae	
γ-Linolenic acid	*Chlorella* sp. NKG042401	9.5
Palmitoleic acid	*Phormidium* sp. NKBG 041105	47
Docosahexaenoic acid (DHA)	*Isochrysis galbana* UTEX LB 2307	15.7
Polysaccharide	*Aphanocapsa halophytia*	45
Glutamate	*Synechococcus* sp. NKBG040607	15.4
Phycocyanin	*Synechococcus* sp. NKBG042902	150
UV-A absorbing biopterin glucoside	*Oscillatoria* sp. NKBG 091600	0.2
Antimicrobial compound	*Chlorella* sp. NKG 0111	—
Plant growth regulator	*Synechococcus* sp. NKBG042902	—
Lactic acid bacteria growth regulator	*Synechococcus* sp. NKBG040607	—
Tyrosinase inhibitor	*Synechococcus* sp. NKBG15041c	107
SOD	*Synechococcus* sp. NKBG042902	—

CULTIVATION AT COMMERCIAL LEVEL

The growth rate and maximum biomass yield of microalgal strains are affected by culture parameters (light, temperature, and pH) and nutritional status (CO_2, nitrogen, and phosphate concentrations). On the other hand, increasing the density of cultures decreases photon availability to individual cells. Light penetration of microalgal cultures is poor, especially at high cell densities, and such poor photon availability decreases specific growth rates. Higher biomass yields could be expected if sufficient photons were provided to high-density cultures of microalgae. Large-scale culture systems have been constructed (classified as open and closed systems) with the greatest attention paid to the light supply. Strains such as *Chlorella, Scenedesmus, Dunaliella, Spirulina, Porphyridium,* and *Haematococcus* have been cultured using these systems to provide for several useful materials.

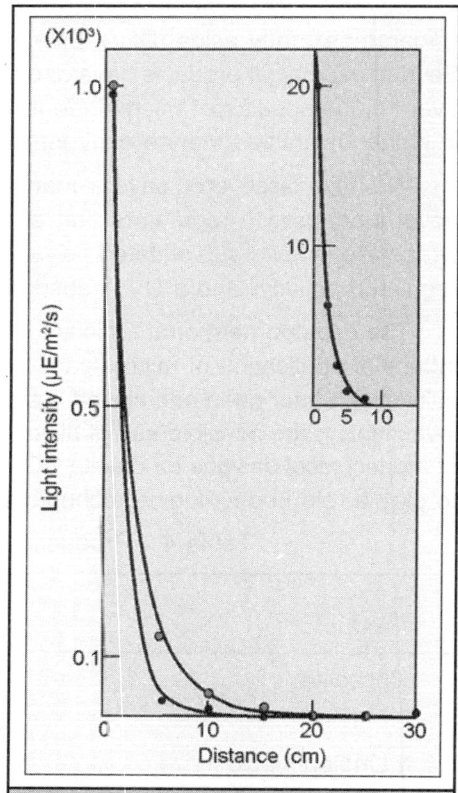

Figure 4.1 The distribution of light intensity as a function of distance from the surface at various cell concentrations; solid circle 2.3 x 10^8 cells/ml; open gray circles, 7.9 x 10^7 cells/ml.

System of Open Culture

Several different types of open culture systems have been proposed. These open culture systems are the simplest method of algal cultivation and offer advantages in construction cost and ease of operation. Open culture systems require a large surface area and shallow depth (ca. 12-15 cm) to improve light utilization. Furthermore, mixing the culture prevents the cells from sinking to the bottom and distributes efficient cell growth with sunlight.

The raceway pound proposed by Oswald has been developed into various types, including raceway pounds employing a paddle wheel, which have most generally been used for outdoor production of microalgae. The raceway pounds for commercial production of microalgae require an area of 1,000-5,000 m^2. Contamination by different algal species and other organisms is a serious problem in open culture systems. *Chlorella, Dunaliella,* and *Spirulina,* which are tolerant to extreme conditions (high nutrients, high salinity, and high pH), are adequate strains for open culture systems.

Vonshak et al. demonstrated that contamination by *Chlorella* in outdoor *Spirulina* culture was prevented by maintaining the culture medium at a high bicarbonate concentration (0.2 M). Grazers sometimes found in *Spirulina* culture were arrested by addition of ammonia (2 mM). The open culture system is easily affected by weather conditions. Rain dilutes salinity, causing contamination. Furthermore, several algae that produce useful chemicals require more restricted conditions for efficient growth and for metabolite production.

System of Closed Culture

Closed systems have been expected to overcome such open culture system disadvantages, and several photobioreactors have been devised. These photobioreactors offer several advantages: easier maintenance of monoalgal cultures because of protection from contamination; avoidance of water loss and subsequent increase of salinity in the culture medium; higher productivity with greater cell densities, reducing harvesting costs; and applicability to varying microalgal species under favorable culture conditions. However, for an efficient and reliable large-scale culture system, several criteria need to be considered, such as light-utilization efficiency, homogeneous mixing (turbulence), low-shear environment, temperature control, and efficient gas transfer. Yield of algal biomass depends on the light path to each cell, and therefore, surface-to-volume ratio is an important factor for efficient light utilization in a photobioreactor.

Tubular reactors have been well studied since their development by Tamiya et al. The diameter of a tubular reactor is now less than 40 cm. Richmond et al. demonstrated that reduction of the tube diameter from 5.0 cm to 2.8 cm enhanced the biomass yield 1.8 times. Narrower tube diameters may allow efficient light utilization as well as a faster flow rate, enhancing the algal productivity and reducing fouling on the inside walls of the tubes. In tubular reactors, flow rates of 30-50 cm/s are generally used, and air-lift is the most effective circulation method of culture, rather than a centrifugal pump, a rotary positive displacement pump, and a peristaltic pump. The main advantages of air-lift systems are their low shear and relative simplicity of construction. Furthermore, the advantage of the tubular system in biomass production per volume using *Anabaena siamensis* was shown where productivities of an open raceway (300-L capacity, height of 15 cm) and a tubular reactor with a 2.8-cm-diameter tube were 86 and 550 mg dry wt/L/day, respectively.

Several modifications in tube arrangement have been carried out for optimizing light penetration. Most of the tubular reactor is laid on the ground, and the lower parts of the tubes receive less light than the upper parts. Torzillo et al. have constructed a two-plane tubular bioreactor for optimizing light availability, in which each tube in the lower plane was placed in the vacant space between two tubes in the upper plane. They showed that this two-plane tubular bioreactor (145 L culture volume) has an effective surface area (ratio of surface area [m^2] to culture volume [m^3] = 49/m) and gave a net volumetric productivity of 1.5 g dry wt/L/day using *Spirulina platensis*. Furthermore, a pilot plant called the Biocoil has been set up. This helical tubular reactor is constructed with a vertical tower coiled up with a tube, increasing the available efficient land surface area. When biomass productivity of various tubular reactors is compared on area basis (surface area), a similar production range of 15-30 g dry wt/m^2/day is obtained for all reactor types.

The effect of temperature on biomass yield is significant in algal culture. The culture in the tubular reactor is often maintained at higher temperature than that in the open raceway. Richmond et al. demonstrated that the *Spirulina* culture using the tubular reactor could be warmed faster than in the open raceway. The culture could be kept at 35–37°C, the optimal range for *Spirulina* growth during most of the light period. However, the closed reactors are sometimes overheated and thus are more suitable for the thermophilic or thermotolerant strains. Temperature control using a heat exchanger and/or evaporative cooling by spraying water onto the surface is sometimes required for the cultivation of general strains. A problem in the closed system is photooxidative damage to the cells caused by accumulation of dissolved oxygen produced by photosynthesis during the light period. In the open system, evolved oxygen is diffused to the atmosphere. By contrast, because oxygen cannot escape from the closed reactors, degasser systems are sometimes required.

The culture part of closed photobioreactors has been constructed with materials such as glass, methyl polymetacrylate, polyethylene, polypropylene, polyvinylchloride, silicone, and stainless steel. These photobioreactors have been designed to utilize external illumination, such as sunlight. Fiber optics also have been employed as internal light sources. Photobioreactors employing fiber optics have the merit of controlling the illumination and duration of the light period. The authors have demonstrated growth of a high-density culture using a photobioreactor employing novel light-diffusing optical fibers (LDOF). Efficient light distribution was achieved as a result of a bundle of LDOFs passing through the cell culture. The ratio of surface area to volume was very high (692/m), and uniform illumination in the cell culture was obtained. Takano et al. cultured a marine cyanobacterium *Synechococcus* sp. in the LDOF photobioreactor, achieving a biomass yield of 2g/L/day. Photobioreactors are less economical but have many superior aspects to the open culture systems mentioned earlier. Thus, they are limited to high-value products. Suitable culture systems have to be adopted according to the target products and available environmental conditions.

RECOMBINANT TECHNOLOGY

Higher productivity of valuable primary or secondary metabolites in microalgae will be possible by high-density cultivation and/or genetically engineered microalgae. Genetic engineering of cyanobacteria (blue–green algae) has been developed in which gene-transfer systems have been established in several cyanobacterial strains, and these have been used as hosts for production of foreign proteins. Recent pharmaceutical interest in unsaturated fatty acids has triggered a search for sources of these valuable compounds. Mono-unsaturated fatty acids such as palmitoleic acid (C16:1), undecylenic acid (C11:1), and tridecenic acid (C13:1) have the potential to prevent several diseases. Several eukaryotic microalgae are known to produce highly unsaturated fatty acids such as eicosapentaenoic acid (EPA; C20:5) and docosahexaenoic acid (DHA; C22:6), which are valuable dietary components. They may be involved in preventing several human diseases. However, wild-type cyanobacteria do not possess the biosynthetic pathway for these fatty acids.

The EPA synthesis gene cluster (ca. 38 kbp) from the marine bacterium *Shewanella putrefaciens* SCRC-2738 was cloned into the marine cyanobacterium *Synechococcus* sp. using a broad-host cosmid vector, pJRD215 (10.2 kbp, Sm^r Km^r). The constructed plasmid (pJRDEPA; ca. 48 kbp) could be transferred by transconjugation to the cyanobacterial host at a frequency of 2.2×10^{-7}. Cyanobacterial transconjugants grown at 29°C produced only 0.12 mg EPA/g dry wt, whereas those grown at 23°C produced 0.56 mg/g dry wt. Furthermore, the content of EPA in cells grown at 23°C increased by 0.64 mg/g dry wt after preincubation for one day at 17°C. *Spirulina* spp. are commercially valuable cyanobacteria. Several *Spirulina* genes have been characterized. The genetic study of *Spirulina* has been focused on the construction of mutants and the cloning of genes. Gene-transfer systems have not been developed because *Spirulina* has high restriction endonuclease activity and is resistant to most antibiotics. Establishment of gene-transfer systems in *Spirulina* will contribute to their industrial application.

UTILIZATION OF CARBON DIOXIDE

Recently, the possible use of biological CO_2 fixation to reduce anthropogenic CO_2 emission has been investigated. For development of on-site CO_2 fixation systems using microalgae, efficient photobioreactors and strains that can fix large quantities of CO_2 are required. Recently, several applied studies aimed at the direct biological utilization of CO_2 in emission gas from power plants have been carried out. In Japan, coal-fired power plants and the steel and cement industries produce large quantities of CO_2. These industries also emit NO_x and SO_x. It is known that NO_x and SO_x reduce biological CO_2 fixation activity, primarily by inhibiting photosynthesis. Therefore, microalgae that can grow under such extreme conditions will be required for direct CO_2 fixation. Identification of microalgal strains capable of growing rapidly in water sparged with emitted gas and under other extreme conditions, such as at high pH, has been carried out. Their ability to grow in saline water (e.g., seawater) and their resistance to other gas constituents (e.g., SO_2 and NO_x) are additional considerations. Several strains have been isolated from environments that have some of the properties described above. Among those strains, the marine alga *Chlorococcum littorale* is characterized by high CO_2 tolerance and high growth rate in the linear growth phase.

Table 4.3 Desired properties of microalgae for CO_2 fixation

1. High pCO_2 tolerance
2. Low pH tolerance
3. Stable and high growth rate in the linear growth phase
4. Capacity to grow at high cell densities
5. Acidic gas (NO_x, SO_x) tolerance
6. Thermotolerance

Reduction of CO_2 Emission

Coal-fired power plants

A designed system for the reduction of CO_2 emission from coal-fired power plants is presented. Microalgal cultivation using CO_2 in gas emitted from power plants has been described. In this system, CO_2-fixed product by microalgal culture is used as biomass fuel, which will substitute for fossil fuel. CO_2 recycling in coal-fired power plants will achieve the reduction of CO_2 emission from power plants.

Cost of microalgae CO_2 mitigation using the designed system has been estimated based on several designed specifications:

1. Plant size: emission gas from a 500 MW power plant
2. Gas condition: 30.85 kg C/S (CO_2) concentration in dry flue gas: 14.2% (vol)
3. Biofixation conditions
   (a) One hundred percent of the CO_2 produced by the power plant is fed to biological systems, and 90% of the CO_2 fed to the system is utilized during daylight summer hours.

(b) Direct biofixation of CO_2 is applied where CO_2 in emission gas from the power plant is directly fed to algal ponds for photosynthesis.

(c) Both 10% (case 1) and 20% (case 2) of photosynthetic efficiency based on visible light are studied as are 4.5% (case 1) and 9.0% (case 2) based on total light.

(d) The average solar radiation should be greater than 200 W/m^2.

Waste concrete → (+ CO_2) → HCO_3^{2-} + Ca^{2+}
Artificial weathering
HCO_3^{2-} + Ca^{2+} → (+ CO_2) → Coccolithophori algae (biomass coccolith)
Photosynthesis Calcification

Figure 4.2 Design of CO_2 fixation by artificial weathering of waste concrete and culture of coccolithophorid algae.

4. Operation of the plant: The power plant is operated at 70% operation factor. That is to say, the plant is operated at 100% of rated capacity for 16.8 h (70% of 24 h) during the day and is shut down for 7.2 h (30% of 24 h) during the night.

5. CO_2 production rate

   (a) Hourly CO_2 production: 407 tCO_2/h

   (b) Annual CO_2 production: 2,495,724 tCO_2/year

6. Algal Strain: NANNP2 (*Nannochloropsis* sp. from the DOE/NREL/ASP culture collection) The expected productivities of algal biomass: 42 g/m^2/day (case 1) and 84 g/m^2/day (case 2)

Two different productivities, 42 and 84 g/m^2/day, were assumed to correspond to about 10 and 20% solar conversion efficiencies, based on visible light, and 4.5 and 9%, based on total light. The lower productivity reflects what could be expected based on present experience. The higher productivity is the theoretical limit. CO2 mitigation costs greatly depend on the productivity of algae and solar radiation. Microalgal strains that can achieve higher photosynthetic efficiency at higher solar radiation are necessary.

A photobioreactor, in which microalgae can convert from CO2 to biomass at higher photosynthetic efficiency, is also required. Further research into microalgae, photobioreactors, the downstream process of microalgal culture, and the burning process of microalgal biomass is still needed. If such research is successful and the economic projections shown in Table 4.4 are verified, this system will represent one of the most promising technologies for future CO_2 mitigation processes.

Cement plants

Coccolithophorids are unicellular planktonic marine algae that produce elaborate structures called coccoliths, comprising scales or plates of $CaCO_3$. In the oceans, huge blooms of coccolithophorid algae occur. These blooms have been recognized as contributing to ocean floor sediment formation, and algae play an important role in the global carbon cycle by recycling $CaCO_3$. We have chosen coccolithophorid algae as model organism to investigate biomineralization and have focused on the ecological significance of $CaCO_3$ recycling. We have proposed CO_2 fixation by artificial weathering of waste concrete and coccolithophorid algae cultures. During artificial weathering of waste concrete suspended in seawater, atmospheric CO_2 can be absorbed and dissolved as bicarbonate ions, which are a major source of coccolith particles. Coccolithophorid algae can use bicarbonate ions to form $CaCO_3$ particles.

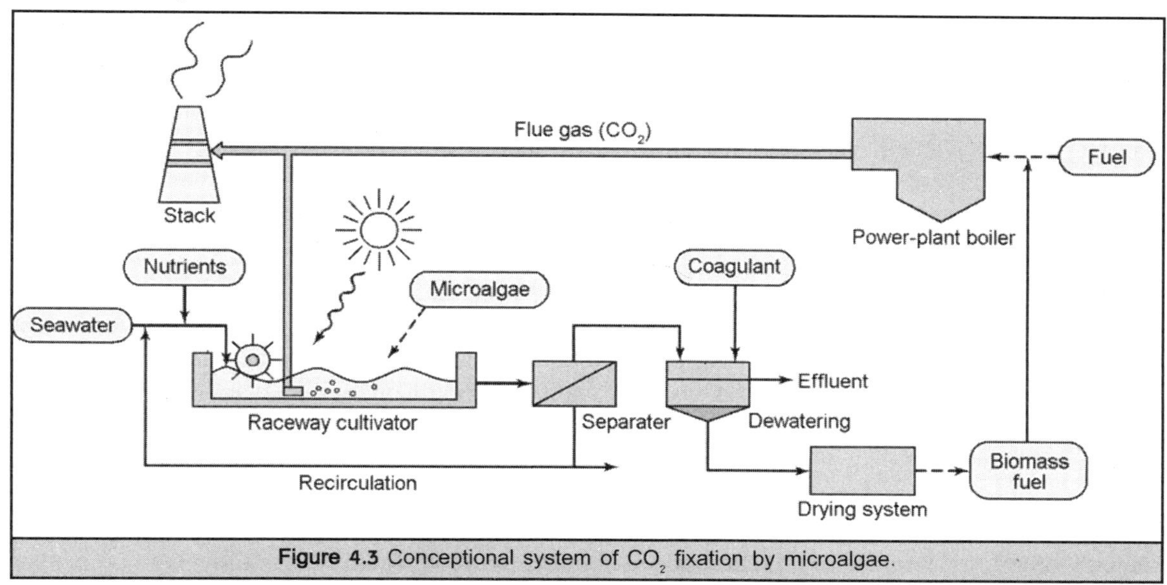

Figure 4.3 Conceptional system of CO_2 fixation by microalgae.

As a result, CO_2 absorbed by artificial weathering can be mineralized and fixed permanently. Artificial weathering of waste concrete is also a useful method to supply bicarbonate ions to cells of the coccolithophorid alga *Emiliania huxleyi*. The CO_2 fixation method by artificial weathering of waste concrete and coccolithophorid algae cultures can be applied to the reduction of CO_2 emission in cement plants. Coccoliths can be used as an alternative to limestone, which is a carbonate source for cement production.

In the cement industry, CO_2 is mainly produced by the decomposition of limestone during the burning of cement clinker. If $CaCO_3$ recycling can be achieved by artificial weathering of waste concrete and coccolithophorid culture, CO_2 emission by the cement industry might be reduced. Moreover, it is proposed that microalgal biomass with fixed CO_2 products maybe stored in concrete. If microalgal biomass can be stored in concrete without the decomposition of the biomass back to CO_2, removal of anthropogenic CO_2 may be achieved.

CO2 absorption by artificial weathering of concrete has been determined. When concrete (calcium content was 15% wt/wt) pieces were weathered by CO_2 and seawater, over 310 mg CO_2/g concrete could be absorbed and dissolved in the form of HCO_3^-. The amount of CO_2 absorbed is directly related to calcium content in concrete, 310 mg CO_2/g concrete is equal to almost 1 g CO_2/g Portland cement (calcium content is almost 46%). On the other hand, amount of emitted CO_2 during a cement production is 708 kg CO_2/ton cement. The amount of absorbed CO_2 by the weathering of waste concrete is greater than that of emitted CO_2 during a cement production. Moreover, CO_2 is absorbed by following coccolithophorid algae cultures.

CONCLUDING REMARKS

The authors reviewed large-scale algal culture systems. Open pond systems have lower productivity of algal biomass, require larger land areas, and involve larger land costs. However, their operation cost is lower. On the other hand, closed systems can achieve high-density culture, and the volume of algal culture can be reduced. The reduction of culture volume decreases the land costs. However, a certain land area is still required for the collection of solar energy, and the operating cost is larger than that of open systems. Moreover, solar radiation, temperature, and other factors regulating algal productivity are significantly affected by location. Suitable culture systems have to be adopted according to the target products and available environmental conditions. Extensive studies of biological CO_2 fixation using microalgal cultures have been pursued for the past 10 years.

A primary goal is the complete removal of CO_2 in discharged gas emitted by such an on-site system. Because of the land-area requirements and the current CO_2 mitigation cost of $264/ton as carbon, it is to be difficult at present to apply microalgal cultures to CO_2 removal. Most nations are seriously considering the increase of atmospheric CO_2 concentration, and intensive efforts to reduce the anthropogenic CO_2 emission will be made. Microalgae culture may be one of the important processes used for such efforts. Regarding technological use of microalgal biomass, several examples, including historical findings, have been presented. Microalgal production of useful chemicals and energy resources have been extensively investigated. Processes that utilize the majority of the resulting microalgal biomass as energy sources would be desired. Such processes may allow the recycling of evolved CO_2 from human energy consumption rather than a one-way emission, as is the case with fossil fuels.

The following six products can be produced from microalgal biomass for use as fuels: hydrogen (through biophotolysis), methane (through anaerobic digestion), ethanol (through yeast or other alcohol fermentation), triglycerides (through extraction of lipids), methyl ester fuels (through transesterification of lipids), and liquid hydrocarbon (from *Botryococcus braunii*). Increasing attention has been paid to resource sustainability in all industries. It must be considered whether resources used for manufacturing products can be sustainable. Developing technologies of microalgal culture will be expected to provide sustainable resources.

5

TECHNIQUE OF AIRLIFT REACTORS

The term *airlift reactor* (ALR) covers a wide range of gas-liquid or gas-liquid-solid pneumatic contacting devices that are characterized by fluid circulation in a defined cyclic pattern through channels built specifically for this purpose. In ALRs, the content is pneumatically agitated by a stream of air or sometimes by other gases. In those cases, the name *gas lift reactors* has been used. In addition to agitation, the gas stream has the important function of facilitating exchange of material between the gas phase and the medium; oxygen is usually transferred to the liquid and in some cases reaction products are removed through exchange with the gas phase. The main difference between ALRs and bubble columns (which are also pneumatically agitated) lies in the type of fluid flow, which depends on the geometry of the system. The bubble column is a simple vessel into which gas is injected, usually at the bottom, and random mixing is produced by the ascending bubbles.

In the ALR, the major patterns of fluid circulation are determined by the design of the reactor, which has a channel for gas–liquid upflow—the riser—and a separate channel for the downflow. The two channels are linked at the bottom and at the top to form a closed loop. The gas is usually injected near the bottom of the riser. The extent to which the gas disengages at the top, in the section termed the *gas separator,* is determined by the design of this section and the operating conditions. The fraction of the gas that does not disengage, but is entrapped by the descending liquid and taken into the down-comer, has a significant influence on the fluid dynamics in the reactor and hence on the overall reactor performance.

PHYSICAL CHARACTERISTICS

Airlift reactors can be divided into two main types of reactors on the basis of their structure: (i) external-loop vessels, in which circulation takes place through separate and distinct conduits; and (ii) baffled (or internal-loop) vessels, in which baffles placed strategically in a single vessel create the channels required for the circulation. The designs of both types can be modified further, leading to variations in the fluid dynamics, in the extent of bubble disengagement from the fluid, and in the flow rates of the various phases. All ALRs, regardless of the basic configuration (external loop or baffled vessel), comprise four distinct sections with different flow characteristics:

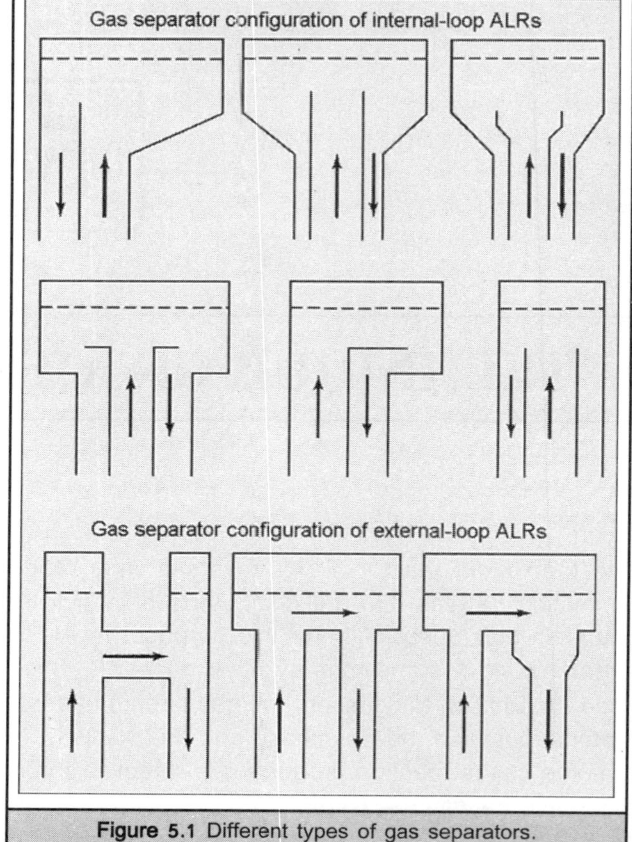

Figure 5.1 Different types of gas separators.

1. *Riser*. The gas is injected at the bottom of this section, and the flow of gas and liquid is predominantly upward.
2. *Downcomer*. This section, which is parallel to the riser, is connected to the riser at the bottom and at the top. The flow of gas and liquid is predominantly downward. The driving force for recirculation is the difference in mean density between the downcomer and the riser; this difference generates the pressure gradient necessary for liquid recirculation.
3. *Base*. In the vast majority of airlift designs, the bottom connection zone between the riser and down-comer is very simple. It is usually believed that the base does not significantly affect the overall behavior of the reactor, but the design of this section can influence gas holdup, liquid velocity, and solid phase flow.
4. *Gas separator*. This section at the top of the reactor connects the riser to the downcomer, facilitating liquid recirculation and gas disengagement. Designs that allow for a gas residence time in the separator that is substantially longer than the time required for the bubbles to disengage will minimize the fraction of gas recirculating through the downcomer.

Momentum, mass transfer, and heat transfer will be different in each section, but the design of each section may influence the performance and characteristics of each of the other sections, since the four regions are interconnected.

APPLICATIONS

For the growth of microorganisms, ALRs are considered to be superior to traditional stirred-tank fermenters despite the fact that the conventional fermenters provide the major requirements for culturing microorganisms: gas– medium interface for the supply of oxygen and the removal of waste gases; means of agitation to ensure proper nutrient distribution and to minimize damage resulting from addition of concentrated acid or base (for pH control); means of heat transfer (for temperature control); and a contamination-free environment. Therefore, the reason for the more successful growth reported in ALRs appears to lie in the difference in the fluid dynamics between ALRs and the more conventional fermenters. In conventional stirred tanks or bubble columns, the energy required for the movement of the fluids is introduced focally, at a single point in the reactor, via a stirrer or a sparger, respectively. Consequently, energy dissipation is very high in the immediate surroundings of the stirrer and decreases away from it toward the walls.

Similarly, shear will be greatest near the stirrer, since the momentum is transferred directly to the fluid in that region, which, in turn, transfers this energy to the slower-moving, more distant elements of the fluid. This results in a wide variation of shear forces; for example, the maximum shear gradient in a stirred tank with a flat-blade turbine has been reported to be approximately 14 times the mean shear gradient. Cells in culture may thus be exposed to contrasting environments in a mechanically stirred vessel, either to minimal shear forces that may generate potentially undesirable gradients in temperature and in substrate, metabolite, and electrolyte concentrations or, alternatively, to highly turbulent zones, with no problems of heat or mass transfer, but with very high shear gradients that may endanger cell integrity or exert some influence on cell morphology and metabolism.

Changes in the morphology of microorganisms associated with high shear forces in the medium have frequently been observed. The nature of the relationship between such morphological changes and the rates of growth and metabolite production is still not properly understood, although it may be of great importance in the design and scale-up of bioreactors. In ALRs, as in bubble columns, the gas is injected at a single point, but in ALRs the direct contribution of gas injection to the dynamics of the system is small; circulation of liquid and gas is facilitated by the difference in gas holdup between the riser and the downcomer, which creates a pressure difference at the bottom of the equipment:

$$\Delta P_b = \rho_L g(\varphi_r - \varphi_d) \quad \text{...(1)}$$

where ΔP_b is the pressure difference, ρ_L is the density of the liquid (the density of the gas is considered to be negligible), g is the gravitational constant, and φ_r and φ_d are the fractional gas holdup of the riser and downcomer, respectively. The pressure difference forces the fluid from the bottom of the downcomer toward the riser, generating fluid circulation in the ALR. Since φ_r and φ_d are both average values integrated along the height of the reactor, it follows that there are no focal points of energy dissipation and that shear distribution is homogeneous throughout the ALR. There is thus a relatively constant environment, with minimization of sharp changes in the mechanical forces acting on suspended particles. Because good mixing is required, shear forces cannot be avoided completely. One of the most critical points is the bottom, where there is a sharp 180^0 turn.

Shear-sensitive mammalian and plant cells in culture should benefit from such an environment. Currently, the research and development of new bioreactors for mammalian cells is indeed focusing on the issue of shear-related damage to suspended cells. Mammalian and plant cells in culture are more susceptible than microorganisms to the reactor conditions. Mammalian cells, which lack the rigid cell wall of microorganisms, have a larger size (one order of magnitude) than microorganisms and are very sensitive to mechanical stress. Plant cells have a rigid cellulose wall, but they are also much larger than microorganisms (usually by about an order of magnitude) and are therefore also sensitive to reactor conditions. Kolmogoroff's model of isotropic turbulence indicates that serious damage may occur at relatively large values of the length scale. The last length is a parameter of the model and indicates the size of the eddy where energy starts to be dissipated by viscous resistance.

Indeed, it has been observed that plant cells, in spite of their rigid wall, are shear-sensitive, and difficulties have been found in stirred-tank cultures. This is especially true when large-scale systems are considered. Although high agitation rates may be detrimental to cell growth, low agitation rates lead to an increase in the number and size of cell aggregates—also an undesirable phenomenon. The aggregates are formed as a result of daughter cells failing to separate after division and as a consequence of the stickiness of the polysaccharides excreted by the cells, especially at the end of the growth phase. An optimal shear rate between these two extremes must be found for each culture. It has recently been shown experimentally that velocity fluctuations related to turbulent shear are relatively homogeneously distributed in an ALR.

The measurements of fluctuating velocity made by Tan et al. show that the liquid turbulence in ALRs is homogeneously distributed in both the riser and the downcomer. It thus seems reasonable to assume that the homogeneity of the stress forces is the main advantage offered by ALRs and that this homogeneity is responsible for the success of shear-sensitive cultures in the ALR type of fermenter. Another advantage of the ALR is the mechanical simplicity of the device. The absence of a shaft and of the associated sealing, which is always a weak element from the point of view of sterility, confers on the ALR an obvious advantage over agitated tanks. This consideration is especially important in processes involving slow-growing cultures, such as animal and plant cells, for which the risk of contamination is large. All the points mentioned above are particularly relevant for sophisticated processes in which the product is usually of high value. But ALRs may be used also for processes involving low-value products, in which case efficiency of energy use may well become the key point for design, as in the use of ALRs for wastewater treatment.

The superiority of ALRs over mechanically agitated contactors in terms of mass transfer rates for a given energy input has been demonstrated by Legrys. Comparison of the efficiency of oxygen transfer, that is, the mass of oxygen absorbed per unit energy invested and unit time, showed that the efficiency of the ALR is among the highest in agitated systems. The ALRs are particularly suited to processes with changing oxygen requirements because aeration efficiency and performance are relatively insensitive to changes in operating conditions. Performance decreases markedly in mechanically stirred systems as the energy input (or oxygen transfer rate) increases, but it is quite constant in ALRs. The efficiency of ALRs decreases relatively slowly as the energy input per unit volume of reactor is increased.

In contrast, in the operation of stirred tanks, the mass transfer rate can be easily increased by increasing the power input, but this improvement is achieved at the cost of a considerable decrease in the efficiency of oxygen transfer. This decrease may constitute a crucial disadvantage in a process like wastewater treatment, where the energy input is an important element in the cost of the final product and flexibility of operating conditions is required because of the constant change of feed composition and flow rate. Energy economy in the ALR may be improved by placing a second sparger in the upper part of the downcomer. If the liquid velocity is greater than the free rising velocity of the bubbles generated, the gas is carried down, resulting in a longer contact time between the bubble and the liquid. This diminishes the energy requirements, since part of the gas is injected against a lower hydrostatic pressure. The advantages described above counterbalance the obvious disadvantage of ALRs, which is the requirement for a minimum liquid volume for proper operation. Indeed, the changes in liquid volume in these reactors are limited to the region of the gas separator, since the liquid height must always be sufficient to allow liquid recirculation in the reactor and must therefore be above the separation between the riser and the downcomer.

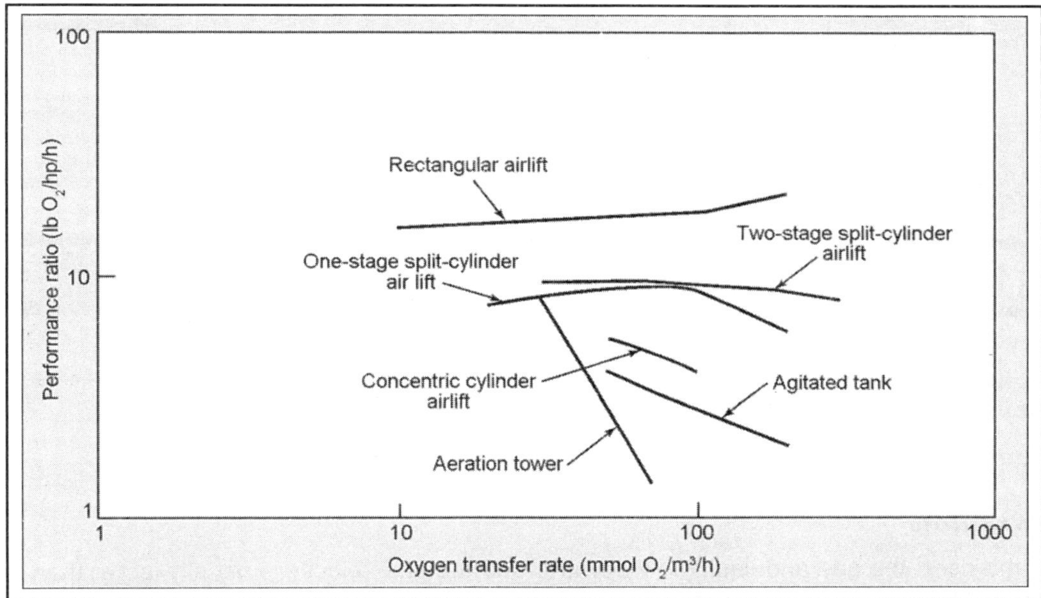

Figure 5.2 Performance ratio as a function of oxygen-transfer rate, showing that aeration efficiency and performance are relatively insensitive to changes in operating conditions in different types of ALRs versus as agitation tank and an aeration tower.

DESIGN VARIABLES

The design variables are the reactor height, the riser-to-downcomer area ratio, the geometrical design of the gas separator, and the bottom clearance (C_b, the distance between the bottom of the reactor and the lower end of the draft tube, which is proportional to the free area for flow in the bottom and represents the resistance to flow in this part of the reactor). The main operating variables are primarily the gas input rate and, to a lesser extent, the top clearance (C_t, the distance between the upper part of the draft tube and the surface of the nonaerated liquid). These two independent variables set the conditions that determine the liquid velocity in the ALR via the mutual influences of pressure drops and holdups. Viscosity is not shown in Figure 5.3 as an independent variable because in the case of gas–liquid mixtures, it is a function of the gas holdup (and of liquid velocity in the case of non-Newtonian liquids), and because in a real process, it will change with time due the changes in the composition of the liquid.

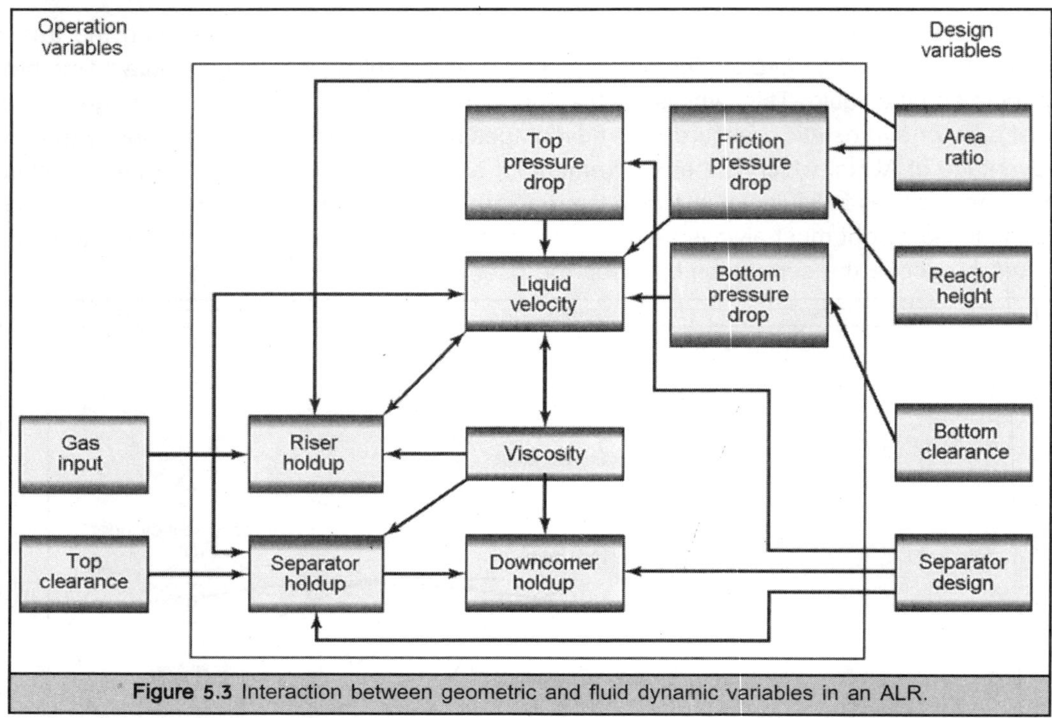

Figure 5.3 Interaction between geometric and fluid dynamic variables in an ALR.

Configuration of Flow

Upward flow

In the riser, the gas and liquid flow upward, and the gas velocity is usually larger than that of the liquid. The only exception is homogeneous flow, in which case both phases flow at the same velocity. This can happen only with very small bubbles, in which case the free-rising velocity of the bubbles is negligible with respect to the liquid velocity.

Although about a dozen different gas-liquid flow configurations have been developed, only two of them are of interest in ALRs:

1. Homogeneous bubbly flow regime, in which the bubbles are relatively small and uniform in diameter and turbulence is low
2. Churn-turbulent regime, in which a wide range of bubble sizes coexist within a very turbulent liquid

The churn-turbulent regime can be produced from homogeneous bubbly flow by increasing the gas flow rate. Another way of obtaining a churn-turbulent flow zone is by starting from slug flow and increasing the liquid turbulence, by increasing either the flow rate or the diameter of the reactor. The slug-flow configuration is important only as a situation to be avoided at all costs, because large bubbles bridging the entire tower cross-section offer very poor capacity for mass transfer.

Downward flow

In the downcomer, the liquid flows downward and may carry bubbles down with it. For bubbles to be entrapped and flow downward, the liquid velocity must be greater than the free-rise velocity of the bubbles. At very low gas flow input, the liquid superficial velocity is low, practically all the bubbles disengage, and clear liquid circulates in the downcomer. As the gas input is increased, the liquid velocity becomes sufficiently high to entrap the smallest bubbles. Upon a further increase in liquid velocity larger bubbles are also entrapped. Under these conditions the presence of bubbles reduces the cross-section available for liquid flow, and the liquid velocity increases in this section.

Bubbles are thus entrapped and carried downward, until the number of bubbles in the cross-section decreases, the liquid velocity diminishes, and the drag forces are not sufficient to overcome the buoyancy. This feedback loop in the downcomer causes stratification of the bubbles, which is evident as a front of static bubbles, from which smaller bubbles occasionally escape downward and larger bubbles, produced by coalescence, escape upward. The bubble front descends, as the gas input to the system is increased, until the bubbles eventually reach the bottom and recirculate to the riser. When this point is reached, the bubble distribution in the downcomer becomes much more uniform. This is the most desirable flow configuration in the downcomer, unless a single pass of gas is required. The correct choice of cross-sectional area ratio of the riser to the downcomer will determine the type of flow.

Device for gas separator

The gas separator is often overlooked in descriptions of experimental ALR devices, although it has considerable influence on the fluid dynamics of the reactors. The geometric design of the gas separator will determine the extent of disengagement of the bubbles entering from the riser. In the case of complete disengagement, clear liquid will be the only phase entering the downcomer. In the general case, a certain fraction of the gas will be entrapped and recirculated. Fresh gas may also be entrapped from the headspace if the fluid is very turbulent near the interface. The extent of this entrapment influences strongly gas holdup and liquid velocity in the whole reactor.

It is quite common to enlarge the separator section to reduce the liquid velocity and to facilitate better disengagement of spent bubbles. Experiments have been reported in which the liquid level in the gas separator was high enough to be represented as two mixed vessels in series. This point will be analyzed further in the section devoted to mixing.

Technique of Holding up Gas

Gas holdup is the volumetric fraction of the gas in the total volume of a gas-liquid-solid dispersion:

$$\varphi_i = \frac{V_G}{V_L + V_G + V_S} \qquad \ldots(2)$$

where the subindexes L, G, and S indicate liquid, gas, and solid, and *i* indicates the region in which the holdup is considered, that is, gas separator (s) the riser (r), the downcomer (d), or the total reactor (T).

The importance of the holdup is twofold: (i) the value of the holdup gives an indication of the potential for mass transfer, since for a given system a larger gas holdup indicates a larger gas-liquid interfacial area; and (ii) the difference in holdup between the riser and the downcomer generates the driving force for liquid circulation. It should be stressed, however, that when referring to gas holdup as the driving force for liquid circulation, only the total volume of the gas is relevant. This is not the case for mass- transfer phenomena, in this case, the interfacial area is of paramount importance, and therefore some information on bubble size distribution is required for a complete understanding of the process. Because gas holdup values vary within a reactor, average values, referring to the whole volume of the reactor, are usually reported.

Values referring to a particular section, such as the riser or the downcomer, are much more valuable, since they provide a basis for determining liquid velocity and mixing. However, such values are less frequently reported. The geometric design of the ALR has a significant influence on the gas holdup. Changes in the ratio A_d/A_r, the cross-sectional areas of the downcomer and the riser, respectively, will change the liquid and gas residence time in each part of the reactor and hence their contributions to the overall holdup. Gas holdup increases with decreasing A_d/A_r.

Internal holding up gas

These take into account liquid properties and geometric differences within a particular design. Most of the correlations take the form:

$$\varphi_r = a(J_G)^{\alpha}\left(\frac{A_d}{A_r}\right)^{\beta}(\mu_{ap})^{\gamma} \qquad \ldots(3)$$

where φ_r is the gas holdup in the riser, J_G is the superficial gas velocity (gas volumetric flow rate per unit of cross-sectional area), μ_{ap} is the effective viscosity of the liquid, and α, β, γ, and a are constants that depend on the geometry of the reactor and the properties of the liquid. The correlation can be used to predict the holdup in a system that is being designed or simulated as a function of the operating variables, the geometry of the system, or the liquid properties. Such correlations are effective for fitting data for the same type of reactor (e.g., a split-vessel reactor) with different area ratios or even different liquid viscosities, but they are mostly reactor-type specific.

The cyclic flow in the ALR complicates the analysis of the system. The riser gas holdup depends strongly on the geometric configuration of the gas-liquid separator and the water level in the gas separator. This has been shown experimentally in a split-vessel rectangular ALR, but the premise can essentially be extended to any internal- loop ALR. Analysis of the system revealed that these factors influence the gas disengagement and hence the gas recirculation in the downcomer. When this influence is taken into account and the holdup is plotted against the true gas superficial velocity, $J_{G,true}$, which is defined as the sum of the gas superficial velocity due to the freshly injected gas, Q_{in}, and to the recirculated gas, Q_d, that is,

$$J_{G,\,true} = \left(\frac{Q_{in} + Q_d}{A_r}\right) \qquad \ldots(4)$$

then all the data for the different gas separators may be represented by a single relationship, such as equation 3. In other words, if the actual gas flow is known, the influence of gas recirculation (which depends on A_d/A_r and the design of the gas separator) has been already taken into account and does not need to be considered again. Nevertheless, this simple approach has a drawback in that the true gas superficial velocity is difficult to measure because the gas recirculation rate is usually not known. A method for evaluation of the extent of the maximum gas recirculation has recently been developed and will be discussed later in this article.

Thus, correlations that take into account all the variables, which may be easily measured, remain the option of choice. A number of correlations have been proposed, and three variables (A_d/A_r, μ_{ap}, and J_G) have been tested by most researchers. The ranges in which these variables were studied varies from source to source. In addition, some other variables (such as bottom clearance, top clearance or gas separator design, and surface tension) have been used by some authors but ignored by others. One example is the disengagement ratio defined by Siegel and Merchuk, which represents the mean horizontal path of a recirculating bubble relative to the external diameter and is equivalent to the parameter obtained by dimensional analysis as:

$$M = \frac{D_S}{4D}$$

where D is the diameter of column and D_s the diameter of gas separator. If this parameter is not taken into account, then studies of the influence of the top clearance are incomplete and difficult to extrapolate to other designs. The same can be said about the filling factor given by the ratio of the gas separator volume to the total volume.

The foregoing discussion thus explains why all the cor- relations coincide for some ranges of these secondary variables while in other ranges they may diverge. In addition, in some cases the number of experiments may not have been sufficient to provide correlations or they may have been ill-balanced from the statistical point of view. The obvious solution to this problem lies in the collection of a large and detailed bank of reliable data that will constitute the basis for correlations with greater accuracy and validity. The safest procedure for the prediction of the gas holdup in an ALR under design is to take data provided by researchers who have made the measurements in that particular type of reactor with the same physicochemical properties of the system.

Gas holdup in the downcomer is lower than that in the riser. The extent of this difference depends mainly on the design of the gas separator. The downcomer gas holdup is linearly dependent on the riser holdup, as a consequence of the continuity of liquid flow in the reactor. Many expressions of this type have been published. At low gas flow rates, φ_d is usually negligible, since most of the bubbles have enough time to disengage from the liquid in the gas separator. This usually happens at the low gas flow rates frequently used for animal cell cultures. The gas holdup in the separator is very close to the mean gas holdup in the whole reactor as long as the top clearance C_t is relatively small (one or two diameters). For larger top clearances, the behavior of the gas separator begins to resemble that of a bubble column, and the overall performance of the reactor is influenced by this change.

External-loop ALR

From the point of view of fluid dynamics, neither the external configuration (shape and architecture) nor the fact that both riser and down- comer are easily accessible is the most important difference between external- and internal-loop reactors. The most important point is that the gas separator of the external-loop ALR is built in such way that gas disengagement is usually much more effective in this type of reactor. In concentric tubes or split vessels, the shortest path that a bubble has to cover from the riser to the downcomer is a straight line across the baffle that separates the two sections. In the case of external-loop ALRs, there is usually a minimum horizontal distance to be covered, which increases the chances of disengagement of the bubbles. In this case, it is worth pointing out that if gas does appear in the downcomer, then most of it will be fresh air entrained in the reactor because of interfacial turbulence or vortices that appear in the gas separator above the entrance to the downcomer.

In many of the studies reported in the literature on holdup in external-loop ALRs, total disengagement is attained. No such data are available for the concentric tubes of split-vessel ALRs, since total disengagement is possible only at very low gas flow rates. Several authors have presented their results of gas holdup as the gas velocity versus the superficial mixture velocity, based on the drift flux model of Zuber and Findlay. These authors derived general expressions for prediction of the gas holdup and for interpretation of experimental data applicable to nonuniform radial distributions of liquid velocity and gas fraction. The drift velocity is defined as the difference between the velocity of the particular phase (U) and the volumetric flux density of the mixture (J) where:

$$J = J_G + J_L \qquad \text{...(5)}$$

The drift velocities of the gas and liquid phases may thus be expressed as:

$$J_G = U_G - J \qquad \text{...(6)}$$

$$J_L = U_L - J \qquad \text{...(7)}$$

Zuber and Findlay derived the relationship, which has been shown to be more than adequate to provide a correlation of gas holdup measurements in tower reactors with high liquid velocities, such as ALRs:

$$U_G = \frac{J_G}{\varphi} = C_0 J + \frac{\frac{1}{A}\int \varphi (U_G - J) dA}{\frac{1}{A}\int \varphi \cdot dA} \qquad \text{...(8)}$$

where A is cross-sectional area, C_0 is distribution parameter, J is superficial velocity, J_G is superficial gas velocity, U_G linear gas velocity, and φ is gas holdup.

Equation 7 describes the relationship between the gas velocity in a two-phase flow and the volumetric flow density of the mixture, J. As stressed by Zuber and Findlay, J has the advantage of being independent on space coordinates for both one-dimensional flow and multidimensional irrotational flows. The distribution parameter C_0 is given by:

$$C_0 = \frac{\frac{1}{A}\int_A \varphi J \cdot dA}{\left[\frac{1}{A}\int_A J \cdot dA\right]\left[\frac{1}{A}\int_A \varphi \cdot dA\right]} \qquad \text{...(9)}$$

The value of C_0 depends mainly on the radial profile of the gas holdup. Zuber and Findlay calculated $C_0 = 1$ for a flat profile and $C_0 = 1.5$ for a parabolic profile. Experimental values have been reported in the range of 1.03–1.2 for upflow and 1.0–1.16 for downflow. Equation 9 shows that this parameter is a function of the profiles of velocities and holdup. The last term of the right-hand side of equation 8 is the weighted mean value of the drift velocity:

$$U_{GI} = U_G - J \qquad \text{...(10)}$$

The drift velocity of a swarm of bubbles can be evaluated by using the expression given by Zuber and Findlay.

$$U_{2J} = 1.53 \cdot \left[\frac{\sigma g \Delta\rho}{\rho_L^2}\right]^{0.25} (1-\varphi)^{1.5} \qquad \text{...(11)}$$

where U_{2J} is the velocity of the swarm of bubbles, g is gravitational acceleration, ρ_L is the density of liquid, $\Delta\rho$ is the density difference, σ is the surface tension, and φ is the gas holdup. This equation is valid for bubble diameters of the order of 0.1 to 2 cm, which covers the population usually observed in ALRs.

It has, however, been shown that a plot of U_G versus J gives a straight line, suggesting that a constant value of the drift velocity satisfactorily represents the two-phase flow in the riser of an external-loop ALR. In this plot, the distribution parameter was C_0 = 1.03, and U_{GS}, the value of the slip velocity of a bubble, was taken as the mean drift velocity. Siegel et al. applied the same model for the study of gas recirculation in a split-vessel ALR and obtained the values of C_0 = 1.11. The slip velocity that they obtained fitting their data to equation 8 was 0.238 m/s. It has been suggested that this simplification holds as long as coalescence is not a predominant factor in the process. It is very important to stress the difference between holdup, φ, and the flowing volumetric concentration (β), which is defined as:

$$\beta = \frac{Q_G}{Q_G + Q_L} = \frac{J_G}{J} \qquad \text{...(12)}$$

Zuber–Findlay's drift flux model allows us to derive the following equation, which establishes a connection between the gas holdup and β.

$$\frac{\beta}{\varphi} = C_0 + \frac{U_{b\infty}}{J} \qquad \text{...(13)}$$

where C_0 is the distribution parameter, J is the superficial velocity, $U_{b\infty}$. is the terminal gas velocity, β is the flowing volumetric concentration, and φ is the gas holdup.

The 45° line indicates that $\varphi = \beta$, an equality that is true only for nonslip flow, where the velocity of the gas is equal to the velocity of the liquid. Such a situation can be visualized for the case of very small bubbles in a relatively fast liquid. In this case, there is no influence of one phase on the motion of the other. As indicated, all the points below the 45° line indicate operation situations in which the liquid is driven by the gas:

$$U_G > U_L;\ \varphi < \beta \qquad \text{...(14)}$$

This happens in the riser of ALRs. For all points above the line the opposite is true:

$$U_G < U_L;\ \varphi > \beta \qquad \text{...(15)}$$

This latter condition reflects the operation of the downcomer.

A number of authors have measured the local holdup profile along the riser of an external-loop ALR. In general, it was found that the holdup increases with height. This finding concurs with the expected expansion of gas bubbles as regions of lower pressure are reached. Common sense indicates that this situation must be limited to a certain range; an increase in bubble size will enhance turbulence and result in an increase in bubble encounters, leading eventually to bubble coalescence. The larger bubbles will rise much faster, resulting in a decrease in holdup. Merchuk and Stein reported a maximum in the holdup profile for the case of a single-orifice gas distributor.

For a multiple-orifice sparger, producing a more homogeneous bubble size distribution, a maximum was not observed within the studied length of the riser, which was 4 m. Literature data from different sources for gas holdup in the riser under conditions of little or no carryover of gas from the separator into the downcomer for different A_d/A_r and top clearance C_t may be represented by the simple exponential:

$$\varphi_r = \alpha J_B^{\beta} \qquad \ldots(16)$$

where the constant α depends on the friction losses in the loop, and β is usually a value between 0.6 and 0.7. The fact that neither the area ratio nor the top clearance affects the gas holdup demonstrates the role of the gas–liquid separator in determining the performance of the reactor in general.

In the absence of gas recirculation, there is no effect on these variables. Moreover, this means that under conditions of no gas entrainment from the separator to the downcomer, it is possible to predict the riser gas holdup as a function of the riser superficial gas velocity alone, which is of great importance for design purposes. It is accepted that liquid velocity has a mild negative effect on gas holdup in the riser. This effect is usually studied by reducing the liquid flow; this is achieved by adding resistance to the liquid loop by means of a valve or other controlled obstruction under conditions of low or nil gas recirculation. Such experiments, which are relatively simple in external-loop ALRs, indicate that the holdup decreases as the liquid velocity is increased from zero (bubble column) to 0.3 m/s (which is close to the bubble free-rise velocity). For higher velocities, the effect of U_L is small. These findings add to our understanding of the fluid dynamics in the column. At liquid velocities that are smaller than the bubble free-rising velocity, the liquid transported in the wake of the bubbles, which must return downward to balance the mass flux, is the cause of the meandering and loops that typically appear in bubble column operation.

As the overall liquid flux increases, the patterns straighten out, the bubbles begin to ascend in a straight pattern, and the holdup goes down. When the liquid velocity is higher than the free-rise velocity of the bubbles, piston flow of bubbles ensues in the tube, and the decrease in holdup for further increases in liquid velocity is due solely to the change in the ratio of gas–liquid volumetric flow rates. When there is gas recirculation, the area ratio A_d/A_r becomes an important variable affecting gas holdup. The effect of A_d/A_r starts in the region in which gas entrainment from the separator to the downcomer occurs. If it is assumed that the riser cross-sectional area A_r remains unchanged and the downcomer cross-sectional area A_d is increased, then it can be expected that the liquid velocity in the riser will increase as a result of the smaller resistance to flow in the loop, which in turn leads to a decrease in the riser gas holdup. An increase in A_d/A_r will result in a decrease in the liquid velocity in the downcomer, which leads to a decrease in the gas recirculation, since fewer bubbles are entrapped in the downcomer. The final outcome of increasing A_d/A_r is thus a decrease is the riser gas holdup.

A similar argument can be applied in the discussion of the effect of reactor height on the riser gas holdup, that is, an increase in the height of the downcomer will result in a higher liquid velocity, which will in turn lead to a decrease, as in the former case, in the holdup in the riser. In contrast, an increase in A_d/A_r will lead to an increase in the extent of bubble entrapment in the downcomer, which will serve to inject some additional gas into the riser. On the other hand, an increase of gas holdup in the downcomer diminishes the driving force for recirculation, as shown in equation 1, and this will moderate the increase of liquid velocity generated by the larger height. This feedback control of the liquid velocity is one of the characteristics particular to ALRs. The differences between the predictions obtained with the different correlations are probably due to the design of the gas separator. The equation given by Popovic and Robinson seems to give an average of the proposed correlations.

Effects of flow of liquid

The effect of rheology on the reactor behavior and performance is of great interest because in most biotechnological processes an increase in biomass provokes changes in the rheology of the fluid, especially in the case of mycelial growth. This effect is enhanced when in addition to the biomass growth, a product of the process is released into the medium in appreciable amounts. A good example of this scenario is the biosynthesis of polysaccharides, which cause an increase in the liquid viscosity. The effect of viscosity on gas holdup in bubble columns has been studied by a number of authors. The main problem to be overcome is that of non-Newtonian flow. If the viscosity is not constant, but changes with changes in the shear rate, then the evaluation of shear rates becomes particularly relevant for the identification of the system. Several authors have confronted this issue. Nishikawa et al. analyzed the problem of heat transfer in a bubble column with non-Newtonian liquids. They found a direct proportionality between the superficial gas velocity and the global shear rate:

$$\gamma = 5000 \cdot J_G \ (J_G > 0.04 \text{m/s}) \qquad \ldots(17)$$

This global shear rate was then used to calculate a global viscosity. In shear-sensitive cultures, the definition of a global shear rate in itself is of great importance.

A number of researchers, Henzler, Kawase and Moo-Young, Schumpe and Deckwer have followed the approach of Nishikawa et al. but have suggested different proportionality constants relating the global shear rate to the superficial gas velocity. This approach is questionable from the rheological point of view because it will predict the same shear rate for a certain superficial gas velocity, no matter which liquid is used. El Tamtamy et al. introduced an improvement by calculating the shear rate from the bubble velocity divided by the bubble diameter. However, accurate evaluation of the latter two parameters is difficult.

Henzler and Kauling suggested relating the shear rate to power input based on dimensional analysis by expressing the shear rate as a function of the power input per unit volume, $(P/[V\rho v])^{1/2}$. Their analysis gives different shear rates for liquids that are rheologically different. The above-described relationships predict different shear rates that vary in up to three orders of magnitude. It is thus generally agreed that the correct solution is still to be found. Recently, a more general approach, known as a global approach, has been proposed by Merchuk and Ben- Zvi (Yona). The shear stress in a bubble column was defined as being equal to the acting force, which can be calculated from the power input divided by the sum of the areas of all the bubbles:

$$\tau = \frac{P}{L_R} S_{ab} \qquad \ldots(18)$$

where L_R is an effective length that represents the mean circulation path of a bubble in the system considered, P is the power input, S_{ab} is the total surface of all of bubbles, and τ is the shear stress.

Assuming ideal gas isothermal expansion, the power input P can be calculated. The interfacial area can be evaluated from correlations or can be obtained by direct measurement if available. A correlation taking into account other variables, like sparger configuration, surface tension, etc., will broaden the range of applications of this approach. If a constitutive equation describing the rheology of the system is available (such as the power law, which has been reported to correspond to many biological systems), equation 17 facilitates the calculation of a global shear force acting on the liquid. The shear rate can be in this case expressed as:

$$\gamma = \left[\frac{\tau}{k}\right]^{1/n}$$

where γ is shear rate and κ is behavior coefficient, and equation 17 can be now used to express γ as:

$$\gamma = \left[\frac{p_1 J_G \ln\left(\frac{p_1}{p_2}\right)}{a L_R^2 \kappa}\right] \qquad \ldots(19)$$

where the subindexes 1 and 2 represent the two extremes of the section considered.

Equation 19 thus gives a global shear rate that is a function of both fluid dynamics and rheology. This approach has been found to be useful for the presentation of results on mass transfer rates in bubble columns.

In contrast to the marked influence of rheology on gas holdup in bubble columns, the data available for ALRs show clearly that the effect of liquid viscosity is less dramatic, but not simpler. At low concentrations of glycerol, a moderate increase of the gas holdup is evident, particularly in the downcomer but also in the riser. These increases are caused by the lower free rise velocity of the bubbles, which increase the gas retention due to the longer residence time. In addition, the entrapment of the bubbles is increased, and this is reflected mainly in φ_d. When the concentration of glycerol becomes too high, a strong decrease of the gas holdup is seen.

This decrease is probably due to the onset of coalescence, which produces larger bubbles that ascend faster in the liquid and easily disengage in the gas separator. The viscosities corresponding to these solutions ranged from 4 mPa s to 14 mPa s. The change in CMC concentration had only a slight effect on the gas holdup for additions in the range 0.01 to 0.05% CMC. Only for solutions with concentrations higher than 0.5% CMC was an appreciable decrease in holdup seen.

Effect of level of liquid

The influence of the liquid level C_t on the gas holdup is exerted as a consequence of changes in the extent of disengagement of the bubbles in the gas separator. This influence is therefore dependent on the geometric design of this section. Whether a bubble will disengage or will be entrapped into the downcomer depends on the interrelationship of several parameters—the free rising velocity of the bubble $U_{b\infty}$. (a function of size and viscosity), the liquid velocity in the downcomer U_{Ld} (a function of the difference in gas holdup between the riser and the downcomer and frictional losses), and the residence time of the bubble in the gas separator (a function of geometric design and liquid height). For a given bubble size, if U_{Ld} is smaller than the $U_{b\infty}$ corresponding to the smallest bubble, then there is no carryover. For smaller bubbles, the balance between the time required to cover the path from the end of the riser to the zone near the entrance of the downcomer and the time needed for disengagement will give the fraction of bubbles recirculated. It should nevertheless be kept in mind that this is a feedback process.

A higher bubble disengagement rate leads to a lower gas holdup in the downcomer, which in turn increases the liquid velocity, enabling larger bubbles to be trapped, until the system eventually reaches a steady state. Due to the above-explained influence of the geometric design, the influence of the liquid height is completely different in internal- and external-loop ALRs. In internal-loop reactors, an increase in C_t increases the zone of the separator in which the gas holdup is higher, and as a result, the gas holdup increases. The extent of this increase depends, as said earlier, on the free-rise velocity of the bubbles. It can be seen that although for water C_t has a small effect, this is not so for the more viscous solution.

In the latter, the lower rising velocity of the bubbles causes a greater number of bubbles to be entrained and carried down by the liquid. Thus, the residence time in the disengagement section becomes very important in determining the fraction of bubbles that recirculate. A lower C_t will give a shorter residence time in the separator, a larger bubble recirculation, and, hence, a larger gas holdup. In external-loop ALRs, the opposite effect is obtained. This is due to the construction of these reactors, in which much of the gas that enters into the downcomer is trapped from the headspace due to the turbulence in this zone. An increase in the liquid height serves to reduce the amount of gas trapped, so that less gas circulates in the downcomer and the liquid velocity increases. The final result is a reduction in the gas holdup, both in the riser and the downcomer.

Recirculation of Gas

The degree to which gas flowing in the riser is entrapped and recirculated through the downcomer is an important variable, since it influences not only the flow configuration in the downcomer, but also the overall performance of the ALR. The liquid velocity depends mainly on the difference in holdup between the riser and the downcomer, and it in turn influences the gas holdup in the riser. Despite the importance of recirculation, very little quantitative data are available on this phenomenon. Siegel et al. evaluated the gas recirculation in a split-vessel ALR by an indirect method based on holdup measurements. From their results, it may be seen that the recirculation rate remains fairly constant for changing gas flow rates in the riser for high values of the last variable. Thus, the recirculation rate is determined largely by the geometric configuration of the gas–liquid separator and the liquid level in the separator. Three zones represent operating conditions giving oscillating, borderline, and straight bubble flow in the downcomer. Oscillating flow patterns produce much larger fractions of gas retained in the downcomer, but they are much more sensitive to J_G. At low superficial gas velocity, the recirculation increases very sharply with J_G.

The bubbles exhibit an oscillating swirling flow pattern, with some larger bubbles escaping toward the top. The borderline condition is defined as oscillatory bubble flow at a low gas flow rate that shifts to straight bubble flow with increasing input gas flow rate. The straight flow operation zone is distinguished by bubble flow in a straight, well-defined flow pattern for all the input gas flow rates studied. If a straight bubble flow pattern is desired, the reactor should be operated at high riser gas flow rates, at which the reactor will shift toward stable operation.

Lubbert et al. attempted to evaluate the recirculation of gas during the cultivation of yeast (*Saccharomyces cerevisiae*) on waste from a starch factory in a 4-m^3 pilot plant. They used microprocessor-aided pseudostochastic tracer input and cross-correlation techniques, which facilitated very reduced tracer feeds due to a high signal-to-noise ratio. The response to a pulse of helium was measured directly at the surface of the liquid in the separator by a quadruple mass spectrometer. The peak obtained showed pronounced shoulders which could be interpreted as superimposition of a second peak that represents the helium tracer one loop after. The fitting of such a model to the experimental data suggested recirculation of 25% of the gas. This figure is in the range of the recirculation rates presented by Siegel et al., considering the differences in the coalescing properties of the liquids used in these two works and the corresponding differences in the downcomer holdup.

Recently, Merchuk and Berzin developed a mathematical model based on the application of the first law of the thermodynamics to each of the regions of an ALR. This model facilitates the evaluation of the maximum liquid recirculation possible in the system. The calculation is based on the premise that the gas recirculated must be compressed from the pressure at the top of the downcomer to the pressure at its bottom. The mathematical expression that gives this maximal gas recirculation is:

$$Q_d = \frac{Q_L(P_2 - P_3) + Q_L\rho_L gh - \frac{1}{2}\rho_L C_d A_d (1-\varphi_d) U_L^3}{P_4 \ln\left(\frac{P_3}{P_2}\right)} \qquad \ldots(20)$$

where Q_d is the gas flow rate in the downcomer, Q_L is the liquid circulation flow rate, P_i is pressure at point i of the reactor (1 is top of the riser, 2 is top of the downcomer, 3 is bottom of the downcomer, 4 is bottom of the riser), C_d is the hydraulic resistance coefficient, A_d is the downcomer cross-sectional area, U_L is the linear liquid velocity, g is the gravitational acceleration, ρ_L is the liquid density, and φ_d is the downcomer gas holdup.

The calculation of Q_d thus requires knowledge of the liquid flow rate, the pressures, and the geometry of the reactor. This equation represents the maximum recycling of gas in the downcomer, which will take place only if all the energy dissipated in the downcomer is invested in gas compression.

Velocity of Liquid in Bioreactor

The liquid velocity is one of the most important parameters in the design of ALRs. It affects the gas holdup in the riser and downcomer, the mixing time, the mean residence time of the gas phase, the interfacial area, and the mass and heat transfer coefficients. Circulation in ALRs is induced by the difference in hydrostatic pressure between the riser and the downcomer as a consequence of a difference in gas holdup. Liquid velocity—like gas holdup—is not an independent variable, because the gas flow rate is the only variable that can be manipulated. The geometric design of the reactor will also influence the liquid velocity, but this remains constant during operation. Experiments have been carried out in devices specially designed to artificially change the resistance to flow, with the aim of studying the effect of the velocity at a fixed rate of aeration. The information emerging from these experiments indicates that an increase in the liquid velocity leads to a decrease in the mean residence time of bubbles in the riser and hence of the gas holdup in the riser. In practice, when the gas flow rate is increased, the higher liquid velocity increases the carryover of bubbles from the gas separator into the downcomer; the carryover dampens the liquid flow by reducing the hydrostatic driving force. As a result, the overall change in liquid velocity is tempered.

Measurement of velocity

Several different methods can be used for measuring the liquid velocity. The most reliable ones are based on the use of tracers in the liquid. If a tracer is injected and two probes are installed in a section of the tube, the velocity of the liquid traveling the distance between probes can be taken directly from the recorded peaks, as the quotient of the distance between the two electrodes and the time required by the tracer to travel from the one to the other. The latter is obtained as the difference of between the first moments of the two peaks. A second method is to calculate the liquid velocity (U_L) from the circulation time (t_c) and holdup (φ) as:

$$U_L = \frac{\text{liquid volume}}{t_c \times A \times (1-\varphi)} \quad \ldots(21)$$

where A is cross-sectional area.

In this case, only one electrode is necessary, φ is the holdup at the point at which the electrode is installed, and the circulation time is obtained from two successive peaks recorded by the electrode.

Estimation of liquid velocity

A number of expressions are available for the estimation of the liquid velocity. Two main methods have been used for the modeling of two-phase flow in ALRs—energy balances and momentum balances. Chakravarty et al. used the energy balance approach to obtain a relationship between superficial gas velocity, holdup, and liquid velocity. Lee et al. calculated U_L by a similar type of model for a series of published data for concentric and external-loop ALRs and from their own results for split vessels. In both the above-mentioned models, constants accounting for friction losses were obtained by adjusting the models to the experimental data. Jones, on the other hand, managed to express the results of his energy balance in a relationship free of empirical constants. His results, however, fit the experimental data only qualitatively, and the fit is satisfactory only for very small diameters. An improvement of this method was suggested by Clark and Jones, who took into account the radial distribution of the gas holdup through the drift flux model. However, the values of the distribution coefficient C_0 needed for satisfactory fitting of the experimental data for lager diameters is far from the range usual in this type of flow. Chisti and Moo-Young extended a model originally proposed by Bello, based on an energy balance over the airlift loop. Their expression for the average superficial liquid velocity is:

$$U_{Lr} = \left[\frac{2gH_d(\varphi_r - \varphi_d)}{\frac{K_t}{(1-\varphi_r)^2} + K_B\left(\frac{A_r^2}{A_d}\right)\frac{1}{(1-\varphi_d)^2}} \right]^{0.5} \qquad \ldots(22)$$

where U_L is the superficial liquid velocity, A_r is the riser cross-sectional area, A_d is the downcomer cross-sectional area, H_d is the downcomer height, K_b and K_t are the hydraulic pressure loss coefficients, φ_r is the riser gas holdup, and φ_d is the downcomer gas holdup.

By choosing suitable values for the friction coefficients in each case, the authors showed that much of the published data on liquid velocity for the different types of ALRs could be satisfactorily correlated by equation 22. Only one coefficient has to be adjusted, since the authors assume that K_t, the friction coefficient at the top of the loop, is negligible in concentric-tube type reactors and that in external-loop reactors K_t can be taken as equal to K_b, the friction coefficient for the bottom of the loop. Equation 22 has thus been adopted by many scientists. Wachi et al. claimed that their derivation of the same equation gives a clearer physical meaning to the adjustable parameters. Equation 22 can also be also derived from a simple momentum balance. Chisti et al. presented an empirical correlation for K_b obtained by comparison of results obtained from several sources:

$$K_b = 11.402 \cdot \left(\frac{A_d}{A_b}\right)^{0.789} \qquad \ldots(23)$$

where A_b is the minimal cross section at the bottom of the airlift reactor and A_d is the downcomer cross-sectional area.

Equation 22 has the particularly that the gas flow rate, which is the main, and often the only, manipulable variable in the operation, is not present directly, but exerts its influence through the gas holdup. Therefore, either experimental data or a valid mathematical expression for the gas holdup in both the riser and the downcomer are required. Chisti and Moo-Young extended this model further in order to facilitate the prediction of liquid circulation in ALRs operating with pseudoplastic fluids, such as mold suspensions. This improvement is very important, since many commercial fermentation processes involve such non-Newtonian liquids. Kemblowski et al. presented a method for the prediction of gas holdup and liquid circulation in external-loop ALRs. In their experiments there was almost no gas recirculation, because of the large size of the gas separators used. Garcia Calvo presented an ingenious model based on energy balances and on an idea originally proposed by Richardson and Higson, and Garcia Calvo and Leton extended the model to bubble columns. The model is based on the assumption that the superficial gas velocity (J_G) in any region can be considered to be the sum of two streams (J' and J'') as follows. The J' stream has a velocity equal to that of the liquid and can therefore be treated by the laws of homogeneous two-phase flow (no slip between the bubbles and the liquid).

The second stream (*J"*) is considered to be responsible for all the energy loses at the gas–liquid interface. The concept in itself is simple and elegant, and it is possible to envisage its application even to the flow in the downcomer, where $U_G < U_L$. In such a case, we would divide the gas flow rate into two parts as follows: One part would be larger than the actual flow rate, i.e., it would have the same velocity as the liquid. In order to arrive at the actual gas flow rate, the second flow rate must have the reverse direction. This type of gas flow can actually be seen under certain conditions, such as when there is coalescence of bubbles and larger bubbles ascend along the walls of the downcomer. Another technique used by several researchers to predict liquid velocity is the momentum balance of the ALR. This method has been used by Blenke in jet-loop reactors and by Hsu and Dudukovic, Kubota et al., Bello, Koide et al., and Merchuk and Stein.

The latter authors presented a simple model for the prediction of the liquid velocity as a function of the gas input in an ALR. They assumed that the pressure drop between the bottom and the top of their external-loop reactor could be expressed as a continuation of the downcomer, using an equivalent length L_E. This length was set as an adjustable parameter describing the pressure loss in the loop. Kubota et al. used a similar approach for the analysis of Imperial Chemical Industries' deep-shaft reactor. They were able to simulate the operation of the reactor and to predict the minimum air supply required to prevent flow reversal. Verlaan et al. used a similar model, in combination with the expression of Zuber and Findlay, to calculate the friction coefficients from experimental data reported by several authors for a wide range of reactor volumes. Koide et al. presented an analysis of the liquid flow in a concentric-tube ARL that was also based on a momentum balance.

The main difference between this model and that used by Merchuk and Stein was that Koide et al. used a convergence–divergence flow model for the bottom and the top of the loop. At the bottom, the effect of flow reversal on the pressure drop was included in the effective width of the gas-liquid flow path under the lower end of the draft tube h_l, which was smaller than the actual gap. Miyahara et al., who studied both the bubble size distribution in an internal-loop ALR and the pressure drop at the top and the bottom of the draft tube, also presented a model facilitating the prediction of the liquid velocity. Other models use the drift-flux model presented in equations 5-7, as:

$$U_{Lr} = \frac{J_{Gr}\left(\frac{1}{\varphi_r} - C_0\right) - U_{GJ}}{C_0(1-\varphi_r)} \qquad \ldots(24)$$

U_{GJ} can be taken from equation 10. The range of variation of C_0 is rather narrow, as shown in the previous section, and therefore it is not difficult to make a judicious guess as to the value of C_0 in an unknown system. The drift flux model has also been used together with energy balances or with the momentum balance. Some studies on liquid measurement present the results in the form of empirical correlations. The usefulness of these correlations depends on the amount of data and the number of parameters taken into account.

In general, the superficial liquid velocity increases with an increase in the superficial gas velocity, but its rate of increase is much lower at high superficial gas velocities. It can be seen that the riser-to-downcomer cross-sectional area ratio and the reactor height are the main parameters that affect the superficial liquid velocity at constant superficial gas velocity. The superficial liquid velocity increases with an increase in A_d/A_r. The effect of the properties of the liquid, such as viscosity, surface tension, and ionic strength, on the superficial liquid velocity are much milder in ALR than in bubble columns. It is expected that increasing the liquid viscosity will decrease the liquid velocity because of frictional losses, but this, in turn, will increase the gas holdup in the riser and consequently increase the driving force for liquid recirculation. Hence, it seems that these two effects balance each other partially and result in a milder effect on the superficial liquid velocity.

The effects of the surface tension and the ionic strength are also exerted via their influence on the gas holdup, as analyzed above. As a rule, it can be recommended that the Zuber-Findlay expression (equation 20) be used when the holdup is known and the liquid circulation velocity is high. For low liquid velocity, a correlation obtained in a piece of equipment as similar as possible to the one under design should be used.

Mixing Behavior

For the design, modeling, and operation of ALRs, a thorough knowledge of mixing behavior is necessary. This is of particular importance during the process of scale-up from laboratory-scale to industrial-scale reactors. The optimum growth rate of a microorganism or the optimum production rate of a specific secondary metabolite usually relates to well-defined environmental conditions, such as pH range, temperature, substrate level, limiting factors, dissolved oxygen, and inhibitor concentration in a specific well- mixed laboratory-scale vessel. Because of the compromises made during scale-up, it is difficult to keep, at different scales of operation, the same hydrodynamic conditions established in the laboratory; mixing on an industrial scale may not be as good as mixing on a laboratory scale. In smaller-scale reactors it is easier to maintain the optimal conditions of pH, temperature, and substrate concentration required for maximum productivity of metabolites in a fermenter. Furthermore, in fermentation systems efficient mixing is required to keep the pH within the limited range, giving maximum growth rates or maximum production of the microorganism during addition of acid or alkali for pH control. Mixing time—or the degree of homogeneity—is also very important in fed-batch fermentation, where a required component, supplied either continuously or intermittently, inhibits the microorganisms or must be kept within a particular concentration range.

A large number of commercially important biological systems are operated in batch or fed-batch mode. In this operation mode, fast distribution of the incoming fluid is required, and the necessity for understanding the dynamics of mixing behavior in these vessels is obvious. Even for batch systems, good control of the operating conditions, such as pH, temperature, and dissolved oxygen, require prior estimation of mixing so that the addition rates can be suitably adjusted. Deviation of the pH or temperature from the permitted range may cause a damage to the microorganism, in addition to its effect on the growth and production rates. Moreover, a knowledge of the mixing characteristics is required for modeling and interpreting mass and heat transfer data. A parameter used frequently to represent mixing in reactors is the mixing time (t_m). It has the disadvantage that it is specific to the reactor design and scale, but it is easy to measure and understand. Mixing time is defined as the time required to achieve the desired degree of homogeneity (usually 90–95%) after the injection of an inert tracer pulse into the reactor. The so-called degree of homogeneity (*I*), is given by:

$$I = \frac{C - C_m}{C_m} \qquad \text{...(25)}$$

where *C* is the maximum local concentration and C_m is the mean concentration of tracer at complete mixing.

A more comprehensive way of analyzing mixing, applicable to continuous systems, is a study of the residence time distribution (RTD). Although ALRs are usually operated in a batch-wise manner, at least in the laboratory, advantage is taken of the fact that the liquid circulates on a definite path to characterize the mixing in the reactor. Hence, a single-pass RTD through the whole reactor or through a specific section is usually measured. Based on the observed RTD, several models have been proposed. These models have the advantage of reducing the information of the RTD to a small number of parameters, which can later be used in design and scale-up. The axial dispersion model, which has the advantage of having a single parameter, is widely accepted for the representation of tower reactors. This model is based on visualization of the mixing process in the tower reactor as a random, diffusion-like eddy movement superimposed on a plug flow. The axial dispersion coefficient D_z is the only parameter in the formulation:

$$\frac{\partial C}{\partial \tau} = D_z \frac{\partial^2 C}{\partial z^2} + U_L \frac{\partial C}{\partial z} \qquad \text{...(26)}$$

where *C* is the concentration of a tracer. The boundary conditions depend on the specific type of tower reactor. This model is attractive, since it has a single parameter, the Bodenstein number (*Bo*), which is used to describe the mixing in the reactor:

$$Bo = \frac{U_L L}{D_z} \qquad \text{...(27)}$$

where *L* is the characteristic length. When the *Bo* number tends to infinity, the mixing conditions are similar to those of a plug-flow reactor, and the reactor can be considered as well-mixed for low *Bo* numbers.

The alternative approach of Buffham and Mason states that the mixing characteristics of a piece of equipment should be expressed as the variance σ^2 of the distribution obtained by injection of a pulse of tracer without adopting any mechanistic model. The relationship between *Bo* and σ^2 depends on the reactor configuration. The approach of Buffham and Mason facilitates the presentation of mixing characteristics free of any modeling assumptions. The variance σ^2 is the second moment of the distribution and carries information on the spread of the distribution around the mean value (first moment). Nevertheless, most of the data on mixing in bioreactors are presented either as t_m or as overall *Bo* numbers, which can be obtained by relatively simple experiments of pulse injection. Single-pass mixing in the ALR is due to mixing in the individual and interrelated sections of the reactor—riser, separator, downcomer, and bottom. Repeated passage mixing is the sum of the mixing in the subsequent passages. The latter is usually reported as the mixing time (t_m), the former as *Bo* or σ^2. Indeed, these parameters are interrelated, and knowledge of *Bo* or t_m is sufficient for calculating, theoretically, the mixing time based on the deviation of the envelope of the maxima in the response curve to a pulse, which is a measure of the degree of in-homogeneity. Verlaan et al. and Lin et al. correlated their results as follows:

$$t_m = MBo \qquad ...(28)$$

where *M* is a constant equal to 0.093 or to 0.089. The coefficient *M* given by Verlaan et al. is in exact agreement with the theoretical relationship derived by Murakami et al. for *Bo* >50 and a degree of in homogeneity, *I* = 0.05.

Equation 28 shows that the circulation path, which enters in the definition of *Bo,* has a linear effect on the mixing time. If the mean circulation time and the axial dispersion coefficients are known, it is possible to theoretically estimate the mixing time using equation 28. Experimental details must, however, be carefully planned to avoid complications. Note that in order to simplify data processing it is important to inject the signal and to measure the response at exactly the same point (often the position of the injection point is not specified despite its effect on the mixing time). In a study of the effect of the injection point on the dynamics of the mixing time, Schigerl et al. concluded that the gas–liquid separator is the best choice for tracer injection for short mixing times. Fields and Slater reported a marked dependence of the respiratory quotients upon the injection point of methanol during unlimited fed-batch gowth of *Methylophilusm ethylotropus* in a concentric-tube ALR. Because of the definition of the overall *Bo,* the values are specific to the reactor for which they were obtained and can be used only as indication of trends and orders of magnitude. As explained above, the ALR is, in fact, a combination of several regions having quite different fluid dynamic characteristics.

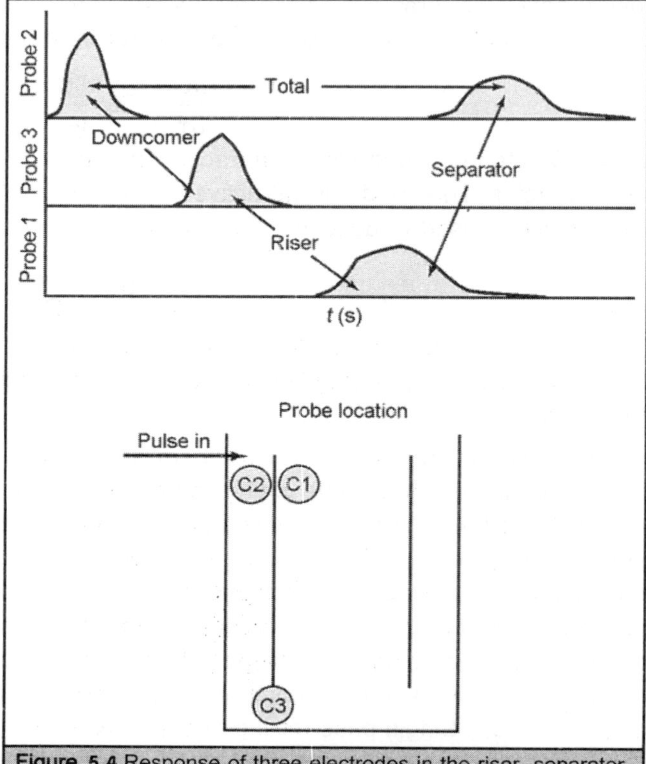

Figure 5.4 Response of three electrodes in the riser, separator, and downcomer to a pulse of electrode.

The overall mixing is the result of the contributions of each of them, and the overall *Bo* represents this combination for a particular reactor. This value thus has some limitations for extrapolation to other configurations or scales. Because each section has different mixing characteristics (it may have a different cross-section, flow configuration, etc.), from a strictly engineering point of view, the mixing in each of the sections of the ALR should be defined and considered separately. One way of obtaining information on the mixing characteristics of each of the regions of the ALR is the simultaneous measurement of the response in the ALR at several points, so that after one single pulse injection the response of each section in the loop can be obtained. This method of measurement has the advantage that multiple measurements are made for the same tracer injection experiment. This enables us to check the consistency of the liquid velocity results obtained, since independent measurements can obtained in the same run.

The results for *Bo* obtained by Verlaan in an external-loop ALR are shown in Figure 5.5 (the higher *Bo* in the downcomer was explained by the fact that data were obtained under conditions of complete gas disengagement so that only liquid flowed). The results indicated that most of the mixing took place in the gas separator, and both riser and downcomer could be considered as plug-flow conduits. The same conclusions were drawn for internal-loop ALRs by Merchuk et al. by analysis of mixing times: the shift in results for mixing times obtained with different probes in response to the same impulse perturbation indicated that the riser and the down-comer were introducing a pure delay in the response, that is, that they behaved as plug-flow sections. Indeed, the simplest model of an ALR, from the point of view of fluid dynamics, is a combination of two plug-flow reactors, representing the riser and the downcomer, and a perfectly stirred tank, representing the gas separator. Merchuk and Yunger showed that this simple model could take into account the mixing in the ALR.

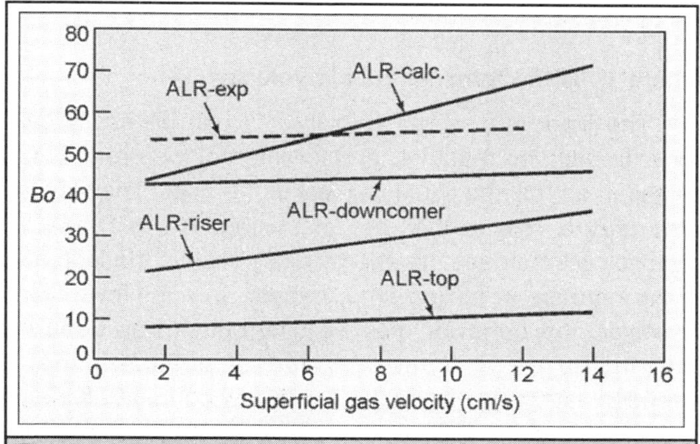

Figure 5.5 Bodenstein numbers as a function of the superficial gas velocity in an external-loop ALR.

The validity of using this simple model as a first approximation is supported by experimental evidence that shows that the mixing time decreases when the separator volume increases. When the volume of the separator is increased substantially without changing the reactor diameter, the gas separator becomes tall and slender and may depart from total mixing behavior. It has been reported that under such conditions the circulation time measured from two successive peaks of a probe is not influenced by changes in the top clearance. This indicates that the gas separator acts as two interconnected well-mixed regions, one of them being the link between riser and downcomer. The extent of mixing in the gas separator of internal-loop ALRs can probably be compared to that in a bubble column of the same dimensions. In the case of external-loop ALRs, the extent of mixing depends on the geometric design. In special designs, like the channel-loop reactor, a different approach must be taken. However, in general, a rule of thumb may be recommended: consider that all the mixing takes place in the gas separator, and calculate the degree of mixing in it using published data for bubble columns. Correlations available for bubble columns can thus be used. Godboole and Shah recommend the use of the correlation proposed by Deckwer et al.:

$$D_Z = 0.678 \cdot D^{1.4} \cdot J_G^{0.3} \qquad \text{...(29)}$$

where D_Z is the dispersion coefficient and D is the column diameter, or the relationship proposed by Joshi:

$$D_Z = 0.33(U_c + J_L)D \qquad \text{...(30)}$$

where J_L is the superficial liquid velocity and U_c is the cell circulation velocity given by:

$$U_C = 1.31\left[gD\left(J_G - \frac{\varphi_G}{1-\varphi_G}J_L - \varphi_G U_{b\infty}\right)\right] \qquad \text{...(31)}$$

where U_b is the terminal bubble velocity.

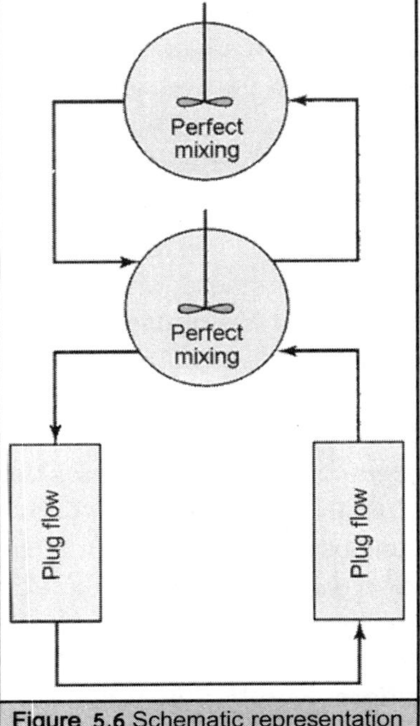

Figure 5.6 Schematic representation of the effect of a tall slender gas separation region in a concentric-tube ALR.

The liquid superficial velocity (J_L) can be easily calculated for external-loop reactors, and for internal loop reactors, J_L may be taken as half the liquid velocity in the riser. The viscosity may also play a role in the rate of mixing. An increase in liquid viscosity will increase the energy dissipation in the loop and result in an increase in mixing time and *Bo*. In non-Newtonian fluids, however, the behavior may be different. It has been reported that in the case of the addition of polymers that confer pseudoplastic behavior to the liquid, low concentrations produce a decrease of the mixing time. This can be explained in terms of the drag reduction due to the presence of polymers in the boundary layer near the walls. Fields et al. found that the mixing time increased for concentrations of the natural polymer, xanthan gum, above the critical concentration. (The critical concentration is a theoretically calculated value at which appreciable overlapping of polymer molecules occurs and which marks the onset of a rapid increase of the apparent viscosity).

Plug Flow Behavior

For all practical purposes, the gas in the riser of an ALR exhibits plug flow behavior. Only for extremely high J_G or hindered liquid circulation will the axial dispersion of the bubbles have some effect on the gas RTD. In the down- comer, the gas flow is almost plug-flow when the bubble recirculation is fully developed. But at the stage at which a stationary phase of suspended bubbles appears at the top of the downcomer, appreciable dispersion will occur. This zone has a large degree of mixing due to coalescence and consequent rise of larger bubbles amid smaller ones, with repeated events of breakup and coalescence. However, this type of operation has no relevance to practical applications. It is an operation mode to be avoided at all costs. Indeed, no data on mixing under these conditions have been reported.

The main question related to the mixing of the gas phase is, in fact, related to gas recirculation. When a particular gas flow has developed in the downcomer, part of the gas is being recirculated. Apulse of gas tracer at the inlet would produce, as a response, a series of pulses, separated one from the other by the gas circulation time. In practice, not many of these pulses would be detected, due to dilution and disengagement of the tracer in each pass through the separator. The only reported study on gas phase mixing in an ALR is that of Frehlich et al. The distribution of the gas residence times in two reactors, one of 0.08 m^3 and the other of 4 m^3, was measured using pseudostochastic tracer signals and a mass spectrometer. The values of *Bo* were calculated from the first and second peaks, indicating the main gas stream and the recirculation. The values obtained in a pilot plant were of the same order.

Shear-sensitive Cells

ALRs are being increasingly used in processes involving shear-sensitive cells (mammalian, insect, and plant cell cultures). This situation has created the need for considering shear stress as one of the parameters relevant in the design of such reactors. Although the a priori evaluation of shear rates is a matter that has been studied for many years in stirred tanks, information on this subject is scanty for pneumatically stirred reactors. The first approach made in this direction in pneumatically agitated vessels was that of Nishikawa et al., who were interested in the problem of heat transfer to a non- Newtonian liquid in a bubble column. This study was extended to mass transfer by Nakanoh and Yoshida, who proposed the expression:

$$\gamma = 5000\, J_G \qquad \ldots(32)$$

This expression has been widely accepted despite the criticism sometimes leveled at it. Some modifications have thus been proposed, which, like the original approach, were based on data for bubble column performance and therefore have limited application for ALRs. Recently, a method facilitating the prediction of the distribution of the energy dissipated in an ALR, based on a simple thermodynamic approach, has been developed. Energy dissipation was considered to occur in the ALR by two main mechanisms, wall friction and bubble-associated dissipation (ideal gas behavior was assumed). The work done by the gas on the liquid (and vice versa) was expressed assuming isothermal expansion of the bubbles. The energy dissipation inside the gas phase was considered negligible. The general energy balance was written as:

$$\Delta(PQ) + \Delta E_p + E_D = W_S \qquad \ldots(33)$$

In this equation, the first term represents the flow work lost by the system under consideration, E_D is the energy dissipated per unit of time, and W_S is the shaft work done by the surroundings on the system under consideration. The schematic representation of the concentric-tube ALR indicates the different points in the reactor considered in the mathematical expressions. The expressions found for the energy dissipated in each zone were the following:

Riser

$$(E_d)_R = Q_L(P_4 - P_5) - Q_L \rho_L gh - Q_r P_4 \ln\left(\frac{P_5}{P_4}\right) \qquad \ldots(34)$$

Gas separator

$$(E_d)_S = Q_L(P_5 - P_2) - Q_{in} P_4 \ln\left(\frac{P_1}{P_5}\right) - Q_d P_4 \ln\left(\frac{P_2}{P_5}\right) \qquad \ldots(35)$$

Downcomer

$$(E_d)_d = Q_L(P_2 - P_3) - Q_L\rho_L gh - Q_d P_4 \ln\left(\frac{P_3}{P_2}\right) \qquad \text{...(36)}$$

Bottom

$$(E_d)_d = Q_L)P_3 - P_4) - Q_d P_4 \ln\left(\frac{P_4}{P_3}\right) \qquad \text{...(37)}$$

Results of this model of an ALR can also be used to estimate the global shear rate in each region of the reactor, according to the global approach presented Merchuk and Ben-Zvi (Yona). The shear stress in the liquid of each region of the reactor can be defined as the energy dissipated divided by the mean path of circulation in the region and by the sum of the areas of all the bubbles. For the region *i* in the ALR

$$\tau_i = \frac{(E_d)_{bulk,i}\, t_i}{a_i h_i^2 A_i} \qquad \text{...(38)}$$

where t_i is the residence time of the liquid, h_i is the effective length, and a_i is the specific interfacial area, in the region *i*.

A global shear rate γ_i can be calculated for each region *i* as

$$\gamma_i = \frac{\tau_i}{\mu} \qquad \text{...(39)}$$

where μ is the effective viscosity of the fluid.

For liquids exhibiting different types of rheological behavior, the corresponding constitutive equation must be used. Such an approach has been used for the interpretation of shear effects on mammalian cells and algal growth.

VOLUMETRIC TRANSFER

The volumetric mass transfer coefficient ($k_L a$) is the rate of gas transfer across the gas-liquid interface per unit of driving force (the driving force is the gas concentration gradient between the liquid and the gas). The mass transfer coefficient $k_L a$ can be seen as the product of two terms: the mass transfer coefficient k_L and the specific interfacial area a. Both terms depend on a series of variables that can be grouped into three categories: (1) static properties of the liquid, such a density, diffusivity, and surface tension; (2) dynamic properties of the liquid (related to liquid flow), such as rheological parameters; and (3) liquid dynamics. In general, the variables in group 1 do not change very drastically. The variables in groups 2 and 3, however, may span wide ranges.

Measurements of Transfer Rate

Methods for the determination of $k_L a$ in a reactor can be grossly classified as steady-state and nonsteady-state methods. In the steady-state methods, the rates of oxygen uptake in steady-state operation are evaluated, either by measurement of inlet and outlet rates of oxygen or by direct analysis of a compound that reacts with the oxygen, as in the case of the sulfite method. One of the problems associated with these procedures is that the changes in oxygen concentration in the gas streams are usually small, and the errors of measurement thus have a substantial influence. When a chemical is added to the system, there is the question of whether the addition has provoked changes in the physicochemical properties, which thus become different from the properties of the original system. Transient methods may be applied to follow the response of the dissolved oxygen concentration in the system after a step-change of oxygen concentration in the inlet gas stream. These methods have the advantage that addition of an alien material is not required and that a single concentration is measured.

The correct use of this method has been analyzed in depth by Linek et al. One important point to take into account is the dynamics of the oxygen electrode. The lag in the response of the oxygen electrode makes it necessary to discern between the electrode response and the real oxygen concentration, especially when close to a sharp change in concentration. A correct analysis should also include the model of the dynamic behavior of the electrode. In order to simplify these procedures, approximations based on truncation of parts of the response curve have been proposed. These methods are based on truncating the first part of the electrode response obtained in a transient experiment. Once the error included in the value of k_La is set, the extent of truncation is fixed, allowing simplification of the analysis of the remainder of the curve. It should be kept in mind that this simplification implies the loss of part of the information, and due care should be given to statistical analysis of the results.

Variations of the method have been proposed to minimize disturbances in the system by introduction of step variations of agitation or pressure. In this way, the method can be applied to bioreactors during real operation of the system. One problem that may appear in the measurement of mass transfer rates, especially when viscous liquids are used, is related to the presence of very small bubbles that are depleted of oxygen very rapidly but do not disengage in the gas separator, thus constituting an inert volume of gas in the reactor. Kawase and Moo-Young analyzed the use of transient absorption of CO_2 for the determination of k_La and concluded that the error due to small bubble retention was much smaller than that in the case of O_2. Whatever the method used for the determination of k_La, the process of data elaboration is basically the same. The measured variable (usually oxygen concentration) is compared to the value predicted by a mathematical model of the process.

The model includes k_La as a parameter, and the value of k_La that gives the best fit is chosen. It is thus obvious that the choice of the model is very important, and poor assumptions on flow characteristics of gas or liquid phases may lead to errors and deviations from the true values. All the models are a simplified (amenable to mathematical treatment) description of the system. It should be kept in mind, therefore, that k_La is not a property of the system, but a parameter of the model adopted. If total mixing is assumed in the model adopted, the mass transfer coefficient obtained will consequently be limited. The assumption of complete mixing is such a common practice that many reports do not even specify explicitly that this has been done, and in many texts the mass transfer coefficient k_La is defined by the equation:

$$\text{OTR} = k_1a(C^* - C) \qquad \ldots(40)$$

It is often forgotten to state that this equation is valid only for perfectly mixed systems.

Strictly, the different sections of ALRs (riser, downcomer, and gas separator) have different flow characteristics, and the mass transfer coefficient may be expected to differ from one region to another. Some researchers have assumed that the contribution of the downcomer to the overall mass transfer is negligible and have reported values of $k_L a$ that are, in fact, the entire mass transfer divided by the volume of the riser. The fact that the values of the mass transfer coefficient are calculated and reported in different ways introduces some ambiguity into the published data; special care should therefore be taken when comparing data or using published mass transfer coefficients for design purposes. The assumption of perfect mixing in the liquid phase may be questionable in the case of tall reactors. One very simple method to verify this assumption is the simultaneous measurement of the response of the concentration to a step-change in the inlet.

In a perfectly mixed system, the location of the probe should be irrelevant. The difference between the figures stems from a different gas separator section, which changes the fluid dynamics in the system. In one case the system behaved as perfectly mixed, because of the large volume of the gas separator and the faster liquid circulation. In the other (closed system in the figure), the responses of the three electrodes were clearly different, calling for a different analysis. The validity of the criterion, originally proposed by Andre et al., that compares the circulation time in the reactor (t_c) with the characteristic time for mass transfer, which is the inverse of the mass transfer coefficient, was thus confirmed:

$$t_c \cdot k_L a < 0.5 \qquad \ldots(41)$$

If this condition is not fulfilled, the bioreactor cannot be considered as a perfectly mixed volume, and more sophisticated ways of analysis are required. In this case, the mathematical model, and, consequently, the analysis of the experimental data become much more complicated. Nevertheless, this approach has been used, and the mass transfer experiments render values of $k_L a$ for each of the regions of the ALR. The graph shows that the mass transfer takes place at the highest rate in the riser. Values for the down- comer are 50% lower, and those for the gas separator are intermediate between the riser and the downcomer. The overall mass transfer rate is the result of the balance between the volumes and rates in the three sections.

Size and Area

As said earlier, the interfacial area per unit volume is an important component of the volumetric mass transfer coefficient. In fact, it is the part of $k_L a$ that is most susceptible to changes in operation variables and fluid properties. The mass transfer coefficient k_L varies only within a limited range, but the interfacial area is the main component responsible for the changes in mass transfer rate due to variations in turbulence, initial bubble size, and liquid properties. The methods for measurement of the interfacial area are based either on rapid chemical reactions or on direct measurement of bubble size. If a mean bubble size can be defined, then the interfacial area can be evaluated with the aid of the holdup measurement, since, in a population of homogeneous bubble size, it can be applied:

$$a = \frac{6\varphi}{d_s} \qquad \ldots(42)$$

where the Sauter mean diameter (d_S) is given by

$$d_S = \frac{\sum n_i d_i^3}{\sum n_i d_i^2} \quad \ldots(43)$$

and n_i is the number of bubbles of diameter d_i. There are very few published data on bubble size applicable to ALRs. For the riser, the correlation presented by Miyahara et al. for the volumetric mean diameter of the bubbles in the riser of a concentric-tube ALR can be used:

$$\bar{d}\left(\frac{g\rho_L}{\sigma d_0}\right)^{1/3} = f(N_W) \quad \ldots(44)$$

where d_0 is the diameter of the sparger orifice and σ the surface tension,

$$N_W = \frac{We}{Fr^{0.5}} \quad \ldots(45)$$

and the function $f(N_W)$ is different for each range of N_W:

$$f(N_W) = 2.9 \qquad N < 1$$

$$f(N_W) = 2.9 N_W^{-0.188} \qquad 1 < N_W < 2$$

$$f(N_W) = 2.9 N_W^{-0.5} \qquad 2 < N_W < 4$$

$$f(N_W) = 3.6 \qquad 4 < N \quad \ldots(46)$$

Data on bubble size in the downcomer of an ALR were published by Popovic and Robinson. No data on interfacial area measured by chemical methods have been published.

Correlative Studies

There are two ways of correlating experimental data from ALRs. First, the hydrodynamic point of view suggests that the movement of the fluid in the reactor determines its overall behavior; the gas superficial velocity is therefore the more appropriate independent variable. Second, the thermodynamic point of view is based on a consideration of energy balance, a more global approach to the system. This will lead to correlation of the phenomena in the system as a function of the energy input. Indeed, it is easier to compare mass transfer coefficients in ALRs with those in conventional reactors when the data are presented as a function of the total power input (both mechanical and pneumatic) per unit volume of the medium. Among the internal-loop reactors, two correlations presented for reactors with a rectangular cross-section. For external-loop ALRs, the correlation by Popovic and Robinson is recommended. In the case of internal-loop ALRs, most of the correlations predict similar values. The correlation of Merchuk et al. can be recommended on the basis that more geometric parameters have been taken into account, and this gives greater generality. Nevertheless, the general considerations already expressed when analyzing other correlations in this chapter are valid here as well.

TEMPERATURE FLUCTUATION

Because of the relatively low reaction rates of processes involving microorganisms and cells, it may—in a very general way—be said that heat-effect problems related to local variations of temperature are not common in bioreactors. Even in the case in which polymeric products are released into the medium and very high viscosity is reached, heat transfer is not the controlling step, because such viscous media will hinder the mass transfer, and heat generation will consequently be limited. In such cases the main point of focus is thus, mass, rather than heat, transfer. Reactions catalyzed by immobilized enzymes, however, may require different considerations, because of higher reaction rates.

There is a far greater body of published data on heat transfer in bubble columns than in ALRs, and some of the basic observations are valid for both types of reactor. The heat transfer rate in bubble columns is much larger than that expected from single-phase flow. This is a result of the bubble-driven turbulence and liquid recirculation, which are characteristic of the flow in pneumatically agitated reactors. Several correlations have been proposed for the prediction of the heat transfer coefficient in these reactors. Recently, Kawase and Moo-Young presented an expression that satisfactorily fits most of the published data. The model is based on Levich's three-zone concept and Kolmogoroff's isotropic turbulence theory and has no empirical adjustable parameters. The model can take into consideration non-Newtonian behavior of the liquid and predicts the enhancement of the heat transfer due to the shear-thinning effect of the fluid. The dimensionless expression may be written as:

$$Nu = 0.075(10.3n^{-0.63})^{\beta} n^{1/3} (Pr*)\frac{1}{3} Fr^{-\beta} (Re*)^{\beta+3(n+1)} \qquad ...(47)$$

with

$$\beta = \frac{4-n}{6(n+1)};\ Pr* = \frac{\kappa D^{1-n}}{\left(\dfrac{\kappa}{C_p}\right) J_G^{1-n}};\ Re* = \frac{D^n J_G^{2-n}}{\left(\dfrac{k}{\rho}\right)} \qquad ...(48)$$

where the symbols defined in "Nomenclature" have been used.

For Newtonian liquids, this equation reduces to:

$$Nu = 0.134^{1/3}\ (Pr)^{1/3}\ Fr^{-1/4}\ (Re)^{3/4} \qquad ...(49)$$

Kawase and Moo-Young compared their model with data and correlations published by various investigators and found satisfactory agreement for both Newtonian and non-Newtonian liquids. For ALRs, the flow in the reactor maybe similar to that in bubble columns if the internal recirculation is high, or it may be even closer to net two-phase flow in pipes, in which case the equations for heat transfer in pipes can be used. Blakebrough et al. studied heat transfer rates in an external-loop airlift fermentor used for culturing *Aspergillus niger*. They concluded that the enhancement of the heat transfer rate could be explained by disturbances in the liquid layer near the surface caused by the presence of the microorganisms.

FLUID DYNAMIC CHARACTERISTICS

The special qualities of the ALR stem, as stated before, from its fluid dynamic characteristics. One such characteristic is the directionality of the liquid flow. Independently of superimposed fluctuations, a clear net flow is present in the reactor, with exception of the gas separator in internal-loop designs. Therefore, it is to be expected that the fluidization capacity of the ALR will be markedly superior to that of a bubble column. Several studies have been conducted on the suspension of solids in ALRs, particularly on the use of this type of device for catalytic processes in the chemical industry, where the solid support is usually heavy. In this regard, a very important point is the minimum gas superficial velocity that leads to complete solid fluidization.

Hysteresis has been observed in some cases; once total fluidization has been attained, the superficial velocity can be reduced to values lower than that required to reach this state. This is due to the high pressure drop related to passage of liquid through a bed of solids, before fluidization, as compared to the drag forces required to maintain the solid in suspension after all the solids are suspended. Contradictory data on the effect of the suspended solids on the reactor performance have been reported. Fan et al. claimed that the overall gas holdup increased due to the presence of the solids, whereas Koide et al. showed the opposite effect on the gas holdup and reported a small decrease in k_La as well. It is possible that these discrepancies are due to the use of different solids. One of the properties of solids that is often overlooked is wettability. Small bubbles may adhere on wettable solids, leading to a change in the apparent density of the particle and thus changing their solid circulation velocity. In the case of suspended solids that take an active part in the process, the mass transfer rate from the liquid to the solid may become the limiting step. The dependence of particle size on the mass transfer to the suspended solids has been studied by several authors.

All the comments made above relate to heavy solids. This situation is not very frequent in biological processes, with the exception of biolixiviation or the special case of microbial desulfuration of coal. In most of the biological processes that may take place in an ALR, the solids are either cells, clumps of cells, or supports that are not much more dense than the medium. Therefore, neither fluidization nor the distribution of solids in the reactor constitutes a problem. Assa and Bar found very small variations in the axial distribution of animal and plant cells suspended in an internal-loop ALR. This is due to the small free-falling velocity of the solids, which is the reason for the difference in loading between the riser and the downcomer when heavy particles are used. Because of the small difference between solid and medium density, the movement of the particles usually present in biological processes is not as dependent on gravity forces as on liquid and bubble movement. In this case, the transport of liquid in the wake of the bubbles maybe considered to be the prevailing transport mechanism.

Snape and Thomas proposed a Monte Carlo algorithm for modeling particle movement by this mechanism in bubbly flow to predict distribution of circulation times in the reactor. Koide et al. conducted a broad study of gas holdup and mass transfer rates in an internal-loop ALR containing Ca-alginate beads, which are often used for cell immobilization. The authors found that solid loading had a negative effect on both parameters, but the particle diameter had no influence in the range studied (1.8 mm $< d_p <$ 3.98 mm). Chang et al. studied the influence of suspended pellets (cylindrical pellets of immobilized penicillin acylase) on the mixing time in an internal-loop ALR. They found mixing time increased when the solid volumetric concentration was raised up to 15%. For higher concentrations the trend inverted, and the mixing time decreased. No such effects were observed with heavier solids. Increases in gas superficial velocity and in top clearance both lead to decreases in mixing time in all cases.

DESIGNING AIRLIFT REACTOR

The problems encountered in the scale-up of bioreactors can be concentrated into two groups. The first includes the cases in which a high power input per unit volume is used on a laboratory scale, but cannot be maintained on an industrial scale due to economic or mechanical limitations. This is not the case for plant or animal cultures, for which a very high specific power input cannot be used because of cell fragility. The second group of problems can be generalized as lack of knowledge about the hydrodynamics of large-volume vessels. The methods available for scale-up of bioreactors have been reviewed by Oosterhuis, Kossen and Oosterhuis, Sweere et al., and Sola and Godia, among others. Because, in general, design from first principles cannot be undertaken because of the lack of basic knowledge about the hydrodynamics of the bioreactors, one possible solution is the semifundamental method, which comprises using approximate simple models for fluid dynamics and integrating them with basic known kinetics and heat and mass transfer rates. It does not happen very often that all this information is available with a degree of certainty that allows safe design of a large system.

Thus, the designer must usually resort to dimensional analysis. This method requires a knowledge of all the variables affecting the process, which can be obtained from a qualitative, but realistic, model. A simplified version of this method is to limit the number of variables to one or two and to follow rules of thumb, which, depending on the specific case, may be constant P/V, constant $k_L a$, etc. The literature shows, however, many cases of inconsistency for this method. For example, design of a scaled-up bioreactor in which the oxygen transfer rate is kept constant can lead to a better performance than expected, as in the example reported by Taguchi et al. for glucoamylase production by *Endomyces* sp., or worse than expected, as in the case of protease production by *Streptomyces* sp.

The method of regime analysis and scale down, proposed by Kossen and Oosterhuis, combines two tools to overcome the problem posed by the complexity of biochemical reactors. Their method is based on considering the regime of the full-scale process as the objective and planning the strategy of process development from this point on. This method is therefore applicable only to conventional and well-studied bioreactors that are to be used in a new bioprocess. Regime analysis is based on the consideration that, generally, biochemical processes involve a series of steps, some being mass or heat transfer by convection, some being diffusive mechanisms (activated or not), and others being chemical reaction steps.

In the latter case, a mass-transfer mechanism is superimposed, since molecules must encounter one another in order to react, and usually a heat effect will accompany the reaction. Depending on whether these steps take place in parallel or in series and on the relaxation time for each step, the rate of the total process is often given as the rate of one single step. But the equilibrium between all the individual rates can be (and usually is) upset by a change in scale. This is to be expected, because a change in scale will not bring a change in the physicochemical or kinetic parameters (scale-insensitive variables), but will affect the overall convective mass and heat rates (scale-sensitive variables). A new equilibrium will be established, and the interplay of all the parameters of the system may lead to a regime in which a different step becomes the step-controlling the process rate. The method of Kossen and Oosterhuis starts with an analysis of the operation of the large-scale system. Once the regime is clarified, a small-scale system is designed in such a way that it simulates the operation regime of the larger one.

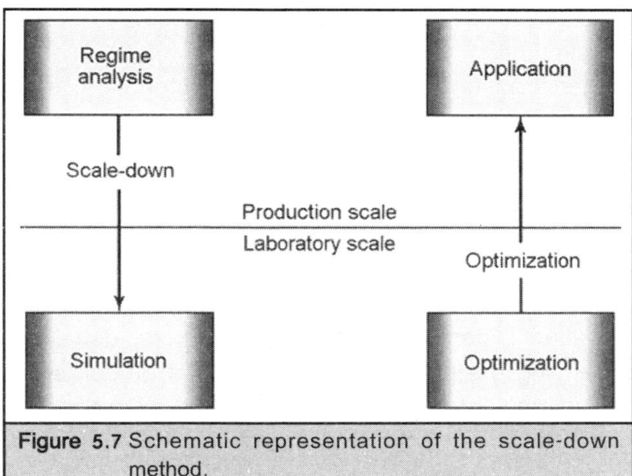

Figure 5.7 Schematic representation of the scale-down method.

Optimization studies can be done on the smaller model, and conclusions will then be extrapolated to the full-scale process. An example of this method is given by Oosterhuis, in which a large-scale stirred tank reactor is simulated on a laboratory scale by two interconnected vessels, one with a small liquid volume, high agitation rate, and high rate of oxygen supply (representing the zone near the impeller-sparger in the reactor), and a second vessel, with a much larger liquid volume, oxygen consumption (simulated by nitrogen sparging), and gentle agitation. The success of this model suggests that large vessels must be carefully analyzed and their internal structure studied for proper modeling and design to be performed. The analysis starts with the definition of the characteristic time constants of the system. In the case of low gas recirculation, the relationship between the total circulation time and the residence time in the downcomer may become very relevant. This was shown by MacNeil and Kristiansen for the scale-up of an external-loop ALR for the production of *Aureobasidium pullulans*.

The poor oxygen supply in the downcomer was reflected on the level of product synthesis. The smaller residence times lead to higher *A. pullulans* production. An analysis of the time constants required for the design of an ALR for plant cell culture has recently been presented. The reported values of the time constants calculated for an external-loop ALR for mixing, mass transfer, and oxygen consumption at two different cell concentrations: 5 and 30 kg/m^3. These graphs exemplify the changes in the controlling mechanism that may take place as the physical properties of the broth change in the course of the bioprocess. Details on bioreactor scale-up may be found in the recent review of Sola. It should be remembered, however, that the reliability of a scale-up will always be limited by the quality of the predictions of gas holdup, liquid velocity, and mass transfer rate.

Essential Modifications

In the design of ALR, several modifications aiming at the improvement of some of the characteristics of the equipment have been proposed. One of the earliest modifications was the two-staged ALR, proposed by Orazem and Erickson. Their design was inspired by the improvement observed by them in the performance obtained with multistage sieve trays over single-stage bubble columns. They claimed that a substantially higher mass transfer coefficient was obtained, as was a better performance in terms of oxygen transferred per unit of energy invested. The combination of a concentric-tube ALR with a marine propeller (1000 rpm) was studied by Pollard et al. This modification, in which the stirrer was located near the bottom of the draft-tube, served to increase the circulation rate, which may be low for viscous liquids. This configuration showed enhanced oxygen transfer and more uniform distribution of the dissolved oxygen concentration throughout the reactor in comparison with that of the unmodified ALR.

The liquid circulation was also intensified. However, the improvement was achieved at the cost of the introduction of a focus of energy input. As has been mentioned before, the absence of such a high shear region is one of the advantages of the ALR. This advantage is, in part at least, lost. It is interesting to note that a bakers' yeast culture grown in the same type of modified reactor did not reveal any improvement. This may be related to the strong shear rate in the vicinity of the propeller. Different types of static mixers, usually located in the riser, have also been used for enhancing the performance of ALRs. Potucek studied the influence of static mixers on the gas holdup and interfacial gas-liquid area in a concentric-tube ALR and found an improvement in interfacial area. Enhancement of $k_L a$ has been also obtained, especially in viscous liquids, as reported by Gaspillo and Goto and Chisti et al. Zhou et al. showed better growth of *Cephalosporum acremonium* in an ALR with static mixers.

As mentioned above, it has been shown repeatedly that the mass transfer rate in a bubble column is usually higher than that in the conventional ALR. It therefore makes sense to try to bring into the ALR some of the characteristics of bubble columns in a controlled fashion. This was done by Bando et al., who tested a perforated draft tube in a concentric-tube ALR. The perforations in the draft tube facilitated communication between the less-well aerated liquid in the downcomer and the better aerated riser. The reported improvements in mass transfer rates were undoubtedly obtained at the cost of a reduction of circulation velocity. One of the advantages of the directionality of flow in ALRs is the improved fluidization capacity. The strengthening of this advantage was the aim of another modification, the helical flow promoter, proposed by Gluz and Merchuk. The helical flow promoter causes the fluid to flow down in the downcomer in a helical pattern. The device comprises several fins or baffles, which have the effect of modifying the flow paths; instead of going in straight lines along the axis, the flow paths move along an helix.

The baffles may be installed in a small section at almost any place along the riser or the downcomer, and the effect is perceived throughout the reactor. One of the best positions for the helical flow promoter is the top of the riser. A helical flow is then generated in the downcomer to produce a swirl at the bottom and a corkscrew-like path in the riser. This has a strong potential for the culture of photosynthetic microorganisms. The helical movement causes secondary flow, which leads to enhanced radial mixing, and therefore more homogeneous distribution of light and heat among liquid elements and suspended particles. With the helical flow promoter, it is thus more likely that all the elements of the fluid get the same exposure to light and heat exchange. One of the most important characteristics of the helical flow promoter is the enhanced capacity for fluidizing solid particles. This is due to the swirls that develop at the bottom of the reactor. Thus, this modification is especially suited to processes operating with cells immobilized on a solid.

The minimal gas flow rate for complete fluidization of solids in an ALR may be reduced drastically by the use of a helical flow promoter. In addition, the mass transfer rate to suspended solids may be enhanced up to 50%, because of the higher relative velocity between the particles and solids.

Mass transfer coefficients obtained in a bubble column, an ALR, and an ALR with HFP were compared, for water and two solutions of CMC: 0.2% and 0.04%. The corresponding effective viscosities for $\gamma = 50\ s^{-1}$ are 16 and 38 mPa s respectively. It is clearly seen that although with water there is a consistent difference in favor of the bubble column, the differences diminish as the CMC concentration increases, and for the highest concentration tested the difference is negligible. This implies that under these conditions the disadvantage of the HFP in mass transfer vanishes, whereas all the advantages mentioned before subsist.

CONCLUDING REMARKS

ALRs are popular in modern bioprocess research and development. These reactors are particularly suitable for biological processes in which a high mass transfer rate is required, but excessive power input may lead to damage of the cells due to shear effects. ALRs also have very appealing characteristics for bioprocesses for low-value products in which efficiency of energy utilization may become the key point for design. Such is the case for ALRs for wastewater treatment. The ALR is particularly effective for solid fluidization, which is important in many biological processes in which the biocatalyst is available in the form of pellets or is immobilized on a solid support. The distinctive characteristics of ALRs are conferred by the fluid dynamics of the liquid–gas or liquid–gas–solid mixtures in it. These characteristics are expressed as gas holdup, liquid velocity, and mixing in each of the zones of the ALR.

It is important to recognize the differences in the fluid dynamic characteristics of these zones: the riser, the gas separator, and the downcomer. Only a correct understanding of behavior and interconnection of these regions can make possible the correct design of a new reactor or the scale-up of a laboratory device up to pilot or industrial size. The purpose of scale-up is to conserve and repeat on a larger scale the fluid dynamics of the reactor. Therefore, one of the most important factors in the design and scaleup of reactors is the influence of the geometric characteristics of the system on the flow of the different phases present. The variables affecting the performance of the reactor are geometric design, operation variables, and fluid properties. Several correlations are available in the technical literature for the prediction of the fluid dynamic characteristics of the reactor (gas holdup, liquid velocity, mixing time of axial dispersion coefficient) and of the mass and heat transfer coefficients. Although attempts have been made to study all the aspects described, no single research group has managed to cover all the variables over a wide range. It is therefore extremely important that the engineer confronting scale-up or de novo design of an ALR analyzes in depth the range of validity of the correlations used for the calculations.

6

INDUSTRIAL DECOLOURATION OF CARBON

Activated carbon is a predominantly amorphous solid that has an extraordinarily large internal surface area and pore volume. These unique characteristics are responsible for its adsorptive properties, which are exploited in many different liquid- and gas-phase applications. Activated carbon is an exceptionally versatile adsorbent because the size and distribution of the pores within the carbon matrix can be controlled to meet the needs of current and emerging markets. Engineering requirements of specific applications are satisfied by producing activated carbons in the form of powders, granules, and shaped products. Through choice of precursor, method of activation, and control of processing conditions, the adsorptive properties of products are tailored for applications as diverse as the purification of potable water and the control of emissions from bioproduct recovery processes. In 1900, two very significant processes in the development and manufacture of activated carbon products were patented. The first commercial products were produced in Europe under these patents: Eponite, from wood in 1909, and Norit, from peat in 1911. Activated carbon was first produced in the United States in 1913 by Westvaco Corp. under the name Filtchar, using a by-product of the papermaking process. Further milestones in development were reached as a result of World War I. In response to the need for protective gas masks, a hard, granular activated carbon was produced from coconut shell in 1915. Following the war, large-scale commercial use of activated carbon was extended to refining of beet sugar and corn syrup and to purification of municipal water supplies. The termination of the supply of coconut char from the Philippines and India during World War II forced the domestic development of granular activated carbon products from coal in 1940. More recent innovations in the manufacture and use of activated carbon products have been driven by the need to recycle resources and to prevent environmental pollution.

CHARACTERISTICS

The structure of activated carbon is best described as a twisted network of defective carbon layer planes, cross-linked by aliphatic bridging groups. X-ray diffraction patterns of activated carbon reveal that it is nongraphitic, remaining amorphous because the randomly cross-linked network inhibits reordering of the structure even when heated to 3,000°C. This property of activated carbon contributes to its most unique feature, namely, the highly developed and accessible internal pore structure. The surface area, dimensions, and distribution of the pores depend on the precursor and on the conditions of carbonization and activation. Pore sizes are classified by the International Union of Pure and Applied Chemistry (IUPAC) as micropores (pore width <2 nm), mesopores (pore width 2-50 nm), and macropores (pore width >50 nm). The surface area of activated carbon is usually determined by application of the Brunauer-Emmett-Teller (BET) model of physical adsorption using nitrogen as the adsorptive.

Typical commercial products have specific surface areas in the range 500-2,000 m^2/g, but values as high as 3,500-5,000 m^2/g have been reported for some activated carbons. In general, however, the effective surface area of a microporous activated carbon is far smaller because the adsorption of nitrogen in micropores does not occur according to the process assumed in the BET model, which results in unrealistically high values for surface area. Adsorption isotherms are usually determined for the appropriate adsorptives to assess the effective surface area of a product in a specific application. Adsorption capacity and rate of adsorption depend on the internal surface area and distribution of pore size and shape but are also influenced by the surface chemistry of the activated carbon. The macroporosity of the carbon is important for the transfer of adsorbate molecules to adsorption sites within the particle. Functional groups are formed during activation by interaction of free radicals on the carbon surface with atoms such as oxygen and nitrogen, both from within the precursor and from the atmosphere.

The functional groups render the surface of activated carbon chemically reactive and influence its adsorptive properties. Activated carbon is generally considered to exhibit a low affinity for water, which is an important property with respect to the adsorption of gases in the presence of moisture. However, the functional groups on the carbon surface can interact with water, rendering the carbon surface more hydrophilic. Surface oxidation, which is an inherent feature of activated carbon production, results in hydroxyl, carbonyl, and carboxylic groups that impart an amphoteric character to the carbon, so that it can be either acidic or basic. The electrokinetic properties of an activated carbon product are, therefore, important with respect to its use as a catalyst support. As well as influencing the adsorption of many molecules, surface oxide groups contribute to the reactivity of activated carbons toward certain solvents in solvent recovery applications. In addition to surface area, pore size distribution, and surface chemistry, other important properties of commercial activated carbon products include pore volume, particle size distribution, apparent or bulk density, particle density, abrasion resistance, hardness, and ash content.

PRODUCTION

Commercial activated carbon products are produced from organic materials that are rich in carbon, particularly coal, lignite, wood, nut shells, peat, pitches, and cokes. The choice of precursor is largely dependent on its availability, cost, and purity, but the manufacturing process and intended application of the product are also important considerations. Manufacturing processes fall into two categories, thermal activation and chemical activation. The effective porosity of activated carbon produced by thermal activation is the result of gasification of the carbon at relatively high temperatures, but the porosity of chemically activated products is generally created by chemical dehydration reactions occurring at significantly lower temperatures.

Method of Thermal Activation

Thermal activation occurs in two stages: thermal decomposition or carbonization of the precursor and controlled gasification or activation of the crude char. During carbonization, elements such as hydrogen and oxygen are eliminated from the precursor to produce a carbon skeleton possessing a latent pore structure. During gasification, the char is exposed to an oxidizing atmosphere that greatly increases the pore volume and surface area of the product through elimination of volatile pyrolysis products and from carbon burn-off. Carbonization and activation of the char are generally carried out in direct-fired rotary kilns or multiple-hearth furnaces, but fluidized-bed reactors have also been used. Materials of construction, notably steel and refractories, are designed to withstand the high-temperature conditions (i.e., >1,000°C) inherent in activation processes. Bituminous coal is pulverized and passed to a briquette press. Binders may be added at this stage before compression of the coal into briquettes. The briquetted coal is then crushed and passed through a screen, from which the on-size material passes to an oxidizing kiln. Here, the coking properties of the coal particles are destroyed by oxidation at moderate temperatures in air. The oxidized coal is then devolatilized in a second rotary kiln at higher temperatures under steam.

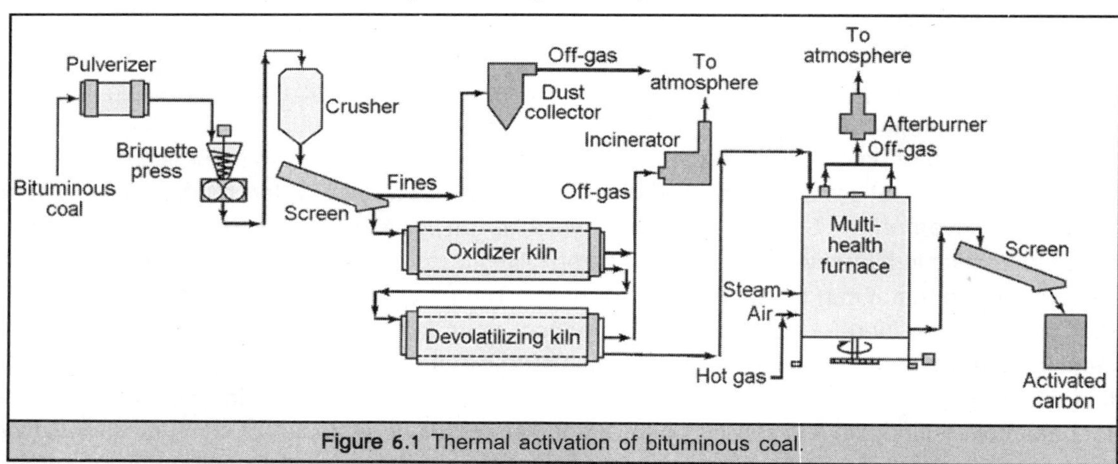

Figure 6.1 Thermal activation of bituminous coal.

To comply with environmental pollution regulations, the kiln off-gases containing dust and volatile matter pass through an incinerator before discharge to the atmosphere. The devolatilized coal particles are transported to a direct-fired multihearth furnace where they are activated by holding the temperature of the furnace at about 1,000°C. Product quality is maintained by controlling coal feed rate and bed temperature. As before, dust particles in the furnace off-gas are combusted in an afterburner before discharge of the gas to the atmosphere. Finally, the granular product is screened to provide the desired particle size. A typical yield of activated carbon is about 30–35% by weight based on the raw coal. The process for the thermal activation of other carbonaceous materials is modified according to the precursor. For example, the production of activated carbon from coconut shell does not require the stages involving briquetting, oxidation, and devolatilization. To obtain a high- activity product, however, it is important that the coconut shell is charred slowly prior to activation of the char. In some processes, the precursor or product is acid-washed to obtain a final product with a low ash content.

Method of Chemical Activation

In contrast to the thermal activation of coal, chemical activation is generally carried out commercially in a single kiln. The precursor, usually wood, is impregnated with a chemical activation agent, typically phosphoric acid, and the blend is heated to a temperature of 450–700°C. Chemical activation agents reduce the formation of tar and other by-products, thereby increasing carbon yield. Sawdust is impregnated with concentrated phosphoric acid and fed to a rotary kiln, where it is dried, carbonized, and activated at a moderate temperature. To comply with environmental pollution regulations, the kiln off-gases are treated before discharge to the atmosphere. The char is washed with water to remove the acid from the carbon, and the carbon is separated from the slurry. The filtrate is then passed to an acid-recovery unit. Some manufacturing plants do not recycle all the acid but use a part of it to manufacture fertilizer in an allied plant. If necessary, the pH of the activated carbon is adjusted, and the product is dried. The dry product is screened and classified into the size range required for specific granular carbon applications. Carbon yields as high as 50% by weight of the wood precursor have been reported.

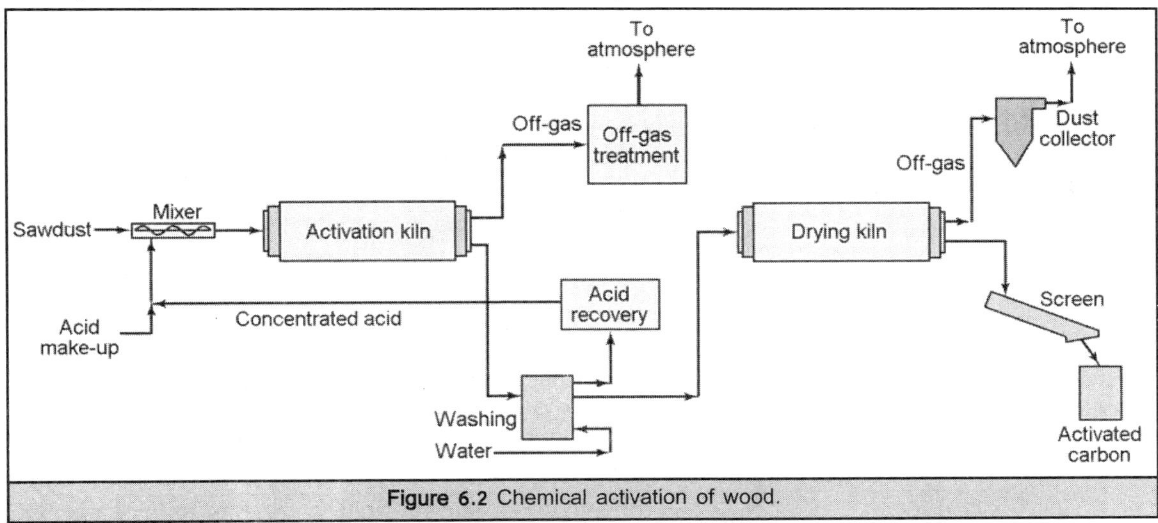

Figure 6.2 Chemical activation of wood.

Other method of activation

Different chemical activation processes have been used to produce carbons with enhanced adsorption characteristics. Activated carbons of exceptionally high surface area (>3,000 m^2/g) have been produced by the chemical activation of carbonaceous materials with potassium hydroxide. Activated carbons are also produced commercially in the form of cloths, fibers, and foams generally by chemical activation of the precursor with a Lewis acid such as aluminum chloride, ferric chloride, or zinc chloride.

Activated Products of Carbon

To meet the engineering requirements of specific applications, activated carbons are produced and classified as granular, powdered, or shaped products. Granular activated carbons are produced directly from granular precursors, such as sawdust and crushed and sized coconut char or coal. The granular product is screened and sized for specific applications. Powdered activated carbons are obtained by grinding granular products. Shaped activated carbon products are generally produced as cylindrical pellets by extrusion of the precursor with a suitable binder before activation of the precursor.

Marketing

Activated carbon products are shipped in bags, drums, and boxes in weights ranging from about 10 to 35 kg. Containers can be lined or covered with plastic and should be stored in a protected area both to prevent weather damage and to minimize contact with organic vapors that could reduce the adsorption performance of the product. Bulk quantities of activated carbon products are shipped in metal bins and bulk bags, typically 1-2 m^3 in volume, and in railcars and tank trucks. Bulk carbon shipments are generally transferred by pneumatic conveyors and stored in tanks. However, in applications such as water treatment where water adsorption does not impact product performance, bulk carbon may be transferred and stored as a slurry in water.

Conditions

Activated carbon producers furnish product bulletins that list specifications, usually expressed as a maximum or minimum value, and typical properties for each grade produced. Standards helpful in setting purchasing specifications for granular and powdered activated carbon products have been published.

COMMERCIAL VALUES

Excluding eastern European countries and China, where production figures have not been published, the world production capacity of activated carbon was estimated to be 375,000 metric tons in 1990. The price of most products was 0.70 to 5.50 \$/kg, but some specialty carbons were more expensive. Forty percent of the production capacity was in the United States, 30% in western Europe, 20% in Japan, and 10% in other Pacific Rim countries.

Table 6.1 World production capacity

Country	*Capacity (10^3 t)*
United States	146
Western Europe	108
Japan	72
Pacific Rim, other	49
Total	375

Production capacity was almost equally split between powdered and nonpowdered activated carbon products. Powdered activated carbon, a less expensive form used in liquid-phase applications, is generally used once and then disposed of. In some cases, however, granular and shaped products are regenerated and reused. In 1990 production capacity for granular and shaped products was split, with about two-thirds for liquid-phase and one-third for gas-phase applications. Over the last decade, production capacity in the United States remained essentially unchanged, but minor fluctuations occurred in response to changes in environmental regulations. A similar reaction was noted worldwide.

The current demand for activated carbon is estimated at 93% of production capacity. The near-term growth in demand is projected to be approximately 5.5%/year. In 1970 the U.S. Congress enacted the Clean Air Act, the Clean Water Act, and the Safe Drinking Water Act. Because activated carbon can often be used to help meet Environmental Protection Agency (EPA) regulations, the U.S. activated carbon industry reacted by increasing its production capacity. A proposed amendment to the Safe Drinking Water Act in 1979 required the use of granular activated carbon systems, but the amendment was not enacted. In response to the projected increase in demand for activated carbon, production capacity remained high until the late 1980s, but when the anticipated need did not materialize, some production facilities were shut down. Currently, because of stricter EPA regulations implementing all three acts in 1990, the industry will increase production capacity by 25% during the next several years.

Western Europe has seven manufacturers of activated carbon. The two largest, Norit and Chemviron (a subsidiary of Calgon), account for 70% of western European production capacity, and Ceca accounts for 13%, Japan is the third-largest producer of activated carbon, manufacturers, but 4 companies share over 50% of the total Japanese capacity. Six Pacific Rim countries account for the balance of the world production capacity of activated carbon, 90% of which is in the Philippines and Sri Lanka. As is the case with other businesses, regional markets for activated carbon products have become international, leading to consolidation of manufacturers. Calgon, Norit, Ceca, and Sutcliffe- Speakman are examples of multinational companies. Activated carbon is a recyclable material that can be regenerated. Thus the economics, especially the market growth of activated carbon, particularly granular and shaped products, is affected by regeneration and industry regeneration capacity.

The decision to regenerate an activated carbon product is dependent on the cost, size of the carbon system, type of adsorbate, and the environmental issue involved. Large carbon systems, such as those used in potable and wastewater treatment, generally require a high-temperature treatment, which is typically carried out in rotary or multihearth furnaces. During regeneration, carbon losses of 1 to 15% typically occur from the treatment and movement of the carbon. However, material loss is compensated for by the addition of new carbon to the adsorber system. In general, regeneration of spent carbon is considerably less expensive than the purchase of new activated carbon. For example, fluidized-bed furnace regeneration of activated carbon used in a 94,600-m^3/ day water treatment system cost only 35% of new material. For this system, regeneration using either infrared or multihearth furnaces was estimated to be more expensive but still significantly less so than the cost of new carbon. Because powdered activated carbon is generally used in relatively small quantities, the spent carbon has often been disposed of in landfills. However, landfill disposal is becoming more restrictive environmentally and more costly.

Thus large consumers of powdered carbon find that regeneration is an attractive alternative. Examples of regeneration systems for powdered activated carbon include the Zimpro/Passavant wet air oxidation process, the multihearth furnace as used in the DuPont PACT process, and the Shirco infrared furnace. Other types of regenerators designed for specific adsorption systems may use solvents and chemicals to remove susceptible adsorbates, steam or heated inert gas to recover volatile organic solvents, and biological systems in which organics adsorbed on the activated carbon during water treatment are continuously degraded.

APPLICATIONS

Activated carbon generally presents no particular health hazard as defined by NIOSH. However, it is a nuisance and mild irritant with respect to inhalation, skin contact, eye exposure, and ingestion. On the other hand, special consideration must be given to the handling of spent carbon that may contain a concentration of toxic compounds. Activated carbon products used for decolorizing food products in liquid form must meet the requirements of the *Food Chemical Codex* as prepared by the Food and Nutrition Board of the National Research Council. According to the National Board of Fire Underwriters, activated carbons normally used for water treatment pose no dust explosion hazard and are not subject to spontaneous combustion when confined to bags, drums, or storage bins. However, activated carbon burns when sufficient heat is applied; the ignition point varies between about 300 and 600°C. Dust-tight electrical systems should be used in areas where activated carbon is present, particularly powdered products. When partially wet activated car bon comes into contact with unprotected metal, galvanic currents can be set up; these result in metal corrosion. Manufacturer material safety data sheets (MSDS) indicate that the oxygen concentration in bulk storage bins or other enclosed vessels can be reduced by wet activated carbon to a level that will not support life. Therefore, self-contained air packs should be used by personnel entering enclosed vessels where activated carbon is present.

LIQUID-PHASE APPLICATIONS

Activated carbons for use in liquid-phase applications differ from gas-phase carbons primarily in pore size distribution. Liquid-phase carbons have significantly more pore volume in the macropore range, which permits liquids to diffuse more rapidly into the mesopores and micropores. The larger pores also promote greater adsorption of large molecules, either impurities or products, in many liquid-phase applications. Specific-grade choice is based on the isotherm and, in some cases, bench or pilot- scale evaluations of candidate carbons. Liquid-phase activated carbon can be applied either as a powder, granular, or shaped form. The average size of powdered carbon particles is 15–25 μm. Granular or shaped carbon particle size is usually 0.3-3.0 mm.

A significant factor in choosing between powdered and nonpowdered carbon is the degree of purification required in the adsorption application. Granular and shaped carbons are usually used in continuous flow through deep beds to remove essentially all contaminants from the liquid being treated. Granular and shaped carbon systems are preferred when a large carbon buffer is needed to withstand significant variations in adsorption conditions, such as in cases where large contaminant spikes may occur. A wider range of impurity removal can be attained by batch application of powdered carbon, and the powdered carbon dose per batch can be controlled to achieve the degree of purification desired. Batch-stirred vessels are most often used in treating material with powdered activated carbon. The type of carbon, contact time, and amount of carbon vary with the desired degree of purification.

Figure 6.3 Multistage countercurrent application of powdered activated carbon.

The efficiency of activated carbon may be improved by applying continuous, countercurrent carbon-liquid flow with multiple stages. Carbon is separated from the liquid at each stage by settling or filtration. Filter aids such as diatomaceous earth are sometimes used to improve filtration. Granular and shaped carbons are generally used in continuous systems where the liquid to be treated is passed through a fixed bed. Compounds are adsorbed by the carbon bed in the adsorption zone. As carbon in the bed becomes saturated with adsorbates, the adsorption zone moves in the direction of flow, and break-through occurs when the leading edge of the adsorption zone reaches the end of the column.

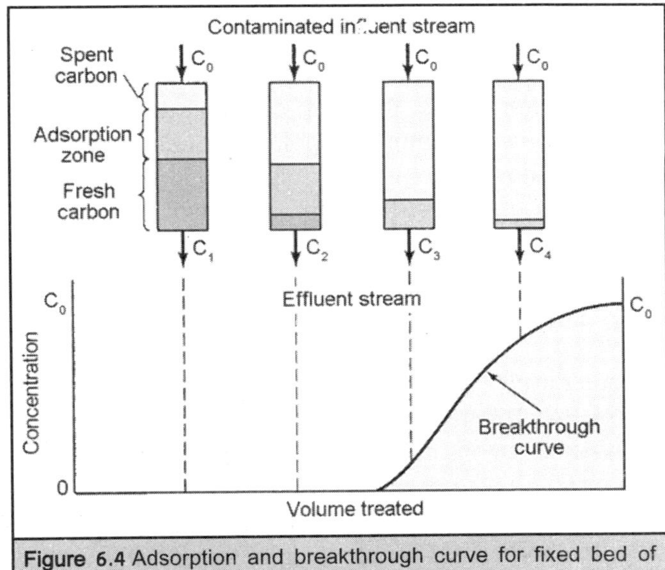

Figure 6.4 Adsorption and breakthrough curve for fixed bed of granular or shaped activated carbon.

Normally at least two columns in series are on line at any given time. When the first column becomes saturated, it is removed from service, and a column containing fresh carbon is added at the discharge end of the series. An alternative approach is the moving bed column. In this design the adsorption zone is contained within a single column by passing liquid upward while continuously or intermittently withdrawing spent carbon at the bottom and adding fresh carbon at the top. The total activated carbon consumption for liquid-phase applications in the United States in 1987 was estimated to be about 76,700 t, which accounted for nearly 80% of the total activated carbon use.

Table 6.2 Liquid-Phase Activated Carbon Consumption (10^3 t).

	Granular-shaped	*Powdered*	*Total*
Potable water	4.5	13.6	18.1
Wastewater			
Industrial	6.4	6.6	13.0
Municipal	0.9	2.0	2.9
Sweetener decolorization	6.8	9.1	15.9
Chemical processing and misc.	4.1	2.3	6.4
Food, beverage, and oils	0.9	3.9	4.8
Pharmaceuticals	2.0	2.3	4.3
Mining	1.6	2.5	4.1
Groundwater	0.9	2.3	3.2
Household uses	1.4	0.9	2.3
Dry cleaning	0.7	0.4	1.1
Electroplating	0.2	0.4	0.6
Total	30.4	46.3	76.7

Drinking Water Treatment

Treatment of drinking water accounts for about 24% of the total activated carbon used in liquid-phase applications. Rivers, lakes, and groundwater from wells, the most common drinking water sources, are often contaminated with bacteria, viruses, natural vegetation decay products, halogenated materials, and volatile organic compounds. Normal water disinfection and filtration treatment steps remove or destroy the bulk of these materials. However, treatment by activated carbon is an important additional step in many plants to remove toxic and other organic materials for safety and palatability.

Groundwater Treatment

Concern over contaminated groundwater sources increased in the 1980s, and in 1984 an Office of Groundwater Protection was created by the EPA. This led to an increase in activated carbon consumption in 1987 for groundwater treatment to about 4% of the total liquid-phase usage, and further growth is expected in the 1990s. There are two ways to apply carbon in groundwater cleanup. One is the conventional method of applying powdered, granular, or shaped carbon to adsorb contaminants directly from the water. The other method utilizes air stripping to transfer the volatile compounds from water to air. The compounds are then recovered by passing the contaminated air through a bed of carbon.

Wastewater Treatment

Wastewater treatment consumes about 21% of the total U.S. liquid-phase activated carbon, and governmental regulations are expected to increase demand over the next several years.

Wastewater may contain suspended solids, hazardous microorganisms, and toxic organic and inorganic contaminants that must be removed or destroyed before discharge to the environment. In tertiary treatment systems, powdered, granular, or shaped carbon can be used to remove residual toxic and other organic compounds after the primary filtration and secondary biological treatment. Powdered carbon is also used in the PACT process by direct addition of the carbon to the secondary biological step.

Purification of Sugar and Syrups

About 21% of the liquid-phase activated carbon is used for purification of sugar (qv) and corn syrup. White sucrose sugar is made from raw juice squeezed from sugarcane or sugar beets. The clarified liquor is decolorized using activated carbon or ion exchange resins. High-fructose corn sweeteners (HFCS) are produced by hydrolysis of corn starch and are then treated with activated carbon to remove undesirable taste and odor compounds and to improve storage life. The demand for HFCS rose sharply in the 1980s primarily because of the switch by soft drink producers away from sucrose.

Biochemical Utility

Activated carbon consumption in a variety of chemical processing applications is about 8% of the total. The activated carbon removes impurities to achieve high quality. For example, organic contaminants are removed from solution in the production of alum, soda ash, and potassium hydroxide. Other applications include the manufacture of dyestuffs, glycols, amines, organic acids, urea, hydrochloric acid, and phosphoric acid.

In Eating and Drinking

Approximately 6% of the liquid-phase activated carbon is used in food, beverage, and cooking oil production. Before being incorporated into edible products, vegetable oils and animal fats are refined to remove particulates, inorganics, and organic contaminants. Activated carbon is one of several agents used in food purification processes. In the production of alcoholic beverages, activated carbon removes haze-causing compounds from beer, taste and odor from vodka, and fusel oil from whiskey. The feed water for soft drink production is often treated with carbon to capture undesirable taste and odor compounds and to remove free chlorine remaining from disinfection treatment. Caffeine is removed from coffee beans by extraction with organic solvents, water, or supercritical carbon dioxide prior to roasting. Activated carbon is used to remove the caffeine from the recovered solvents.

In Pharmacy

Pharmaceuticals account for 6% of the liquid-phase activated carbon consumption. Many antibiotics, vitamins, and steroids are isolated from fermentation broths by adsorption onto carbon followed by solvent extraction and distillation. Other uses in pharmaceutical production include removal of chromogenic impurities, process water purification, and removal of impurities from intravenous solutions prior to packaging.

In Mining

The mining industry accounts for only 4% of liquid-phase activated carbon use, but this figure may grow as low-grade ores become more common. Gold, for example, is recovered on activated carbon as a cyanide complex in the carbon-in-pulp extraction process. Activated carbon serves as a catalyst in the detoxification of cyanides contained in wastewater from cyanide stripping operations. Problems caused by excess flotation agent concentrations in flotation baths are commonly cured by adding powdered activated carbon.

Other Uses

Several relatively low-volume activated carbon uses comprise the remaining 6% of liquid-phase carbon consumption. Small carbon filters are used in households for purification of tap water. Oils, dyes, and other organics are adsorbed on activated carbon in dry cleaning recovery and recycling systems. Electroplating solutions are treated with carbon to remove organics that can produce imperfections when the thin metal layer is deposited on the substrate. Medical applications include removal of toxins from the blood of patients with artificial kidneys, and oral ingestion into the stomach to recover poisons or toxic materials. Activated carbon also is used as a support for metal catalysts in low-volume production of high-value specialty products such as pharmaceuticals, fragrance chemicals, and pesticides.

GAS-PHASE APPLICATIONS

Gas-phase applications of activated carbon include separation, gas storage, and catalysis. Although only 20% of activated carbon production is used for gas-phase applications, these products are generally more expensive than liquid-phase carbons and account for about 40% of the total dollar value of shipments. Most of the activated carbon used in gas-phase applications is granular or shaped.

Table 6.3 Gas-Phase activated carbon

Application	*Consumption (10^3 t)*
Solvent recovery	4.5
Automotive/gasoline recovery	4.1
Industrial off-gas control	3.2
Catalysis	2.7
Pressure swing separation	1.1
Air conditioning	0.5
Gas mask	0.5
Cigarette filters	0.5
Nuclear	0.3
Total	17.4

Separation processes comprise the main gas-phase applications of activated carbon. These usually exploit the differences in the adsorptive behavior of gases and vapors on activated carbon on the basis of molecular weight and size. For example, organic molecules with a molecular weight greater than about 40 are readily removed from air by activated carbon.

Preservation of Volatile Compounds

Most of the activated carbon used in gas-phase applications is employed to prevent the release of volatile organic compounds into the atmosphere. Much of this use has been in response to environmental regulations, but recovery and recycling of solvents from a range of industrial processes such as printing, coating, and extrusion of fibers also provides substantial economic benefits. The structure of activated carbons used for solvent recovery has been predominantly microporous. Micropores provide the strong adsorption forces needed to capture small vapor molecules such as acetone at low concentrations in process air. In recent years, however, more mesoporous carbons, specifically made for solvent recovery, have become available and are giving good service, especially for the adsorption of heavier vapors such as cumene and cyclohexanone that are difficult to remove from micropores during regeneration. Regeneration of the carbon is performed on a cyclic basis by purging it with steam or heated nitrogen.

Gasoline Vapor Control

A principal application of activated carbon is in the capture of gasoline vapors that escape from vents in automotive fuel systems. Under EPA regulations, all U.S. motor vehicles produced since the early 1970s have been equipped with evaporative emission control systems. Most other auto-producing countries now have similar controls. Fuel vapors vented when the fuel tank or carburetor are heated are captured in a canister containing 0.5–2 L of activated carbon. Regeneration of the carbon is then accomplished by using intake manifold vacuum to draw air through the canister. The air carries desorbed vapor into the engine where it is burned during normal operation. Activated carbon systems have also been proposed for capturing vapors emitted during vehicle refueling, and activated carbon is used at many gasoline terminals to capture vapor displaced when tank trucks are filled.

Typically, the adsorption vessels contain around 15 m^3 of activated carbon and are regenerated by application of a vacuum. The vapor that is pumped off is recovered in an absorber by contact with liquid gasoline. Similar equipment is used in the transfer of fuel from barges. The type of carbon pore structure required for these applications is substantially different from that used in solvent recovery. Because the regeneration conditions are very mild, only the weaker adsorption forces can be overcome, and therefore the most effective pores are in the mesopore size range. A large adsorption capacity in these pores is possible because vapor concentrations are high, typically 10–60%.

Other Applications

Other applications that depend on physical adsorption include the control of krypton and xenon radionuclides from nuclear power plants. The gases are not captured entirely, but their passage is delayed long enough to allow radioactive decay of the short-lived species. Highly microporous coconut-based activated carbon is used for this service.

Regulating Biochemical Activities

Pick-up of gases to prevent emissions can also depend on the chemical properties of activated carbon or of impregnants. Emergency protection against radioiodine emissions from nuclear power reactors is provided by isotope exchange over activated carbon impregnated with potassium iodide. Oxidation reactions catalyzed by the carbon surface are the basis for several emission control strategies. Sulfur dioxide can be removed from industrial off-gases and power plant flue gas because it is oxidized to sulfur trioxide, which reacts with water to form nonvolatile sulfuric acid. Hydrogen sulfide can be removed from such sources as Claus plant tail gas because it is converted to sulfur in the presence of oxygen. Nitric oxide can be removed from flue gas because it is oxidized to nitrogen dioxide. Ammonia is added and reacts catalytically on the carbon surface with the nitrogen dioxide to form nitrogen.

Preventing from Atmospheric Contaminants

Activated carbon is widely used to filter breathing air to protect against a variety of toxic or noxious vapors, including war gases, industrial chemicals, solvents, and odorous compounds. Activated carbons for this purpose are highly microporous and thus maximize the adsorption forces that hold adsorbate molecules on the surface. Although activated carbon can give protection against most organic gases, it is especially effective against high molecular weight vapors, including chemical warfare agents such as mustard gas or the nerve agents that are toxic at parts per million concentrations. The activated carbon is employed in individual canisters or pads, as in gas masks, or in large filters in forced air ventilation systems. In air-conditioning systems, adsorption on activated carbon can be used to control the buildup of odors or toxic gases such as radon in recirculated air. Inorganic vapors are usually not strongly adsorbed on activated carbon by physical forces, but protection against many toxic agents is achieved by using activated carbon impregnated with specific reactants or decomposition catalysts. For example, a combination of chromium and copper impregnants is used against hydrogen cyanide, cyanogen, and cyanogen chloride, whereas silver assists in the removal of arsine.

All of these are potential chemical warfare agents; the Whetlerite carbon, which was developed in the early 1940s and is still used in military protective filters, contains these impregnants. Recent work has shown that chromium, which loses effectiveness with age and is itself toxic, can be replaced with a combination of molybdenum and triethylenediamine. Oxides of iron and zinc on activated carbon have been used in cigarette filters to absorb hydrogen cyanide and hydrogen sulfide. Mercury vapor in air can be removed by activated carbon impregnated with sulfur. Activated carbon impregnated with sodium or potassium hydroxide has long been used to control odors of hydrogen sulfide and organic mercaptans in sewage treatment plants. Alkali-impregnated carbon is also effective against sulfur dioxide, hydrogen sulfide, and chlorine at low concentrations. Such impregnated carbon is used extensively to protect sensitive electronic equipment against corrosion by these gases in industrial environments.

Isolation of Gas Streams

Differences in adsorptivity between gases provides a means for separating components in industrial process gas streams. Activated carbon in fixed beds has been used to separate aromatic compounds from lighter vapors in petroleum refining process streams and to recover gasoline components from natural and manufactured gas. Molecular sieve activated carbons are specially made with restricted openings leading to micropores. These adsorbents are finding increasing use in separations utilizing pressure swing adsorption, in which adsorption is enhanced by operation at high pressure and desorption occurs upon depressurization. Larger molecules are restricted from entrance into the pores of these carbons and, therefore, are not retained as strongly as smaller molecules.

The target product can be either the adsorbed or unadsorbed gases. Examples include separation of oxygen from air and recovery of methane from inorganic gases in biogas production. Hydrogen can be removed from gases produced in the catalytic cracking of gasoline, and carbon monoxide can be separated from fuel gases. Use of pressure swing techniques for gas separation is an area of growing interest in engineering research. The hypersorption process developed in the late 1940s used a bed of activated carbon moving countercurrent to gas flow to separate light hydrocarbons from each other and from hydrogen in refinery operations. The application is of interest because of its scale, treating up to 20,000 m^3/h of gas, but the plants were shut down within a few years, probably because of problems related to attrition of the rapidly circulating activated carbon. It should be noted, however, that in recent years moving-bed and fluid-bed adsorption equipment using activated carbon has been successfully employed for solvent recovery.

Packing

Adsorption forces acting on gas molecules held in micropores significantly densify the adsorbed material. As a result, activated carbon has long been considered a medium for lowering the pressure required to store weakly adsorbed compressed gases. Recent work with modern high-capacity carbons has been directed toward fueling passenger cars with natural gas, but storage volume targets have not yet been attained. Natural gas storage on activated carbon is now used commercially in portable welding cylinders. These can be refilled easily at about 2,000 kPa and hold as much gas as a conventional cylinder pressurized to 6,000 kPa (59 atm).

Altering the Rate of Reaction

Catalytic properties of the activated carbon surface are useful in both inorganic and organic synthesis. For example, the fumigant sulfuryl fluoride is made by reaction of sulfur dioxide with hydrogen fluoride and fluorine over activated carbon. Activated carbon also catalyzes the addition of halogens across a carbon-carbon double bond in the production of a variety of organic halides and is used in the production of phosgene from carbon monoxide and chlorine.

7

BIOINTERACTION EXPLOSION

The strong, specific, and well-characterized interaction between the Fc part of immunoglobulin G (IgG) and *Staphylococcus aureus* protein A (SpA) was the first biointeraction exploited for the creation of an affinity fusion system allowing affinity chromatography purification of expressed recombinant gene products. To date, hundreds of proteins have been produced as SpA fusions in a wide variety of host cells. During the past decade a large number of additional affinity fusion systems have been presented. Affinity fusions are primarily used to simplify recovery of a fused target gene product but are also widely employed for detection and immobilization purposes. Most affinity tags are easy to use as detection moieties to monitor the production or distribution of a produced fusion protein, and certain affinity tags are particularly suited to obtain efficient and directed immobilization of a target protein. This article describes the use of two affinity tags, the IgG-binding SpA domains and the serum albumin–binding region of streptococcal protein G (SpG), which both carry a number of features that make them highly suited as fusion partners in different applications.

The wide spectrum of applications for affinity fusions in biotechnology is illustrated by selected examples in which SpA or SpG fusion proteins have been used to (i) improve the stability to proteolysis of produced recombinant proteins, (ii) allow integrated downstream processing schemes by careful genetic design, (iii) improve in vitro renaturation schemes for expressed gene products, (iv) enable affinity-assisted folding in vivo of target proteins, (v) prolong the in vivo half-life of therapeutic proteins, (vi) facilitate subunit vaccine development and functional cDNA analysis, and (vii) select novel receptor variants with new specificities by the use of phage display technology.

ESTABLISHING AFFINITY

A large number of different affinity fusion systems have been described involving fusion partners that range in size from one amino acid to large multisubunit proteins capable of selective interaction with a ligand immobilized onto a chromatography matrix. Different types of interactions have been utilized, such as enzymes–substrates, bacterial receptors–serum proteins, polyhistidines–metal ions, and antibodies–antigens. The degree of specificity in the interaction and conditions for the affinity purification differ from system to system. The environment tolerable by the target protein is an important factor for deciding which affinity fusion system to choose. Also, other factors, including costs for the affinity matrix and buffers and the possibilities for column sanitation and reuse, are important to consider. Numerous gene fusion systems for affinity purification have been described in the literature, all with different characteristics.

The polyhistidyl tags, which suffer from a somewhat lower specificity, have the advantage of allowing affinity purification under denaturing conditions. The in vivo biotinylated "Bio-tag" is particularly suited for immobilization purposes because of the strong interaction (association constant; $K_A \approx 10^{15}\ M^{-1}$) between biotin and streptavidin, but is inconvenient to use for affinity purification purposes because efficient elution is difficult to accomplish. One of the most extensively used affinity tags is the glutathione S-transferase (GST) from the parasitic helminth *Shistosoma japonicum*. GST fusion proteins can be purified using glutathione as immobilized ligand, also following renaturation from inclusion bodies after solubilization with 6 M guanidinium hydrochloride.

Table 7.1 Commonly used affinity fusion systems

Fusion partner	*Size*	*Ligand*
Protein A	31 kDa	hIgG
Z	7 kDa	hIgG
ABP	5-25 kDa	HSA
Poly His	≈1 kDa	Me^{2+} chelator
GST	25 kDa	Glutathione
MBP	41 kDa	Amylose
FLAG	1 kDa	MAbs M1/M2
PinPoint	13 kDa	Streptavidin/avidin
Bio	13 aa	Streptavidin/avidin

A possible complication associated with the GST fusion system is the use of reduced glutathione (a reducing agent) for elution, which can affect target proteins containing disulfides. Although a multitude of systems have been described, no single affinity fusion strategy is ideal for all expression or purification situations. For example, if secretion of the product into the periplasm or culture medium is desired, fusion to affinity tails derived from normally intracellular proteins (e.g., the β-galactosidase system) is not applicable, and if the purification must be performed under denaturating conditions, protein ligands such as monoclonal antibodies (mAbs; e.g., anti-FLAG M1 and M2 mAbs) are unsuitable. Therefore, to obtain general expression vectors for affinity gene fusion strategies applicable in several situations, a combination of affinity fusion domains could be introduced into a single fusion partner. Such composite fusion partners, consisting of several independent affinity domains, could potentially be used in different detection, purification, and immobilization situations, employing the affinity function that is most suitable for the situation. However, the included affinity domains should be carefully chosen not to interfere functionally with each other, and each moiety should be able to withstand the purification conditions dictated by the affinity domain requiring the harshest affinity chromatography conditions.

A further advantage using combined affinity tags is that different methods for the detection and immobilization of the fusion protein also can be envisioned. Recently, a new affinity fusion partner was described in which this strategy was utilized. An affinity-tag combination was designed consisting of the in vivo biotinylated peptide (Bio), a hexahistidyl peptide (His_6), and the SpG-derived albumin-binding protein (ABP) that could allow for flexible use in different detection, purification, and immobilization situations. Fusions to the Bio-His_6-ABP affinity tag could be used for easy detection using commercial streptavidin conjugates, for purification under both native and denaturating conditions, and for directed immobilization purposes with an option to choose among three strong affinity interactions. In addition, the ABP part of the affinity tag is highly soluble and effectively refolds after denaturation, which can be an advantage during refolding of the target protein fused to the Bio-His_6-ABP affinity tag. Such composite affinity tags have the potential of becoming widely used in the future because they offer flexibility. This article does not focus on the description of different characteristics of various affinity-tag systems, as up-to-date reviews are available, but instead describes various application areas in which affinity fusions to the IgG-binding domains of SpA and the serum albumin- binding regions of SpG have been used.

BIOINTERACTION BETWEEN A AND G PROTEINS

SpA is an immunoglobulin-binding surface receptor found in the Gram-positive bacterium *Staphylococcus aureus*. SpA has found extensive use in immunological and biotechnological research. SpA binds to the constant (Fc) part of certain immunoglobulins, but the exact biological significance of this property is not clarified. Functional and structural analysis of SpA has revealed the presence of a signal peptide, S, which is processed during secretion, five highly homologous domains (E, D, A, B and C) all capable of binding to IgG, followed by a cell wall-attaching structure designated XM. Here, X represents a charged and repetitive region, postulated to interact with the peptidoglycan cell wall, whereas M is a region common for Gram-positive cell surface-bound receptors containing an $LPX_{aa}TG$ motif, linked to a C-terminal hydrophobic region ending with a charged tail. It has been demonstrated that the complete M region is required for cell surface anchoring and that the cell wall sorting is accompanied by proteolytic cleavage at the SpA C-terminus and subsequent covalent linking of the surface receptor to the cell wall. SpA binds to IgG from most mammalian species, including humans.

Of the four subclasses of human IgG, SpA binds to IgG1, IgG2, and IgG4 but shows only weak interaction with IgG3. Several characteristics of the IgG-binding domains of SpA have made them suitable as fusion partners for the production and purification of recombinant proteins: (i) SpA is proteolytically stable. (ii) The structure of the IgG-binding domains, each being a three-helix bundle, appears to be favorable for independent folding of the fusion partner and the fused product because the N- and C-termini of each IgG-binding domain are solvent exposed. (iii) It has been demonstrated to be feasible to introduce sequences accessible for site-specific cleavage of SpA fusion proteins so as to release the target gene product. (iv) The high solubility of SpA enables the production of soluble fusion proteins to very high concentrations within the *Escherichia coli* cell. To create a "second-generation" affinity tag, an engineered IgG-binding domain, Z, was designed based on domain B of SpA. This domain naturally lacks methionines, making it resistant to CNBr cleavage. In addition, an NG dipeptide sequence present in all native domains was changed to NA by altering the glycine codon to a codon for alanine, rendering the Z-domain also resistant to cleavage by hydroxylamine.

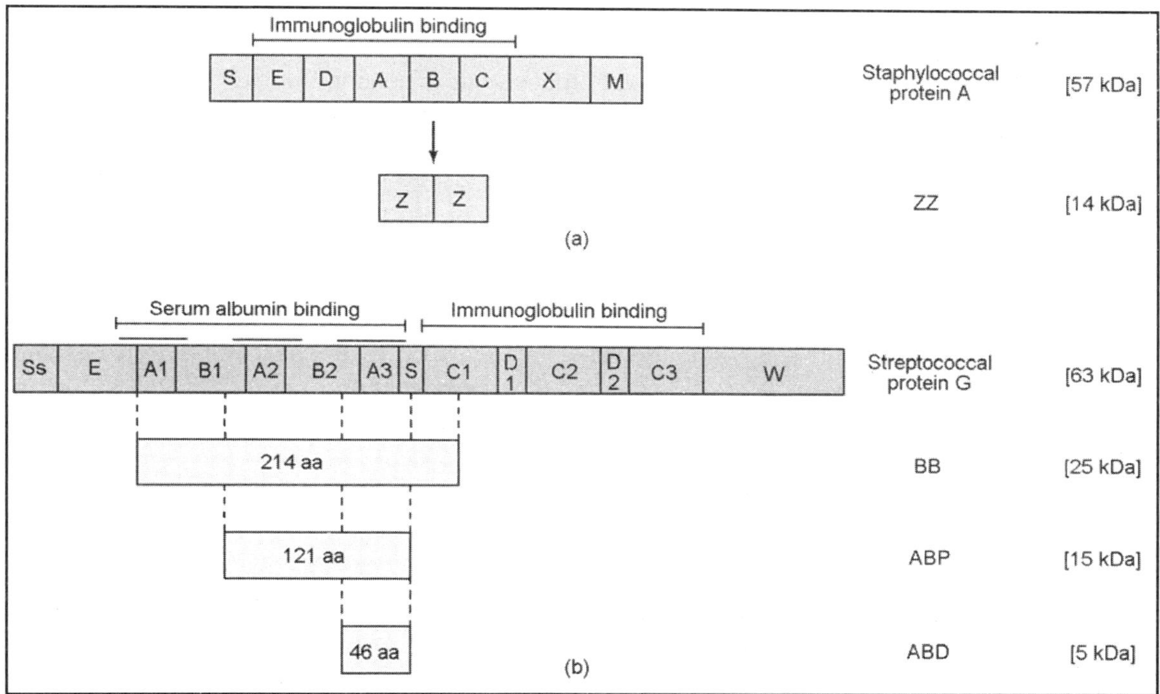

Figure 7.1 Schematic presentation of staphylococcal protein A (SpA) and streptococcal protein G (SpG). (a) The complete SpA gene product and the 7-kDa B-domain analog Z, represented in the most widely used divalent form. (b) The SpG gene product, with the separate albumin- and immunoglobulin-binding regions indicated, and a presentation of some of the most commonly used albumin-binding affinity tags.

Analysis of the interaction between IgG and Z domains, polymerized to different multiplicities, demonstrated that a dimeric form, ZZ, was the optimal fusion partner by its strong IgG-binding and efficient secretion. A different affinity-tag system, which has a number of features in common with the SpA system but also some additional advantageous properties, is based on the serum albumin-binding region of streptococcal protein G (SpG). The regions of SpG mediating binding activities to serum albumin and IgG have been shown to be structurally separated. SpG binds to serum albumins of different mammalian species, including humans, mouse, and rat. The complete region comprises three serum albumin-binding motifs, each being approximately 46 amino acids in size. One of these postulated minimal motifs has been produced and tested for its serum albumin binding and is considered to be a suitable affinity tag in a mono- or divalent fashion. However, tags comprising one, two, two-and-a-half, or three serum albumin-binding motifs have all been successfully produced and used in different applications. The affinity tags, denoted BB, ABP (for albumin-binding protein), and ABD (for albumin-binding domain), are proteolytically stable, highly soluble, and possible to produce at high yields. The 46–amino acid albumin-binding domain (ABD) was recently subjected to a heteronuclear nuclear magnetic resonance (NMR) analysis, concluding that it constitutes a three-helix bundle, remarkably similar to that of the IgG-binding domains of SpA.

Multitude of Proteins

To date, a multitude of proteins have been produced as fusions to the IgG-binding domains of SpA (native domains or the engineered Z-domains), in host cells from all five kingdoms of living organisms: (i) Gram-negative bacteria, such as *E. coli* and *Salmonella typhimurium*, as well as Gram-positive bacteria such as *Staphylococcus aureus, Streptococcus sanguis*, and *Bacillus subtilis;* (ii) the yeast *Saccharomyces cerevisiae;* (iii) plants; (iv) insect cells; and (v) mammaiian cells. The SpG-derived expression systems have found several applications (as are described later), some completely new, and some as a complement to the SpA/ZZ systems. The albumin-binding tags have not yet been evaluated for production in all host organisms but have been successfully expressed for different purposes in Gram-negative bacteria such as *E. coli*, Gram-positive bacteria such as *Staphylococcus xylosus* and *Staphylococcus carnosus*, and mammalian cells such as Chinese hamster ovary (CHO) cells.

The SpA fusion proteins, which bind to the Fc region of IgG, are commonly affinity purified using IgG-Sepharose and the serum albumin-binding SpG fusions are normally purified on human serum albumin (HSA) Sepharose. However, other matrices, such as protein-coated paramagnetic beads, have also been utilized. SpA (ZZ) and SpG (BB, ABP, or ABD) fusion proteins can be easily detected during expression, without the need of specific antisera, by blotting techniques after polyacrylamide gel electrophoresis (PAGE) by taking advantage of their IgG-(Fc) and HSA-binding properties, respectively. SpA fusions are efficiently stained using a commercially available complex of rabbit anti-horseradish peroxidase–IgG and horseradish peroxidase. SpG fusions are stained using biotinylated HSA and conjugated streptavidin–alkaline phosphatase.

System of Expression Vector

A number of expression vectors based on gene fusions to the SpA gene, often in the form of the divalent SpA analogue ZZ, or all five native IgG-binding domains, have been developed for recombinant protein production in *E. coli*. The promoter and secretion signals of SpA are functional also in the Gram-negative *E. coli*. Protein A fusions expressed in *E. coli* can thus be efficiently secreted to the periplasm of the bacteria, and in many cases also to the culture medium from which they can be easily purified by IgG affinity chromatography. In the vectors designed for secretion, the expressed SpA-fusion products are in most examples transcribed from the SpA promoter, which is constitutive and efficiently recognized in *E. coli*. Analogous expression vectors for production of secreted fusions also to the serum albumin–binding regions of SpG have been developed, taking advantage of the same SpA promoter and secretion signals. Fusion proteins produced in this way can be efficiently affinity purified using human serum albumin (HSA) columns. There exists no general strategy applicable to all recombinant gene products to guarantee secretion.

Instead, the inherent properties of the target protein largely dictate if a secretion route is possible to use. However, a large fraction of the proteins produced as SpA fusions are soluble and efficiently secreted to the periplasmic space and culture medium of *E. coli*. This may at least be partially caused by the high solubility of the ZZ domains, which most likely increases the overall solubility of the fusion proteins.

There are a number of advantages connected with a secretion strategy. First, the gene product will be less exposed to cytoplasmic proteases, which might enable production of also labile proteins. Second, disulfide bond formation, which is occurring in the nonreducing environment outside the cytoplasm, could improve on the folding of certain gene products. Third, the recovery of the recombinant protein is simplified because a large degree of the purification from host cell proteins has been achieved through the secretion. Hansson and coworkers described the *E. coli* production of a fusion protein ZZ-M5, a malaria subunit vaccine candidate, using a secretion strategy. More than 65% of the recombinant gene product was found to be secreted to the culture medium, from which it could be recovered in a single step by expanded-bed anion exchange adsorption. As the ZZ-M5 protein was to be used in a preclinical malaria vaccine trial in *Aotus* monkeys, a polishing step was included in which the IgG-binding capacity of the fusion protein was employed. After affinity chromatography on IgG-Sepharose, contaminating DNA and endotoxin levels were well below the demands set by regulatory authorities.

The overall yield of the process, performed in pilot scale, exceeded 90%, resulting in 550 mg product per liter of culture. As mentioned earlier, there is no guarantee that a certain protein can be secreted even if an expression system with secretion signals is chosen for the production. Proteins with a strong tendency to precipitate intracellularly and proteins containing hydrophobic transmembrane regions maybe extremely difficult to secrete. Intracellular expression of such proteins could be a more attractive alternative because of recent advances in in vitro renaturation of recombinant proteins from intracellular precipitates, that is, inclusion bodies. The mechanism for inclusion body formation is not fully elucidated, but expression driven from exceptionally strong promoters such as the T7 promoter, leading to very high expression levels, seems to increase the tendency for intracellular aggregation of the produced recombinant protein. Production by the inclusion body strategy has the main advantages that the recombinant product normally is protected from proteolysis and that it can be produced in large quantities. Levels up to 50% of total cell protein content have been reported.

Vectors for intracellular production of SpA fusions have thus been constructed in which the transcription is under the control of inducible promoters, such as *lac*UV5, λ_{PR}, *trp*, *trc*, or T7, enabling a more controlled expression. Also, for intracellular production of BB and ABP fusion proteins, inducible vector systems have been used, employing *trc*, *trp*, or T7 promoters, respectively. The most robust and in addition the most tightly controlled of these systems seems to be the T7 RNA polymerase–regulated promoter system. Using this system, Larsson and coworkers recently described the expression of five different mouse cDNAs as fusions to ZZ and ABP, respectively. Expression levels ranged from 4 to 500 mg gene product per liter of culture. Intracellularly produced and still soluble fusion proteins can be released by sonication or high-pressure homogenization of cells before purification by IgG or HSA affinity. However, precipitated gene products need to be solubilized before the affinity purification procedure. It was recently demonstrated that affinity purification indeed can be useful also for the recovery of proteins with a strong tendency to aggregate during renaturation from inclusion bodies following standard protocols. Murby and coworkers devised a recovery scheme for the production of such ZZ and BB fusions, applied on various fragments of the fusion glycoprotein (F) from the human respiratory syncytial virus (RSV).

Earlier attempts to produce these labile and precipitation-prone polypeptides in *E. coli* had failed, but several different F fusion proteins could by this novel strategy be produced and recovered as full-length products with substantial yields (20-50 mg/L). Because it was demonstrated that the IgG and HSA affinity ligands both were resistant to 0.5 M guanidine hydrochloride, efficient recovery from inclusion bodies of the ZZ-F and BB-F fusions could be achieved by affinity chromatography in the presence of the chaotropic agent throughout the purification process. The described strategy has so far been successfully evaluated for a number of proteins of low solubility and should be of interest for efficient recovery of other heterologous proteins that form inclusion bodies when expressed in a bacterial host. It is, of course, almost impossible to give general information concerning the production levels for the described expression systems because production levels are strongly dependent on inherent properties of the target gene products. It is however evident that both the secretion strategy and intracellular protein production in *E. coli* using the SpA and SpG expression systems can give high expression levels. For the secreted production, expression levels of 0.5 to 0.8 g product per liter of culture have been reported, and for intracellular production levels up to 3 g gene product per liter of culture have been obtained. The SpA and SpG expression systems are thus for many gene products a good choice because high expression levels can be obtained and efficient affinity purification of produced fusion proteins can be easily achieved.

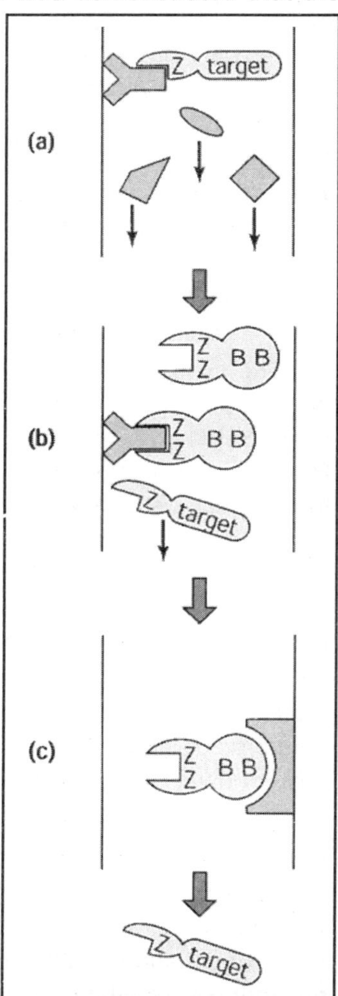

Figure 7.2 The concept of competitive elution, here used for elution of a monovalent Z-fusion by the competitor ZZ-BB.

Interactive Drawback

One possible drawback associated with the SpA and SpG affinity purification systems has been the need for elution by low pH from the affinity column. The elution of fusion proteins at pH 3, used routinely, has for some products been destructive, yielding biologically inactive products. By a recently presented concept, the low-pH elution can be circumvented by competitive elution based on an engineered competitor protein that can be efficiently removed from the eluate mix after elution. The principle for the competitive elution strategy was described by Nilsson and coworkers. The target protein in the presented examples was produced as a fusion to a monovalent Z-domain of SpA. A sample (e.g., crude cell lysate) containing the recombinant fusion Z-target is passed through an IgG-Sepharose column. The recombinant fusion protein is thus captured, enabling extensive washing to remove contaminants. A stoichiometric excess of an engineered bifunctional competitor fusion protein ZZ-BB was added in which BB constitutes a second affinity tail capable of selective binding (association constant; $K_A \approx 1.4 \times 10^9$ M^{-1}) to HSA. The competitor ZZ-BB, having an affinity to human IgG more than 10-fold higher than that to the monovalent Z, is used for competitive elution of the Z-fusion at neutral pH. Specific removal of ZZ-BB from the effluent mix is accomplished by a passage through a second affinity column (HSA) to which the ZZ-BB fusion is bound. The described strategy has been used to produce the Klenow fragment of *E. coli* DNA polymerase I as a Z-Klenow fusion, and the recovered product retained full enzymatic activity.

Stabilization to Proteolysis of Labile Gene Products

When expressing mammalian proteins in a heterologous host such as *E. coli*, proteolytic degradation is one of the more common problems encountered. A review describing the different strategies available to minimize proteolytic degradation of recombinant gene products was recently presented. Fusion of a proteolytically labile target protein to stable fusion partners, such as the ZZ and ABP tags, has been shown to have a certain, but limited, stabilizing effect. However, a much more pronounced stabilizing effect has been obtained when employing a dual affinity fusion concept, where the protein of interest, X, is fused between the two different affinity tag proteins, ZZ and BB. The concept was applied to the production of proteolytically sensitive human insulin-like growth factor II (IGF-II) in *E. coli*. The concept first of all allows the use of two consecutive affinity chromatography steps, but in addition, dramatic stabilization effects can be obtained as compared with more conventional expression routes. The dual affinity fusion approach was found to have a stabilizing effect on three small and labile recombinant proteins, namely human proinsulin, a domain from rat protein disulfide isomerase, and a fragment from a human T-cell receptor, when expressed in *E. coli*. The concept of dual affinity fusions has thereafter been employed also with affinity tags other than the SpA and SpG tags.

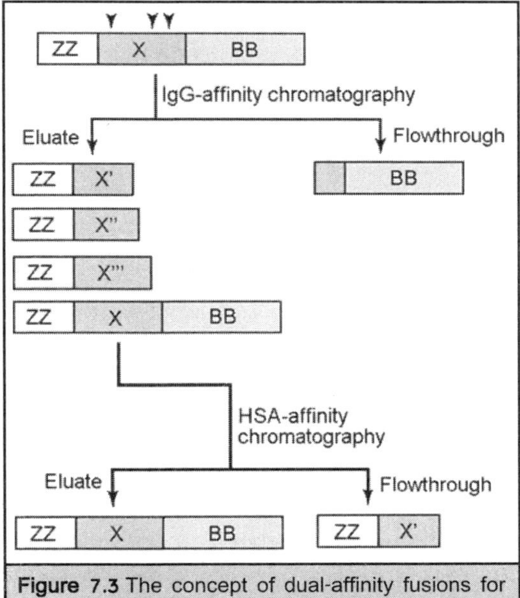

Figure 7.3 The concept of dual-affinity fusions for recovery of proteolytically liable target proteins.

Interactive Strategies

An important part of modern biotechnology is to develop simplified schemes for fermentation and downstream processing of recombinant proteins by the integration of unit operations. Genetic product design can be used in several ways to obtain such schemes, that is, by influencing the yield and localization of the product as well as to adapt the gene product by fusion technology to specific unit operations suitable for large-scale downstream processing. The recently presented expanded-bed adsorption procedure constitutes a nice example of process integration.

Efficient recovery of a secreted recombinant fusion protein ZZ-M5 was achieved directly from a crude fermentor broth without prior cell removal. The fusion protein had a relatively low isoelectric point (pI), which allowed anionic exchange adsorption at pH 5.5 at which most *E. coli* host proteins are not adsorbed. This strategy allowed an integration of the cell separation step with ion exchange adsorption of the gene product with simultaneous volume reduction, which resulted in a highly condensed but still efficient recovery process. The two-step purification process, ion exchange chromatography in an expanded-bed format followed by IgG affinity chromatography for polishing, demonstrated an overall yield of more than 90%. An alternative strategy to simplify downstream processing was investigated by Kohler and coworkers, where recombinant technology was applied to alter the partitioning properties of a ZZ-fusion protein, which thereby could be efficiently recovered by aqueous two-phase extraction as the primary purification step.

An attractive approach to simplify recovery of products, which may be produced with modified primary sequences, e.g., industrial enzymes, would be to alter the partitioning properties by substitution of one or more surface amino acid residues, which are not involved in the biological activity, for tryptophans, which are known to influence partitioning characteristics. This type of protein engineering will most probably be used in the future to reduce production costs for bulk proteins. A third example of how unit operations can be integrated by careful design of the gene product and the recovery process was recently demonstrated for the production of native human insulin and its C-peptide in *E. coli*. Human proinsulin was genetically fused to ZZ, and recovered from inclusion bodies via solubilization and refolding of the intact fusion protein, followed by IgG affinity chromatography.

Native insulin, consisting of cystein-bridged A- and B-chains, as well as the clinically relevant C-peptide, could be released by a single-step trypsin treatment, cleaving off the ZZ-tail simultaneously to the processing of the C-peptide. The careful genetic design of a fusion protein as an intermediate product, which could be efficiently converted to the final products, allowed the design of a pilot-scale process involving few unit operations that thus might result in a cost-efficient bioprocess. A further improved bioprocess for production of the proinsulin C-peptide was presented in which the C-peptide-encoding gene fragment was heptamerized to increase yields. A fusion protein containing seven copies of the C-peptide was thus produced, and after affinity purification, enzymatic digestion was employed to release native C-peptide.

Recovery of Native Proteins

The recovery of biologically active or native proteins by in vitro refolding from insoluble inclusion bodies can in some cases be hampered by the aggregation of the product during the procedure, leading to low overall yields. To make the procedure more efficient, several improved protocols have been developed, including the addition of different "folding enhancers". Alternatively, the protein itself can be engineered to facilitate the refolding. The presence of hydrophilic peptide extensions during the refolding can dramatically improve the folding yield, probably by conferring a higher overall solubility of the protein. Samuelsson and coworkers showed that the reshuffling of misfolded disulfides in recombinant insulin-like growth factor I (IGF-I) was greatly facilitated by fusion to the highly soluble ZZ fusion partner. Compared to unfused IGF-I, the fusion ZZ-IGF-I could be successfully refolded at a 100-fold higher concentration (1-2 mg/mL), without forming precipitates.

Further Improvement

A completely new concept to improve the fraction of correctly folded recombinant IGF-I was recently presented. It was demonstrated that coexpression of a specific binding protein, IGF binding protein 1 (IGFBP-1), significantly increased the relative yield of IGF-I having native disulfide bridges when expressed in a secreted form in *E. coli*. In addition, a glutathione redox buffer was added to the growth medium to enhance formation and breakage of disulfide bonds in the periplasm of the bacteria. In the presented example, both IGF-I and IGFBP-1 were produced as affinity fusions, to Z and BB, respectively, which facilitated the purification of *in vivo*–assembled heterodimers by alternative purification methods.

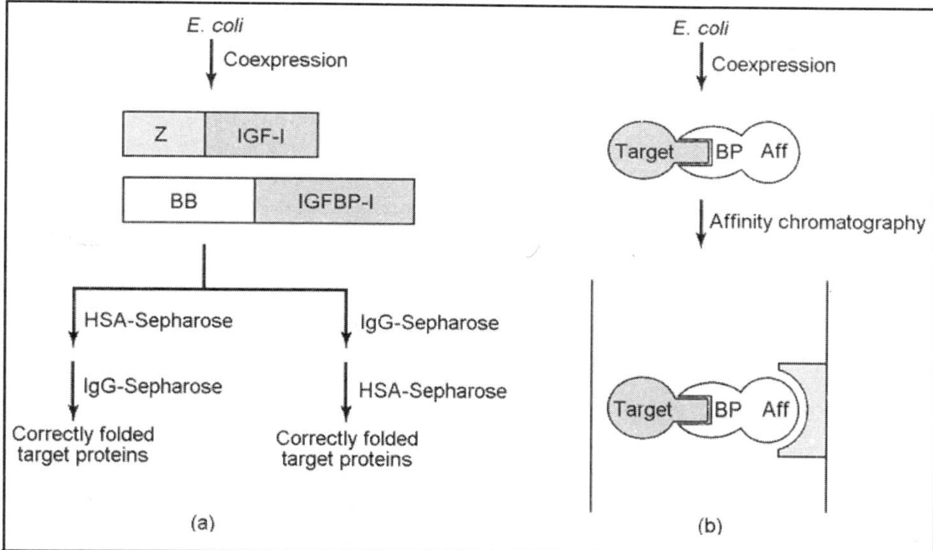

Figure 7.4 Affinity-assisted in vivo folding and its use for recovery of correctly folded gene products. (a) Schematic description of the example presented by Samuelsson and coworkers where a fusion protein Z-IGF-I, with correctly folded IGF-I, could be recovered using alternative affinity chromatography strategies taking advantage of the interaction between IGF-I and IGF-BP-I. (b) The basic principle might be suitable for general use both to increase the yield of correctly folded target protein and to achieve recovery of only correctly folded target protein via the interaction of a specific binding protein (BP), which binds only to the correctly folded form of the target protein.

A further development of the strategy would be to express the target protein as a nonfused gene product and the specific binding protein in an affinity-tagged configuration. Correctly folded target protein would thus be affinity captured as a heterodimer via the tagged binding protein. This would employ the benefits of high specificity for affinity chromatography without introducing requirements of proteolytic processing to recover the native target protein.

Site-Specific Cleavage of Fusion Proteins

When gene fusion strategies are used for the production of native proteins, efficient means for site-specific cleavage of the fusion protein and subsequent removal of the affinity fusion partner are needed. In addition, if employing enzymatic strategies for cleavage, strategies for the removal of the proteolytic enzyme itself must be developed. Special considerations must be taken if the target protein itself also has to be further processed to give the desired final product. Careful upstream design of the fusion protein construct using genetic strategies can greatly facilitate the subsequent purification of the target protein and also allow for integrated systems involving coprocessing of the protein and efficient removal of the affinity fusion partner as well as the protease used for cleavage. Enzymes for use in biotechnological applications should preferably be highly specific proteases, easy to produce by recombinant means in large scale. Interesting candidate enzymes can be found in human picorna viruses, whose maturation relies on the site-specific cleavage of a large polyprotein precursor to yield the viral components.

Some of these proteases are functionally produced at high levels in bacterial expression systems, and this could facilitate the production of variants constructed by protein engineering. Recently, a new general strategy was described where the 3C protease of rhinovirus was fused to an affinity tag and subsequently used for specific cleavage of a target protein. This enabled removal of the affinity-tagged protease and the cleaved affinity fusion partner by simple affinity chromatography steps. It was in addition possible to add any desired amount of protease to ensure efficient cleavage because the protease could be easily removed after cleavage. Recently, a different affinity-tagged protease, consisting of coxsackie virus 3C ($3C^{pro}$) fused to the serum albumin–binding ABP at the N-terminus and His_6 at the C-terminus, was produced in *E. coli* and used for the production of a truncated *Taq* DNA polymerase (ΔTaq). The heat-stable polymerase was produced as an ABP-ΔTaq fusion protein having a $3C^{pro}$ cleavage site introduced between the two protein moieties.

After affinity purification of ABP-ΔTaq, the fusion was efficiently cleaved using the affinity-tagged protease ABP-$3C^{pro}$, which allowed for subsequent recovery of nonfused, fully active ΔTaq via passage of the cleavage mixture over an HSA-column. Note that the affinity-tagged protease, ABP-$3C^{pro}$, is removed together with eventual remains of unprocessed ABP-ΔTaq fusion protein. The use of affinity fusions for recombinant production of therapeutic target gene products has probably been hampered by the difficulties with efficiently cleaving off the affinity tag and thereafter removing not only the tag but also the used protease. The increasing availability of affinity-tagged, highly specific proteases for cleavage of fusion proteins to enable straightforward methods for recovery of unfused authentic gene products will probably influence researchers to further increase their use of fusion strategies for production of target proteins.

Prolonging Therapeutic Gene Products

Gene fusion technology might also prove to be useful in prolonging the in vivo half-life of pharmaceutical proteins by administering the therapeutic protein as fused to a protein with extended half-life. The principle was demonstrated to be efficacious by Capon and coworkers, who showed that by fusing a soluble portion of CD4 (the target receptor for the human immunodeficiency virus particle) to the Fc part of IgG, the serum half-life of CD4 in rabbits increased 200-fold. Nygren and coworkers described an alternative strategy where in vivo stabilization was achieved by fusion of the unstable CD4 protein to BB, leading to the formation of complexes between the fusion protein and serum albumin.

The serum half-lives of different fusion proteins containing the albumin-binding receptor were tested in mice. In these studies, the half-life of the CD4 fusion in vivo was shown to be comparable to the CD4-IgG molecule described earlier. Interestingly, the albumin-binding receptor in macaque monkeys was found to have a serum half-life similar to that of serum albumin itself. Recently, the same strategy was employed to increase the serum half-life of human soluble complement receptor type 1 (sCR1) in rats, by fusion to different serum albumin–binding fragments of SpG. These examples suggest that gene fusions to BB, ABP, or the minimal binding motif ABD of SpG might constitute useful strategies to improve the in vivo stability of pharmaceutical proteins. It may also be possible to combine these approaches with new methods for drug targeting or slow drug release.

Method of Immunodetection

Bifunctional SpA fusion proteins have found extensive use for various analytical purposes, for example, in enzyme immunoassays. Protein A fusion proteins can after expression be efficiently purified using the IgG-binding activity, and in most assays the protein A moiety binds antibodies to be quantified and the activity of the fused gene product is monitored. Thus, the fusion protein can replace more traditional reagents, such as secondary antibodies conjugated to an enzyme. The fusion between catechol dioxygenase and protein A produced in *E. coli* was used in an enzyme immunoassay to quantify antibodies. The principle of a bioluminescent immunoassay using the fusion between protein A and bacterial luciferase has also been demonstrated, and recently a bifunctional fusion between domain D of SpA and luciferase from sea-firefly was used in a sandwich enzyme-linked immunosorbent assay (ELISA).

Furthermore, an SpA-streptokinase fusion was shown to be equally useful as a commercial SpA-horseradish peroxidase conjugate for use in immunoassays. A fusion between SpA and LacG, 6-phospho-β-galactosidase from *S. aureus*, was shown to be able to replace secondary antibodies in an ELISA. In a similar approach, an SpA-alkaline phosphatase fusion was used in ELISA and a fusion between maltose-binding protein and protein A was used for a solid-phase immunoassay, in which the fusion protein was immobilized to determine IgG concentrations. Using sandwich techniques, DNA fragments also can be involved in various purification and detection strategies. Taking advantage of the lac repressor (LacI) from *E. coli*, binding to the *lac* operator sequence, a LacI-SpA fusion protein was produced for simplified analysis of polymerase chain reaction (PCR) amplified DNA fragments generated for diagnosis of pathogens. In addition, this LacI-SpA fusion protein has been used for mild reversible recovery of protein antigens or whole cells using paramagnetic beads with a coupled DNA fragment containing the *lacO* recognition sequence.

An antigen detection system, termed immuno-PCR, has been described in which the principle of immunodetection was combined with the sensitivity of PCR. A streptavidin-SpA chimera was used to attach a biotinylated DNA fragment to antigen–IgG complexes immobilized on microtiter plate wells. Although the method gives a sensitivity greater than any existing antigen–antibody detection system, the sensitivity of the PCR assay could have the drawback of causing false positives resulting from nonspecific binding of IgG, fusion protein, or DNA. Nevertheless, SpA fusion proteins have through the IgG Fc binding activity found extensive use in a vast number of immunodetection techniques.

"Hot-Start Procedure PCR"

Immobilization of enzymes has proven to be a useful procedure for many applications because of the altered characteristics of the immobilized enzyme as compared with the enzyme's soluble counterpart. One practical application utilizing such differential activities was recently described in which a heat-mediated release of an affinity- immobilized recombinant *Taq* DNA polymerase (TagTaq) was used to create a hot-start PCR procedure. *Taq* DNA polymerase was produced in *E. coli* as fused to the multifunctional Bio–His6–ABP tag for affinity purification and immobilization. HSA-affinity immobilization of the fusion employing the ABP moiety resulted in an apparent deactivation of the *Taq* DNA polymerase. However, the ABP–HSA interaction was shown to be disrupted at elevated temperatures (>70°C), resulting in a restored DNA polymerase activity of released *Taq* DNA polymerase fusion protein.

This concept for controlling the activity of the *Taq* DNA polymerase was demonstrated to be of practical use to achieve a so-called hot-start PCR procedure, where it is desired to suppress the DNA polymerase activity until temperatures are reached at which nonspecific annealing of PCR primers is eliminated. The affinity immobilization concept for hot-start PCR was successfully used in full multiple-cycle PCR amplification and was shown to eliminate artefactual primer-dimer products in the amplification of the human oncogene K-*ras* gene in contrast to standard amplification protocols. Noteworthy, the ABP affinity fusion partner was still functional after 30 PCR cycles, allowing post-PCR reimmobilization of the *Taq* DNA polymerase fusion protein onto fresh HSA-Sepharose, facilitating the removal of the DNA polymerase from reaction mixtures after PCR. Further studies are needed to investigate if this concept for modulating biological activity can also be applied to other widely used enzymes, for example, reverse transcriptase.

Vaccine Development

Immunogenic proteins and peptides that are produced as fusion proteins have been frequently used for immunization purposes and represent a strategy to construct subunit vaccines against infectious diseases. An important aspect regarding production of antigens by recombinant DNA technology is the purification of the gene product from contaminating host components. The approach to express peptides or proteins as one part of a fusion protein, where the other part constitutes an affinity handle, allows rapid and efficient affinity purification of the gene product. As a powerful strategy in subunit vaccine development, a dual expression system was devised, combining the two similar affinity-fusion expression systems (ZZ and BB) for the parallel expression of immunogenic peptides and proteins. After affinity purification, one fusion protein is used for immunization while the corresponding second fusion protein is used to analyze the induced antibody response to the fused peptide. This strategy eliminates any background originating from antibodies reactive with the affinity tag of the fusion used for immunization. In addition, the second fusion protein can, after immobilization, be used as a ligand for affinity purification of peptide-specific antibodies.

The dual expression system has been used to produce immunogenic structures derived from different *Plasmodium falciparum* malaria blood-stage antigens, and the expressed fusion proteins have been shown to be immunogenic in mice, rabbits, and monkeys and to induce antigen-specific antibody and T-cell responses. The dual expression strategy has thus shown to be of significant value in subunit vaccine research. Besides the use of BB and ABP fusion proteins in the dual expression concept for analysis of elicited immune responses, the serum albumin–binding region of SpG has been shown to have inherent immunopotentiating properties when used as a carrier protein genetically fused to immunogen used for immunization.

Recently, it was demonstrated that a BB-fusion protein (BB-M3), containing a malaria peptide, induced significant antibody responses in mice strains that were nonresponders to the malaria peptide (M3) alone, suggesting that BB has the ability to provide T-cell help for antibody production. Furthermore, a fusion protein BB-G2N, comprising a 101–amino acid sequence from human RSV, was shown to induce protective immunity in mice to RSV challenge. It was shown that, by inclusion of the BB part, a more potent G2N-specific B-cell memory response was evoked. This indicates that the SpG-derived BB can function both as an affinity tag to facilitate purification and as a carrier protein with immunopotentiating properties. To date, it is not fully elucidated whether this capacity results from strong T-cell epitopes or is related to the serum albumin–binding activity resulting in a prolonged exposure of the immunogen to the immune system, or is due to a combination of both.

Role in Functional cDNA Analysis

The principle of dual expression systems was recently also implemented for functional analysis of cDNA-encoded proteins. Selected cDNAs were then expressed as ZZ fusion proteins and used for immunization to elicit antibodies, whereas the corresponding ABP fusion protein, having the cDNA-encoded portion in common, was used for purification of target protein–specific antibodies. Such highly enriched antisera should be suitable for functional analysis of the cDNA-encoded protein, for example, in Western blots and immunohistology on tissue sections or within cells. The principle was evaluated by expressing five cDNAs, isolated from prepubertal mouse testis by a differential cDNA library screening strategy. All five clones could be expressed intracellularly in *E. coli* as fusion proteins with high production levels, and affinity purification yielded essentially full-length products.

Characterization of affinity-purified antibodies revealed that there existed no cross-reactivity between the two fusion systems, and that such antibodies indeed were valuable for various analyses to elucidate the functions of the cDNA-encoded proteins. For example, in the presented study a previously unknown cDNA, with no sequence homology to genes in the public databases, could be connected with a gene product found in nuclear bodies within spermatogonial cells. To make it possible to use the dual expression system for functional analysis of cDNA, where robustness of the expression systems is crucial, completely new expression vectors were constructed. Intracellular expression was used and transcription was tightly controlled by the T7 promoter system. This method should ensure that a higher fraction of investigated cDNA-encoded proteins are successfully produced. The different ongoing genome projects have generated enormous amounts of partially sequenced cDNAs, and there exists a significant need for various alternative systems to take care of this information to elucidate the function of the corresponding proteins. By automation of some of the unit operations, the system described by Larsson and coworkers might have the potential to make large-scale functional analysis of cDNA libraries possible.

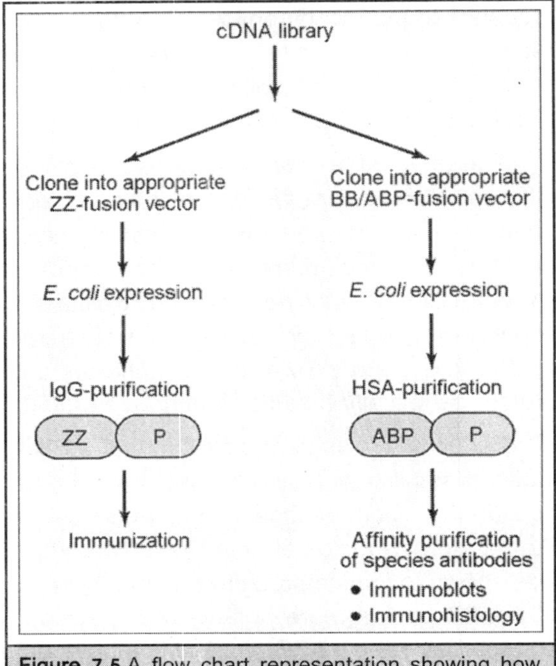

Figure 7.5 A flow chart representation showing how the dual expression concept has been used in functional cDNA research.

Display of Heterologous Proteins

The display of heterologous proteins on the outer surface of bacteria has become an emerging topic in different fields of research within biotechnology, microbiology, and immunology. Genetic insertion of a target sequence into the genes for various outer membrane proteins has constituted the general strategy to enable secretion and subsequent surface anchoring of the recombinant target gene products. For Gram-negative bacteria, represented by *E. coli* and *Salmonella* spp., a number of different types of heterologous proteins have been surface displayed including antigenic determinants for the purpose of vaccine development, enzymes, metal-chelating histidyl peptides, and single-chain antibodies, as well as entire peptide libraries. Recently, surface display on Gram-positive bacteria has gained interest in the context of vaccine development. Approaches based on the attenuated mycobacteria, commensal streptococci, and nonpathogenic food-fermenting staphylococci have been presented.

To achieve surface display of various target proteins, the staphylococcal systems take advantage of the cell surface anchoring regions (XM) of SpA. The surface-exposed proteins are thus genetically fused between a functional signal peptide, responsible for the translocation from the cell, and the XM-regions of SpA. Heterologous antigenic determinants of various origin (bacteria, protozoa, and human virus) have been expressed as anchored on recombinant staphylococci, which are being investigated as a bacterial vaccine delivery system. Even active single-chain antibody fragments have been surface displayed on recombinant staphylococci. Such recombinant bacteria could be envisioned as inexpensive tools in diagnostic tests.

Phage Technology

An example of emerging interest in which gene fusions have made new technology possible is the display of libraries of peptides or proteins on filamentous phage surfaces. Phage particles decorated in such manner can be enriched in vitro from a background of irrelevant phages by biopanning techniques, employing an immobilized ligand capable of binding to the displayed protein. Moreover, inside its protein coat the captured phage carries the genetic information of the displayed protein, which allows later identification by DNA sequencing. Thus, the technology provides a link between genotype and phenotype and allows the selection of peptides or proteins of desired functions to be expressed on the surface of the filamentous phage.

This technology, as pioneered by Smith, has emerged into a powerful technology to select, rather than screen, for rare phages capable of binding to a defined target molecule from libraries comprising as many as 10^{10} phage species. Such phage libraries are obtained by the fusion to phage pIII or pVIII genes, of cassettes encoding randomized peptides (usually 6-to 10-mers), antibody repertoires, or multiple variants of a protein. Thus, it is possible to rapidly and efficiently select high-affinity ligands or effector molecules that can be used for analytical or preparative applications. The Z-domain from SpA has several features including proteolytic stability, high solubility, small size, and compact and robust structure, which makes it ideal as a fusion partner for the production of recombinant proteins. The residues responsible for the IgG Fc–binding activity have been identified from the crystal structure of the complex between its parent B-domain and human IgG Fc, showing that these residues are situated on the outer surfaces of two of the helices making up the domain structure and do not contribute to the packing of the core.

The Fc-binding surface covers an area of approximately 600 $Å^2$, similar in size to the surfaces involved in many antigen–antibody interactions. Taken together, these data suggest that if random mutagenesis of this binding surface was performed, novel domains could be obtained with the possibility of finding variants capable of binding molecules other than Fc of IgG. In addition, these domains would theoretically share the overall stability and α-helical structure of the wild-type domain. Recently, an attempt to obtain such ligands with completely new affinities was initiated, employing phage display technology. A genetic library was created in which 13 surface residues (involved in the IgG Fc binding) of the Z-domain were randomly and simultaneously substituted to any of the 20 possible amino acids. This Z-library has, after genetic fusion to the gene for coat protein 3 of filamentous phage M13, resulted in a phage library adapted for selection of novel specificities by biopanning. The library has been subjected to such affinity selection against different molecules for investigation as a source of novel binders. So far, novel Z-variants, so-called affibodies, have successfully been selected to diverse targets, such as *Taq* DNA polymerase, human insulin, and a human apolipoprotein variant. This concept suggests that the described strategy might prove to be beneficial in generating a "next generation" of ligands or artificial antibodies to various target molecules, to be used for purification, detection, immobilization, and perhaps even therapeutic purposes.

8

INDUSTRIAL BIOTRANSFORMATIONS

The use of biotransformations in food, pharmaceutical, and chemical industries has increased steadily since the early 1970s. Applications range from synthesis on the milligram scale for precious oligosaccharides, to the multiton scale for bulk chemicals using either whole cells or isolated enzymes. This process has been documented in numerous reviews and textbooks. When new processes are being developed, the special properties of the biocatalyst must be considered, as well as the basic elements of chemical engineering. There are strong interactions among the individual aspects, which may be considered for process optimization. The aim of this article is to outline the basic strategy for the development of a biocatalytic step. For some selected examples the beneficial role of (bio)chemical engineering is demonstrated. A more detailed description can be found in textbooks of (bio)chemical engineering.

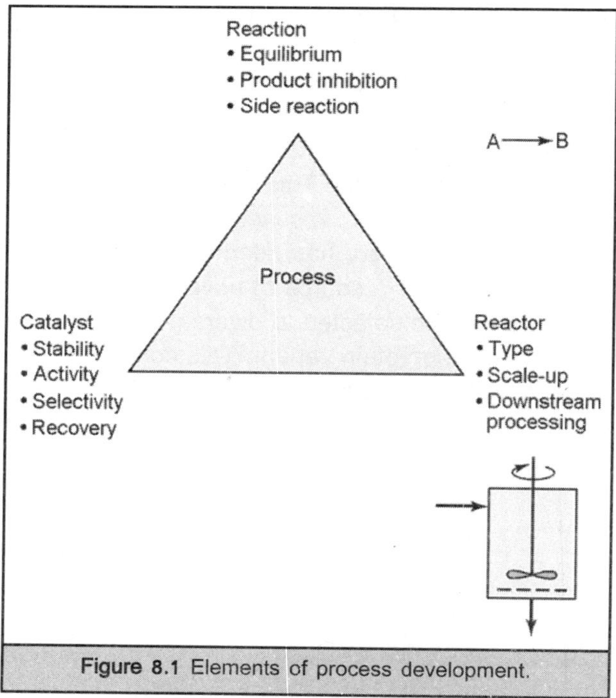

Figure 8.1 Elements of process development.

MECHANICAL SEPARATION

Whole microorganisms or their components, enzymes, may be used for biotransformations. Both approaches have specific advantages and requirements. Commonly, the word *biotransformation* is used to describe the transformation of a substrate or precursor into a certain product in only a few steps. Fermentation processes, for example, production of amino acids from glucose and ammonia or production of (recombinant) proteins, are normally not regarded as biotransformation. Some special information about process characteristics is given as well. When isolated enzymes are used for biotransformations, the purity must only be such that no side reactions occur. There is no need to purify the enzyme to homogeneity, as is required for biochemical characterization. On the contrary, many enzymes show reduced stability with increased purity.

In some cases, the addition of inert proteins, such as bovine serum albumin, may be used as a simple way to increase the stability of the production enzyme. In others, enzyme stability may be increased by immobilization on a support. Covalent attachment fixes the conformation of the enzyme often limiting denaturation. However, the increase of stability must be balanced against the losses during the immobilization step, as well as the additional work and material costs. A method introduced quite recently for the stabilization of enzymes and their immobilization is the formation of cross-linked enzyme crystals (CLECs). Reactors for immobilized enzymes are discussed elsewhere in this volume. However, the engineering approach, as outlined in the following sections is the same for all forms of biocatalysts. Specific requirements and properties such as oxygen delivery, mass transport limitations, or particle size must be considered for individual cases.

Table 8.1 Comparison of biotransformations using whole cells vs. isolated enzymes

Whole cells	*Isolated enzymes*
Large amounts of catalyst easy available	Additional step for obtaining the catalyst required
No stability problem when growing cells are used	Catalyst stability strongly influenced by reaction conditions
Limited volumetric catalyst activity	High volumetric catalyst activity possible
Multistep processes	Multistep processes difficult
Cofactor recycling solved	Cofactor recycling so far only for NADH and ATP
Side reactions may occur	Few to no side reactions
Toxicity of substrate/product may be complex	Enzyme stability may be influenced by substrate or product, but can be enhanced by choosing the appropriate reactor
Oxygen supply may control scale-up	Nonnatural substrates and reaction conditions possible
Normally no cosolvent, possibly leading to low concentrations of hydrophobic compounds	Cosolvents possible, but often reducing enzyme stability
Transport in and out of the cell may be limiting	Mass transport limitations only with immobilized enzymes

TYPES OF REACTORS

There are three basic types of reactors, which have several variations: (i) Stirred-tank reactors (STRs) or batch reactors; (ii) Continuously operated stirred-tank reactors (CSTRs); and (iii) Plug-flow reactors (PRFs). Unlike the stirred-tank reactor, which is operated batchwise, the latter two are operated continuously. By knowing the main characteristics of these fundamental reactors and some of their variations, for example the cascade reactor, it is possible to choose the appropriate reactor for a specific application. This is especially important when dealing with a kinetically or

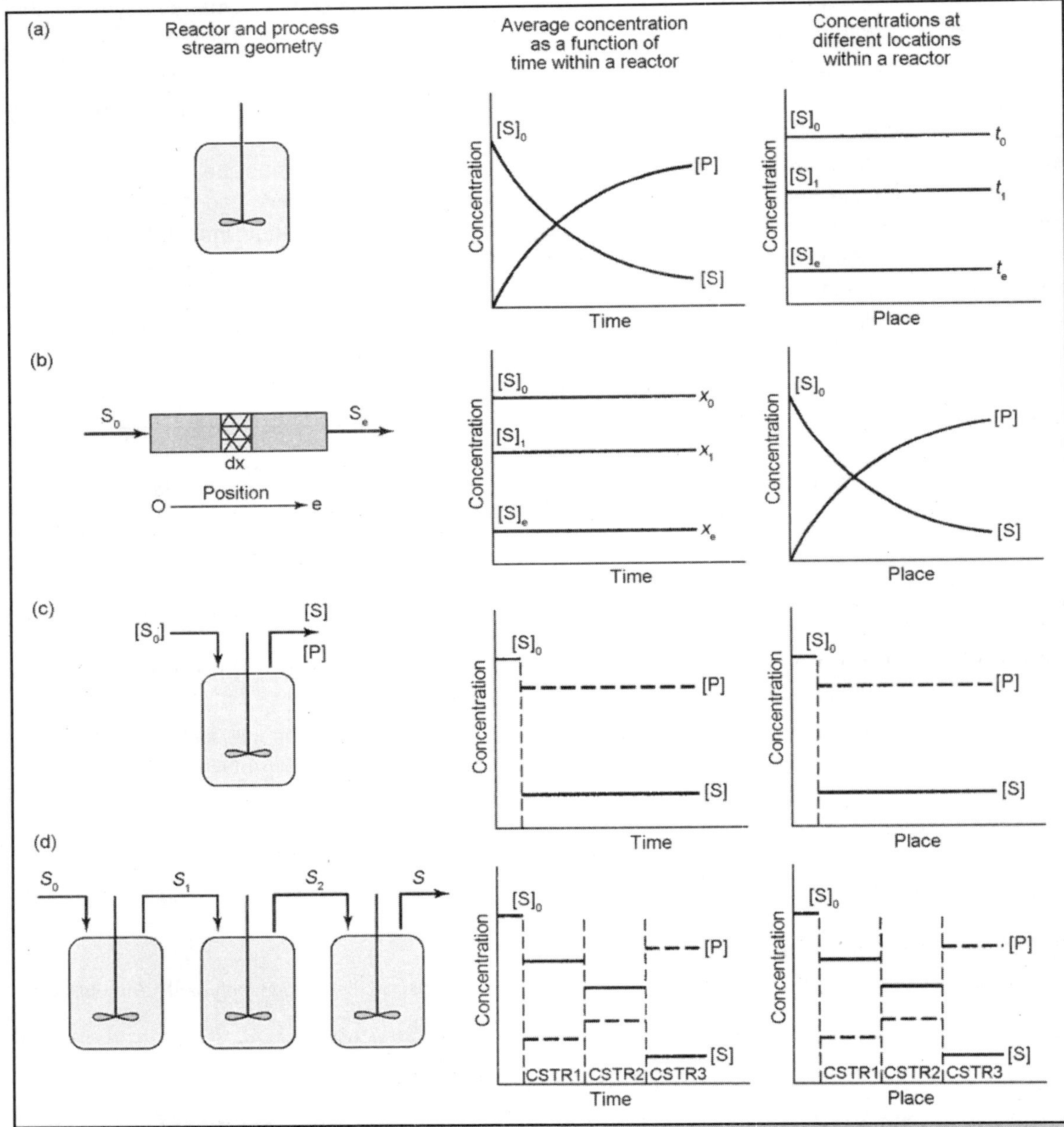

Figure 8.2 Characterization of different reactors. (a) Stirred-tank reactor; (b) plug-flow reactor; (c) continuously operated stirred-tank reactor; (d) Cascade.

thermodynamically limited system, as demonstrated later. In the following analyses, only the basic terms are explained.

The stirred-tank reactor is operated in an instationary way. Assuming ideal mixing characteristics, the instantaneous concentration of every species is the same in every volume element. With advancing conversion the substrate concentration is decreased and the product concentration is increased. A steady state is reached only at equilibrium conversion.

In the plug-flow reactor the product concentration increases slowly over the length of the reactor. Therefore, the average reaction velocity is faster than in the continuously operated stirred-tank reactor. In each single-volume element in the reactor, the concentration is constant in the steady state. In other words, the dimension of time is exchanged with the dimension of place in comparison to the stirred-tank reactor.

The continuously operated stirred-tank reactor works under product outflow conditions, meaning that the concentrations in every volume element are the same as those at the outlet of the reactor. If the steady state is reached, the concentrations are independent of time and place. The conversion is controlled by the catalyst concentration and the residence time, τ:

$$\tau = \frac{V}{F}$$

where $\tau[h]$ is the residence time, $V[L]$ is the total reactor volume, and F is the $[L/h]$ substrate feed rate.

One very common variation of the CSTR is the cascade of n CSTRs. With an increasing number, n, of reaction vessels, the cascade approximates the plug-flow reactor. The product concentration is increased stepwise from vessel to vessel.

Mathematical Balances

The performance of the different reactor types concerning the reaction can be simulated mathematically. This is also the verification of the kinetic model of the reaction because it should describe the course of the concentration for each compound with only a small error. The main part of the simulation model is the coupled system of differential equations of first order, which are the mass balances of all substrates and products. The change in the concentration of one compound in time and in a volume element (called the accumulation) is the sum of convection, reaction, and diffusion.

$$\text{accumulation} = \text{convection} + \text{reaction} + \text{diffusion} \qquad \ldots(1)$$

The convection term describes the change of the concentration of one compound in the reactor as the difference of the influx into the reactor and the efflux. The reaction term describes, by use of the kinetic model, the change of the concentration of one compound as result of the reaction. The reaction velocity, V_s, is the sum of the individual reaction velocities describing consumption of a substrate or formation of a product. Diffusion occurs only in the case where nonideal mixing is stated.

With regard to the chosen reactor type the mass balance can be simplified by assuming ideal mixing or ideal plug- flow behavior.

Stirred tank reactor

Diffusion = 0 — At a defined time the concentrations are the same in every volume element.

Convection = 0 — There is no influx or efflux of substrates or products to a single volume element in time.

The mass balance is simplified to

$$-\frac{d[S]}{dt} = v_s \rightarrow t = [s]_0 \int_0^{\text{desired conversion}} 1/v \, dx \qquad \ldots(2)$$

Continuously operated stirred-tank reactor

Diffusion = 0 — See previous.

Accumulation = 0 — In the steady state the concentrations will not change with time.

The mass balance is simplified to:

$$0 = -\frac{d[S]}{dt} = \frac{[S]_0 - [S]}{\tau} + v_s \qquad \ldots(3)$$

Plug flow reactor

Diffusion = 0 — Diffusion is neglected in an ideal plug-flow reactor.

Accumulation = 0 — In the steady state the concentrations will not change with time.

The mass balance is the same as for the stirred-tank reactor. In the integrated form the time, *t*, has to be replaced by the residence time, τ.

Kinetics

The fundamental equation (equation 4) for the description of enzyme kinetics was derived by Michaelis and Menten.

$$v = [E]\frac{A_{max} \cdot [A]}{K_m + [A]} \qquad \ldots(4)$$

The enzyme activity is expressed as a function of the substrate concentration for an irreversible one-substrate reaction without inhibition. As this is rarely found in nature, equation 5 gives the rate for a reversible reaction A $\leftrightarrow$ B. In this form the product inhibition is considered as a competition of A and B for the active sites, as expressed by the two K_m values. This form might be extended to cover substrate surplus inhibition as well as multisubstrate reactions by adding additional terms to the numerator and denumerator.

$$v = [E] \cdot \frac{\dfrac{A_{max,A} \cdot [A]}{K_{m,A}} - \dfrac{A_{max,B} \cdot [B]}{K_{m,B}}}{1 + \dfrac{[A]}{K_{m,A}} + \dfrac{[B]}{K_{m,B}}} \qquad \ldots(5)$$

The determination of the kinetic parameters can be performed either by initial reaction velocity measurements or by batch experiments at different substrate concentration, such that substrate limitation is encountered. The kinetic parameters are determined by fitting the kinetic model to the experimental data by use of nonlinear regression; in the case of the batch experiments, numerical integration of the reaction velocity equations is required.

STEPWISE METHODOLOGY

The main focus is on the use of isolated enzymes, but for whole-cell processes a similar scheme must be followed. In practice, some of the investigations listed here stepwise will be performed simultaneously. For example, first experiments should hint to possible problems from side reactions. Other factors may be considered as well.

1. Buffer compounds may be important for enzyme stability, but must be removed during product isolation. Avoiding them may simplify product isolation and save costs.
2. Any reaction exhibiting substrate surplus inhibition should not be carried out in a batch-reactor setup, since this results in longer reaction times due to the high substrate concentration at start-up, which lowers the initial reaction velocity. In such a case, a continuously operated stirred-tank reactor that is operated at a high conversion in the steady state is preferred, resulting in a low substrate concentration. Another possibility is the use of a fed batch reactor, which means that the substrate is added in small portions as a function of time, also resulting in a small substrate concentration. One example of such a system is the enzymatic synthesis of L-phenylalanine by reductive amination, which can be found in the literature.
3. If product inhibition occurs, either a stirred-tank reactor, a plug flow reactor, or a cascade of *n* continuously operated stirred-tank reactors should be chosen. In all these reactors the product concentration increases over time. Another possibility is the use of a differential reactor with integrated product separation (30). These two possibilities will be discussed in more detail in the section titled "Examples".
4. The biocatalyst may be recirculated in and between reactors by various means to enhance catalyst concentration. Whole cells are often concentrated by microfiltration membranes or centrifugation. Isolated enzymes are much smaller, but can be retained by an ultrafiltration membrane. This type of equipment is commercially available for the processing of volumes of a few microliters up to the cubicmeter scale. If the enzyme is immobilized on a support, its size determines the separation step. For example, penicillin amidase is immobilized on a support suitable for separation using a stainless steel mesh in a stirred tank reactor. Also, whole cells might be immobilized on a support, most often simply by adsorption, or by entrapment in alginate gel or similar material.
5. The optimization process shown here and discussed later is based on a kinetic model of the whole reaction system, including side reactions. This model, together with the mass balance for the reactor, can be used for simulation. By a variation of feed concentrations, catalyst concentration, and residence time, conditions can be found that result in high conversion and high space–time yield. When enzyme deactivation is known, the product-specific catalyst consumption may be calculated as well. Alternatively, an empirical optimization can be performed using a simplex method or a generic algorithm. The latter have been successfully applied to

fermentation processes, but only tested in preliminary studies for optimization of multienzyme systems. These methods are especially powerful for optimization of processes where a direct connection between input and output values cannot be given.

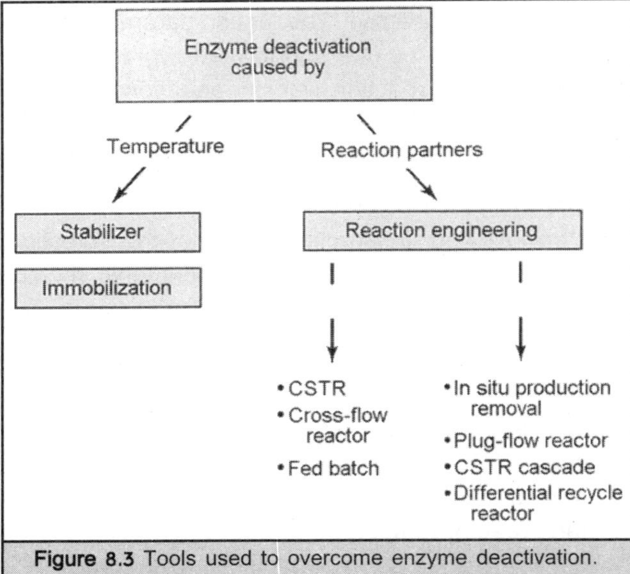

Figure 8.3 Tools used to overcome enzyme deactivation.

Figure 8.3 summarizes some measures that can be taken to overcome or modulate against enzyme deactivation caused either by thermal condition or by the reaction partners. Whereas in the first case lowering of the temperature, application of stabilizers, or immobilization on a support may be helpful, the deactivation caused by the reaction partners, may be overcome with the help of the tools of reaction engineering. For the deactivation caused by substrates or products, different tools can be used.

EXAMPLES

By choosing the appropriate fundamental reactor type on the basis of the individual reaction kinetics, the production of the fine chemicals is possible even in processes that exhibit a strong inhibition by one of the reaction compounds.

Product Inhibition

Product inhibition is the consequence of the competition of a substrate and a product for complexation in the active site of a catalyst. This kind of inhibition is not only observed with enzymes, but is also found for reactions that are catalyzed by homogeneous soluble and heterogeneous chemocatalysts. Whereas the reaction velocity is reduced by product bonding in the case of an irreversible reaction, the situation is more complex for reversible reactions.

In terms of the mathematics, product inhibition is expressed as an additional term in the denominator of the Michaelis–Menten equation (equation 6). This form of the equation describes competitive inhibition. An explicit discussion of other types of inhibition can be found in the textbooks of enzyme kinetics. By transformation of equation 6, equation 7 is can be obtained, which is similar in its structure to the common reaction velocity equation (equation 8), derived for chemical systems by Hougen and Watson in 1947.

$$\nu = [E] \cdot \frac{A_{max} \cdot [\text{substrate}]}{K_m^{\text{educt}} \cdot \left(1 + \frac{[\text{product}]}{K_i^{\text{product}}}\right) + [\text{substrate}]} \quad \ldots(6)$$

where v(kat/L) is reaction velocity (1 kat = 1 mol/s), [E](g/L) is enzyme concentration, [substrate](mol/L) is concentration of substrate, A_{max} (kat/g, U/mg) is maximal mass specific activity of the reaction, $K_m^{substrate}$ (mol/L) is Michaelis–Menten constant of substrate, and $K_i^{product}$ (mol/L) is competitive product inhibition.

$$\nu = [E] \cdot \frac{\frac{A_{max}}{K_M^{substrate}} \cdot [\text{substrate}]}{1 + \frac{[\text{substrate}]}{K_M^{substrate}} + \frac{[\text{product}]}{K_i^{product}}} \qquad \ldots(7)$$

$$\nu = \frac{(\text{kinetic term}) \cdot (\text{potential term})}{(\text{adsorption term})^n} \qquad \ldots(8)$$

where n is the number of centers, which are part of a velocity-limiting elementary reaction.

The decrease of the reaction velocity with increasing product concentration is described by the ratio of K_m to Ki. Three different ratios of K_m to Ki are plotted, which relate to strong (curve A), moderate (curve B), and weak (curve C) product inhibition. In 1985, Lee and Whitesides suggested, using the example of an oxido-reductase-catalyzed alcohol oxidation, that the feasibility of a process can be judged by the ratio of $K_m^{substrate}$ to $K_i^{product}$:

1. If $K_m^{substrate}$: $K_i^{product} \leq 1$, it is, in principle, possible that the reaction can proceed with practicable conversions.
2. If K_m^{educt}: $K_i^{product} > 1$, the reaction is unusable with regard to preparative syntheses.

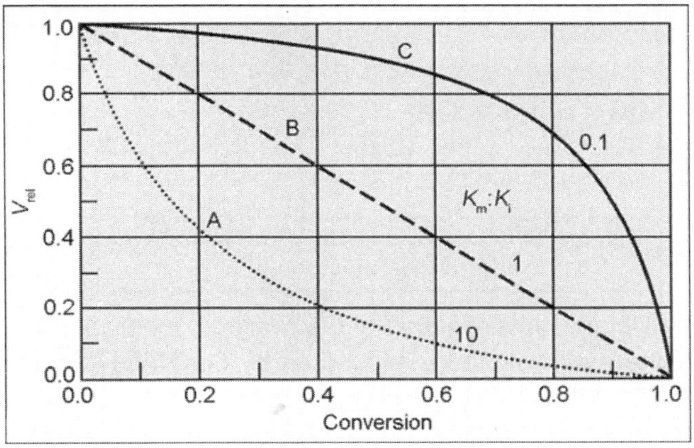

Figure 8.4 Influence on reaction velocity as function of conversion. Curve A, strong product inhibition with the ratio of Michaelis-Menten constant K_m, and the product inhibition constant, K_i; K_m:K_i = 10. Curve B, moderate product inhibition with K_m:K_i = 1. Curve C, weak product inhibition with K_m:K_i = 0.1.

This clearly is a simplified view. It is also important to note, that the kinetic parameters of a given reaction cannot be changed.

Role of Building Blocks

Chiral diols are important building blocks for the synthesis of chiral ligands, agrochemicals, and pharmaceutical intermediates. Using the example of the oxidative resolution of 1-phenyl-1,2-ethanediol, it is demonstrated that product inhibition can be overcome in the case of hydrophobic products. The enantioselective oxidation of the (*R*)-enantiomer to ω-hydroxyacetophenone (1-oxo-1-phenyl-2-hydroxyethane) is catalyzed by glycerol dehydrogenase (GDH) from *Enterobacter aerogenes*. Because GDH needs β-nicotinamide adenine dinucleotide (NAD^+) as a cofactor, a cofactor regeneration carried out by lactate dehydrogenase (LDH) was incorporated. LDH reduces pyruvate to (L)-lactate enantioselectively. The cofactor regeneration must proceed faster than the racemic resolution in order not to become the limiting step. The remaining (*S*)-enantiomer of the diol is the desired product for further synthesis.

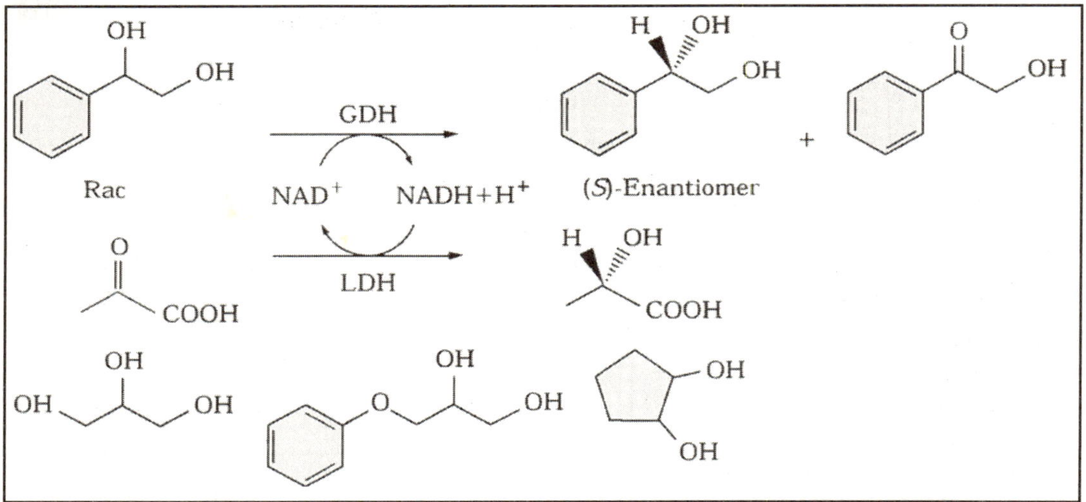

Figure 8.5 Enzymatic resolution of racemic 1-phenylethan-1,2-diol. GDH, glycerin dehydrogenase; LDH, lactate dehydrogenase.

Detailed investigation of the reaction conditions and kinetics revealed strong product inhibition by the hydroxyketone, as well as by the NADH. This resulted in a ratio of $K_m^{substrate} : K_i^{product}$ = 43.2 for the ω-hydroxyacetophenone. The inhibition by NADH was not taken into account, because NADH is converted in situ to NAD by LDH. Utilizing the above-determined criteria, the reaction as investigated here would never proceed efficiently. Additionally, it was observed that the hydroxyketone reacted in a chemical side reaction with NAD^+. Thus, ω-hydroxyacetophenone not only reduced the GDH activity due to its inhibition, it furthermore reduced the available NAD concentration. Therefore, it is very important that the ω-hydroxyacetophenone concentration is kept as low as possible. One possibility is an integrated extraction of the inhibiting compound. Best results were obtained using hexane as the extracting solvent.

In contrast to phenylethanediol, which was not extracted, up to 98% of the hydroxyketone was extracted after eight extraction steps. The reactor consists of three different loops. The membrane reactor (loop I) integrates the stirred-reaction vessel and the membrane module. A hydrophilic ultrafiltration membrane is used to retain the glycerol dehydrogenase and the lactate dehydrogenase. The consequence is the decoupling of the residence times of the catalysts and the reactants. Prior to the membrane reactor, an extraction module is positioned in a closed-loop arrangement (loop II). The aqueous phase is continuously extracted by means of liquid–liquid extraction using a hollow-fiber module consisting of Celgard microporous, hydrophobic hollow-fiber membranes. The module was circulated countercurrently by a hexane stream. The best ratio of the flow rates of the aqueous phase to the organic phase was determined to be 1:6. To reduce the amount of organic solvent used, a distillation (not shown) is incorporated in the extraction loop III. The oxidation product ω-hydroxyacetophenone is collected in the bottom of the distillation.

The entire reactor behaves as a batch reactor with recirculation, meaning that only a low conversion per cycle is reached, corresponding to a small amount of oxidation product, which is directly extracted. The consequence is a continuous low concentration of ω-hydroxyacetophenone in the membrane reactor. This reactor represents a modification of the continuously operated bimembrane reactor that was constructed for the conversion of sparingly soluble substrates. When no extraction is incorporated, an enantiomeric excess of only 58% and a conversion of 37% was reached after 170 h. The reason for this unsatisfactory result was the accumulation of the reaction product, which lowers the reaction velocity of the GDH and also reduces the concentration of the NAD^+ by the chemical side reaction. With integrated product extraction a conversion of 50% and an enantiomeric excess >99% is reached after 64 h. This process example demonstrates that by means of reaction engineering, kinetically unfavorable reaction can still be used on a preparative scale.

Effects of CSTR Cascade

D-Mannose can be found as building block of numerous glycoproteins and glycolipids, for example, in the common core structure of N-glycans. For the enzymatic in vitro synthesis of oligosaccharides, guanosine-5′-diphospho-α-D-mannose (GDP-Man) is used as donor for mannosyl transferases. Recombinant mannosyltransferases have been used for the synthesis of the trisaccharide Manβ1- 4GlcNAcβ1-4GlcNAc. The enzymatic synthesis of GDP-Man starting from

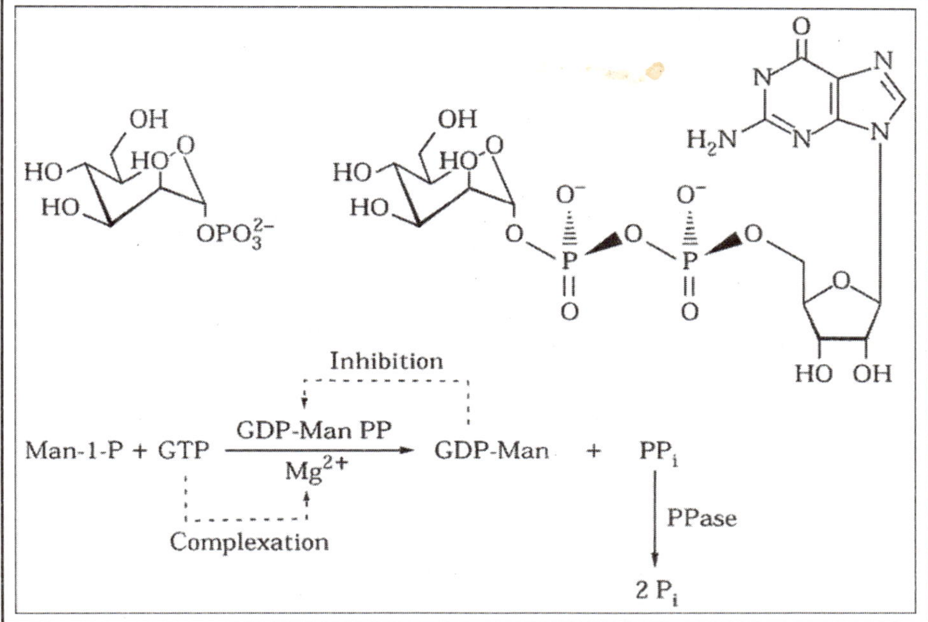

Figure 8.6 Enzymatic synthesis of GDP-Man starting from Man-1phosphate and GTP; GDP-Man PP = GDP-Man pyrophosphorylase, PPase = inorganic pyrophosphate.

mannose-1-phosphate (Man-1-P) and guanosine-5′-triphosphate (GTP) using a recombinant GDP-Man pyrophosphorylase has been recently described. Both reaction products show a very strong inhibition of GDP-Man pyrophosphorylase.

The inhibition by pyrophosphate can be overcome by cleaving it to phosphate using an inorganic pyrophosphatase. The product inhibition by the product GDP-Man itself can be reduced in a CSTR-cascade, as shown below. Additionally, the cofactor Mg^{2+}, required by the GDP-Man pyrophosphorylase, is complexed by the substrate GTP. Its influence on enzyme activity and the preferable use of a continuously operated reactor to establish a constant Mg^{2+} to GTP ratio, together with a complete set of kinetic data for the pyrophosphorylase, has been recently published. For the ratio $K_m^{Man\text{-}1\text{-}P}/K_i^{GDP\text{-}MAN}$, a value of 4.4 is obtained. According to the classification outlined in the section "The Problem of Product Inhibition", this reaction would be of only limited use. Unfortunately, the substrates and products are quite similar with respect to size, hydrophilicity, or charge, prohibiting a selective removal of the inhibiting product. However, simulations on the basis of the kinetic model and various reactor configurations revealed that a cascade of two CSTR in series reduces the required amount of enzyme by >50%.

A two-stage cascade is a good compromise: plug-flow behavior is approximated to a large extent, and the experimental setup does not require too much effort. In a two-stage cascade 98% conversion is reached for 4.5 (U/mL)min, whereas a single CSTR requires 9.0 (U/mL)min. That means that at a constant enzyme concentration the residence time can be cut in half, resulting in a doubling of the space–time yield. When the enzyme deactivation is only a function of time, by an increase of the space–time yield at constant enzyme concentration, the product-specific enzyme consumption can be decreased. The experimental verification revealed good agreement between calculated and experimental values. At a conversion at 98%, a space–time yield of 28 g/(L day) and an enzyme consumption of 900 $U/kg_{product}$ was obtained. For a single CSTR, an enzyme consumption of 2,000 $U/kg_{product}$ was reached due to the higher enzyme concentration required to reach the desired conversion under otherwise identical conditions. The relatively low space–time yield is caused by the low substrate concentrations. Complex interactions between K_m values of the substrates and the K_i of GDP-Man require simulation to show that the space–time yield can be increased 10-fold by a corresponding increase of the substrate concentration. In order to keep a high conversion at >98%, the enzyme concentration must be increased at the same time, but only 2.5-fold. This clearly demonstrates the potential of process optimization based on a model and a complete set of kinetic data.

9

PRODUCTION OF COMPLEX PROTEINS

Continuously propagatable mammalian cells, in general, and Chinese hamster ovary (CHO) cells, in particular, have become preferred host systems for the production of complex proteins for clinical applications. Nowadays, in pharmaoriented biotech companies, more protein species are being produced in CHO cells than in *E. coli* and in *Saccharomyces*, which were production hosts of first choice during the early years of applied DNA technology. These microbial organisms are capable of producing large amounts of recombinant protein. However, the decision to employ somewhat more complex and costly in-vitro cultures of recombinant mammalian cells has been driven by the need to obtain proteins with superior biological activity and more complex biochemical structures. Biological activity and, last but not least, the pharmacokinetic characteristics of recombinant proteins, when administered in therapy, frequently depend on a number of complex protein modifications, that is, proper folding and disulfide bridge formation, oligomerization, proteolytic processing, phosphorylation, and the addition of specific and complex carbohydrate groups.

Most of these and other protein processing steps can be executed with high efficiency by mammalian cells. CHO cells have become dominant in industry, and not only for historic reasons. Of course, they were chosen as the host substrate for the first recombinant protein from mammalian cells, human tissue plasminogen activator (Activase tPA), achieving market approval in the United States and other global markets. CHO cells are unique in many ways and exhibit a number of advantages, ranging from ease of introduction of exogenous DNA to the capacity for growth and production at very large scales. This article attempts to summarize and highlight arguments, issues, advantages, questions, and ongoing research, in a cursory way, for the industrial production of high-value proteins derived from CHO cells. In doing so, the development path followed to establish a high-yield process will be followed, from the construction and use of appropriate plasmid vectors to the planning and design of the cell culture manufacturing process. Issues raised and discussed herein are not meant to be comprehensive, yet it is hoped that the most critical points impacting development efforts will be addressed.

RECOMBINANT TECHNOLOGY

The following gives a brief introduction into the underlying principles for use of CHO cells as recipients of protein-encoding DNA vectors, and the most popular way to improve productivity with these cells: experimentally induced gene amplification. A mutant cell line lacking enzyme activity for dihydrofolate reductase, developed by Urlaub and Chasin, can be cultivated in the presence of glycine, hypoxanthine, and thymidine (GHT). When these cells are transfected with a functional expression vector for dihydrofolate reductase (DHFR vector), cell lines that have acquired a functional DHFR gene can be selected and expanded in media lacking GHT. A second (or third or more) expression cassette for a product of interest (e.g., a protein with therapeutic value) can be included in the DHFR plasmid or cotransfected as part of one or more additional vectors. Screening a number of "clones," or more correctly, cell lines derived from colonies, will lead to the identification of cells coexpressing the product of interest. In a second and often time-consuming phase of cell culture work, expression can be augmented by exposing recombinant cells to the drug methotrexate (MTX), a folate derivative that blocks DHFR activity completely and irreversibly.

Usually, within a period of 2–3 weeks, from a population of cells a few colonies emerge (while the majority of cells die) that are resistant to the drug due to elevated expression of DHFR. It was found that these cells frequently contain chromosomally integrated plasmid sequences in a higher copy number than observed in cells prior to exposure to MTX. Treatment with stepwise elevated concentrations of MTX can be repeated several times and may result in the isolation of cells that contain dramatically increased copy numbers of the transferred genes. The same principle for MTX-driven amplification applies to other immortalised cell lines as well; however, DHFR-mediated DNA transfer and methotrexate-driven amplification works best with cells that lack an endogenous DHFR gene (activity) prior to transfection.

REGULATORY AGENCIES

CHO cells, originally derived from the ovary of a female Chinese hamster, emerged as a spontaneously immortalized line, more than 40 years ago in the hands of Kao and Puck. The cells have been the object of many studies since then and are very well characterized with respect to a variety of aspects, including karyotype, chromosome structures, gene mapping, general culture conditions, cell physiology, and media requirements. For the purpose of genetic and physiological studies on the activity of dihydrofolate reductase and its genomic alleles, in the late 1970s, Urlaub and Chasin mutagenized the original cell line with the help of radiation and chemicals. The goal was to create a mutant cell line lacking functional activity of both alleles of the DHFR gene locus.

The resulting cell line was called CHO-DUX (sometimes also referred to as CHO-DUKX-B11) and, although not initially intended for this purpose, proved to be an ideal substrate for a number of pioneering gene transfer experiments. The "climate" for permission by regulatory agencies, in particular the FDA in the United States, was not favorable in the early 1980s for using "tumor" cell lines, such as immortalized CHO cells, for the production of recombinant proteins. Discussions about risks associated with the use of mammalian cells were controversial, and had been initiated more than two decades earlier when a first generation of "classical" biological products (i.e., vaccines and, later, natural interferons) were developed on the basis of primary monkey kidney cells, human diploid cells, and later, transformed mammalian cells. Manufacturers of recombinant proteins for clinical applications and regulatory agencies were in agreement that it was very important to minimize eventual risks associated with the use of recombinant mammalian cell hosts.

Risks were seen in tumor principles, carried by the DNA of the host and in adventitious agents (viruses, mycoplasma, etc.), which could infect the host cell lines. Also, consistency and quality of the recombinant proteins were discussed in the context of risk assessment and risk control. The result of a decade-long discussion was that very stringent manufacturing controls and monitoring procedures were enforced as a prerequisite for manufacture of proteins from such cells. A balance had to be found between the almost assured clinical benefit of some of those first recombinant products and the perceived (real or imaginary) risks associated with the unavoidable necessity to produce them in "tumor" cells.

The first product, recombinant tissue plasminogen activator (Activase rtPA), proved to be a good candidate to achieve such a balance because the benefit, saving lives of heart-attack patients, outweighed by far anticipated risks, for example, the debated risk of exposing patients to DNA sequences that may transfer a tumor-inducing principle. However, approval was achieved only after a huge amount of data was provided to the regulatory agencies showing that (i) consistently, only a very small quantity of CHO DNA (less then 10 pg/dose recently related to less than 100 pg/dose) was present in the final product, (ii) the product itself could be produced with a very high degree of reproducibility, and (iii) it was produced with a purity not achieved before in any biological derivation from mammalian cell culture.

Air Borne Contamination

Prevention of airborne and other bacterial or fungal infections of cell culture and recovery systems can be assured, to a very high degree of confidence, by the use of a piping and vessel system that maintains absolute containment of the sterile medium fluids. The equipment used must be of a nature that allows cleaning and sterilization by clean-in-place (CIP) and sterilize-in-place (SIP) procedures (usually using high-quality stainless steel). Most cell culture processes require complex media containing amino acids, vitamins, protein hormones, and fetal bovine serum. Some of these components cannot be autoclaved. To exclude, therefore, the introduction of bacterial and fungal contamination through these raw materials, prior testing and, in addition, filtration through membranes of 0.2-μm-or even 0.1-μm-pore-size into presterilized containers is employed. Rigorous testing cell banks, done by analyzing cells of a number of representative cryovials, which represent the bank when frozen in liquid nitrogen, assures that the production cell line itself is not contaminated with viruses, bacteria, mycoplasma, and fungi. Sterility testing of the cell line must be done in appropriate media lacking antibiotic or antifungal compounds, for obvious reasons. Virus testing is done in suitable cell systems that are validated for each of the individual virus species.

Bovine Spongiformous Encephalopathy

The use of sera or other products derived from bovine sources represents a potential risk of transfer of the causative agent for bovine spongiformous encephalopathy (BSE) to patients receiving recombinant proteins. Regulatory agencies therefore request detailed information on the origin and processing by the vendor of bovine derived products. For example, sera obtained from countries in which BSE was diagnosed, even in a small number of animals, are considered unacceptable by regulatory agencies. Companies have, in most cases anticipating these regulations, assured their supply of bovine- derived process materials from countries like New Zealand or the United States where BSE has not yet been reported.

Recently, due to the high degree of concern raised in Europe, but also in the United States, very stringent process regulations have been discussed. These regulations would restrict to a very large extent, or prevent altogether, the use of bovine-derived process additives, not taking into account the processing of the materials by the vendor or the receiving company. The high degree of concern regarding BSE is based on the possibility that (i) a very small proteinaceous compound could be responsible for the disease, (ii) transfer to the human population as Creutzfeld–Jacob disease has been considered possible, and (iii) detection of the causative agent is extremely difficult, even with the most sophisticated methods available today. It appears that the complexity of the involved issues will require further discussion in the future and must occur at an international level.

Viral Contaminants

Why were hamster cells chosen as a host for making human recombinant proteins? A strong argument, favoring nonhuman cell lines over human cell lines for the production of proteins, is the fact that certain life-threatening human viruses cannot be propagated, or grow poorly, in nonhuman cell lines. CHO cells have been found to not support the replication of pathogenic viruses such as polio, herpes, hepatitis B, HIV, measles, adenoviruses, rubella, and influenza. Thus, the risk of a viral adventitious agent, belonging to this group of viruses, involuntarily carried along with the product of interest can be considered extremely low. Wiebe et al. tested a total of 44 human pathogenic viruses for replication in CHO cells and found only seven (Reo 1,2,3; mumps; and parainfluenza 1,2,3) that were able to infect these cells. Exclusion of these few virus species, as well as some other viruses that can be propagated on CHO cells, such as the minute virus of the mouse (MVM), can be assured to a very high degree of certainty through testing.

Tests can be performed with all materials that enter the manufacturing process and would support the viability of the virus in question. Tests are also obligatory with fluids that contain the product of interest and that represent, for example, toward the end of the cell culture production cycle, a manufactured product unit, defined as a "batch." Sterility filtration of fluids containing a variety of raw materials, some of which may have been exposed to viruses, does not prevent the introduction of viral contaminants into the process of cell culture–based manufacturing of proteins. Only recently have membranes with pore sizes small enough to exclude passage of virus particles become available for industrial-scale operations. However, these membranes cannot be introduced into existing processes without complex consequences on regulatory issues. Testing therefore still appears to be the most efficient method to exclude viruses that may reside within the host cell line itself, or that could be introduced by biologically derived raw materials needed for the cell culture process. Specialized service companies have developed, in close collaboration with the pharmaceutical client companies, batteries of validated test procedures.

These utilize cell culture systems in which supernatants or lysates of the production cells are cocultivated with the corresponding virus-sensitive substrates. Because cells in master seed banks and in working cell banks are of the highest importance for the cell culture production process, samples from these are the first to be considered for the rather expensive and time-consuming testing exercise. Another and complementary approach to virus safety is the design of virus kill and removal steps of the recovery process for the protein of interest. These include physical and chemical principles separating (theoretical) viral contaminants from the product or inactivating them. Again, appropriate testing procedures and the demonstration of inactivation and removal of model viruses, as discussed in the important paper by Wiebe et al. was considered a major provision for the achieved safety of recombinant products from hamster cells.

Purity of Products

A very rich collection of methods has become available for the analysis of purified proteins within the last two decades. In addition, most of these methods have been recently optimized and fine-tuned to very high sensitivities and resolution. When employed as routine analytical procedures during the manufacturing process, they can assure high quality and consistency of protein products. Still, the manufacturer of a pharmaceutical protein has one major concern: will a product that has demonstrated its efficacy in extensive and expensive clinical trials remain the same in its essential characteristics when employing a defined manufacturing process for its production over many years? It seems surprising, but due to the large size and complexity of proteins under study for clinical use today, their structural features and their functional principles within the human body may not be fully understood by the manufacturer when completing successful clinical trials. This is especially true for the newer generation of pharmaceutical proteins that are larger in size and often contain multiple polypeptides and/or multiple specialized domains exhibiting secondary modifications such as glycosylaton, phosphorylation, and sialylation.

Subtle changes, sometimes difficult to detect due to inherent heterogeneities of structures of protein populations, may result in loss or modification of activity and could pose risks to the patient. In order to reduce this possibility, batteries of in-process controls and tests are an inherent part of the production of clinical proteins. In the following section, sensitive and robust techniques and methods are outlined and discussed. The objective is to (i) prevent the occurrence of even small changes in the procedures employed to produce the product of interest and (ii) enable the detection and exclusion from the final vial protein variants differing in a major way from the one tested in clinical trials.

GMP protocol

The first level of protection against inadvertent changes of the product rests in effective management of the production process over time. The challenge is the same as that of any mature industry: to produce large amounts of material at competitive cost while ensuring that product consistency and quality are maintained. Obviously, manufacturing teams and their supervisors undertake every effort to reproduce the manufacturing process to the utmost detail in every production run. This is achieved by defining and describing each of the various steps in the form of detailed protocols. Almost every aspect of the procedure is documented. These documents establish, in the form of cGMP (current good manufacturing practice) protocols, the basis for the overall procedure.

A "sign off" procedure by supervisors represents an integral part, assuring that the operating personnel for the manufacturing process are in fact controlling, assessing, and executing it according to the established protocols. At critical points in the overall process, the signatures are prerequisites allowing the progression of the process to the consecutive steps. The time period of cell line and process development leading up to the establishment of the manufacturing process is very important for the definition of critical "checkpoints." During this period, knowledge about parameters and steps is acquired that can result in product changes. Once certain limits of variations of process conditions have been identified (within which no change was observed) the cGMP protocol is drafted and finalized. Specific events, defined in precise terms as part of the manufacturing protocols, can trigger a more elaborate investigation. Supervisors and managers can even order an interruption of the manufacturing process. In extreme cases, crude product batches are withheld from further processing and are discarded.

Quality control

Quality control (QC) is an integral part of the manufacturing process of recombinant products. A comprehensive approach, utilizing independently validated techniques, is applied to assess the quality and identity of the product from various angles. It is the goal of QC efforts to assure that products made over years of manufacturing will meet the stated specifications in terms of identity, quantity, activity, and purity. In principle, it is no longer very difficult to produce large quantities of highly purified recombinant proteins. However, methods to produce these proteins are still considered part of a young technology, because its basis is the manipulation of genetic material of various sources in the laboratory. Those manipulations involve the creation of plasmid vectors, based on fragments of genetic information that are linked together as a first step. In a second step, the transfer of such DNA for transgene-dependent protein synthesis to a complex organism such as *E. coli*, yeast, or, as discussed here, into the chromosomes of mammalian cells cultivated in vitro occurs.

A major concern has been the fidelity (amino acid sequence identity) of the final product, particularly in view of the high degree of ignorance with respect to gene transfer mechanisms in higher eucaryotes and its consequences. Also, it appears that transfected DNA may have, possibly during the transfer to the host cell, a somewhat elevated propensity for mutation. A discussion of issues concerning structure and fate of DNA transferred to mammalian cells will be discussed later. Nevertheless, based on a history of experience with this technology for more than 10 years, it can be stated that this young technology is in fact a very reliable one. It has been suggested that rigorous and extensive nucleic acid based tests, most notably a complete sequence assessment of the genomically integrated DNA (or transcribed RNA from recombinant cells) should be performed. It appears that such rather complex, expensive, and time-consuming tests are not necessary. An entire session was devoted to nucleic acid considerations during a conference organized by the International Association of Biological Standardization, held in Annecy, France in 1994. Berthold specifically addressed genetic stability in CHO cells. As already pointed out, following the rapid development of techniques allowing manipulation and transfer of DNA molecules to various hosts, powerful and reliable new methods have emerged in analytical protein chemistry. These have increased the capacity to characterize purified protein preparations to an extremely high degree of sensitivity and resolution.

These methods represent efficient tools to assess identity (including the amino acid sequence), quantity, potency, and purity of the product at the level where it is most needed—just prior to administration (delivery) to the patient. The results from QC assessments from thousands of lots of approved products are, as far as this author knows very encouraging with respect to (amino acid sequence) identity: no lot has been called into question based on a QC finding that has indicated a newly emerged mutant gene. Still, this statement does not negate the possibility that mutants may occasionally emerge. They will, over and over again, within the large population of cells in bioreactors. A 1,000-L reactor usually contains more than 10^{12} cells. However, there is no reason to believe that mutations would occur at a frequency significantly different from the frequency of mutations occurring within the human body (1 bp change in 10^9 bp for each generation).

Of course, such mutations will be "scattered," and as individual entities are extremely rare. Therefore, individual and specific variants are not likely to emerge in the population of protein molecules as a sizeable subclass of identical molecules. A telling example of the power of in-process controls and the associated biochemical assays to identify variants in a population of molecules is given in an article by Sliwkowski and coauthors. It describes the detection of an amino acid exchange mutant of a recombinant monoclonal antibody (anti-Her2). This variant was detected early during process-development efforts when using a methotrexate-amplified clone of CHO cells. The variant represented about 10% of the total population of antibody molecules. The origin of this mutant remains mysterious, but it seems it was the result of an early event during the development of the cell line that was used in the process.

BIOTECHNOLOGICAL APPROACH

Role of Calcium Phosphate in DNA Transfer

Graham and van der Eb showed more than 20 years ago that the transfer of DNA into mammalian cells in culture can be achieved by exposing cells to microprecipitates of DNA and calcium phosphate. This simple method (termed calcium phosphate transfection) has become very popular. Typically, a glycerol shock is part of the transfection procedure. The shock is thought to promote the transport of individual DNA–calcium phosphate complexes across the cell membrane, perhaps by inducing endocytosis or phagocytosis. When optimized transfection conditions are used, a high percentage of cells take up DNA. However, in addition to the movement across the outer membrane of the cells, the transport of DNA from the cytoplasm to the nucleus seems to be a significant barrier. In 1982 Loyter et al. reported that only 1–5% of transfected cells have detectable calcium phosphate complexes within the nucleus. The stable integration of transfected DNA into chromosomes of mammalian cells was first shown by Axel and collaborators. The number of emerging colonies upon transfection by the calcium phosphate technique has been reported to be low: usually, only between 0.0 1% and 0.5% of the cells gave rise to recombinant colonies. More recent investigations and appropriate improvements of crucial physicochemical parameters of the calcium phosphate transfection methodology in several laboratories have increased transfection efficiencies for stable recombinant cells to about 5–10% (and those of transient transfections to levels of 50% or higher) of the population of cells exposed to exogenous DNA. Calcium phosphate transfection still seems to be the most frequently used method for the generation of recombinant CHO cells. Yet, it appears that the transfer of a sufficient quantity of DNA into the nucleus and the integration of at least a few functional plasmid molecules into transcriptionally active regions of the receiving genome is by no means an efficient process. Halfa million or more plasmid molecules are usually provided per cell in a typical calcium phosphate transfection, the result of which is stable gene transfer in a small percentage of the exposed cells.

It has been shown that the association of the DNA with calcium phosphate protects against nuclease attack during the transport to the nucleus. Eventually these complexes become dissociated and nuclear endo- and exonucleases will have access to the "naked" DNA. Moreover, it has been shown that supercoiled plasmid DNA molecules will be converted into relaxed (single-stand cut), open circular, and finally linear molecules (double-strand cut) within the nucleus after 1–2 h. This is, of course, an essential step for integration into the linear DNA backbone of a chromosome. The mechanism by which the DNA is transported across the nuclear membrane (nuclear pores are not large enough for diffusive transport), and finally to the site of chromosomal integration, are not yet known. On the receiving side, at least one strand of the chromosomal DNA must be opened while a linear plasmid molecule is in sufficiently close proximity. Nuclear ligases could then mediate covalent linkages. Very little is known about the mechanism and about the factors affecting site specificity for integration. Site specificity is a very relevant factor in this context. It is believed that it has influence on the transcription rate of the integrated DNA, even if the DNA carries a powerful (viral) promotor, such as the SV40 early promoter or a CMV promoter.

The current assumption is that exogenous DNA will integrate at "random" within the genome, even if the transfection cocktails contain DNA with homology to sequence segments of the receiving genome. This is an observation from numerous gene-targeting experiments in mammalian cells. It was found that only 1 in 1,000 to 1 in 10,000 events would result in targeted integration. With respect to the actual sites of integration in CHO cells, there is only limited and sporadic data available at present. Only a small proportion of recombinant cell lines should contain the transferred DNA within transcriptionally active regions of chromosomes if integration would truly occur at random subsequent to transfection. In reality, we cannot know whether DNA will end up more frequently than not in transcriptionally inactive regions of the genome. We have no opportunity to analyze cells that did not overcome the selection step. The selection step is, as we know from the low transfection efficiencies in stable transfections, a very significant barrier. Among the emerging cell lines there is usually extensive heterogeneity in productivity for recombinant proteins.

The expression levels from primary CHO clones for proteins like rtPA, a secreted version of the CD4 receptor (rCD4-truncated), and a chimeric TNFR-FC fusion protein had a range of more than two orders of magnitude, observations which this author believes can be generalized for all transfections in mammalian cells. As a consequence, the identification of high-producer cell lines is a tedious and labor-intensive exercise and requires the screening of hundreds of individual cell lines. In the author's experience, it is usually necessary to invest 2–4 months of laboratory work into this task. Only then has one enough confidence in having appropriately assessed the upper range of sustainable expression for a novel protein in CHO cells. Nuclear enzymes such as endo- and exonucleases, but also ligases and possibly recombinogenic enzymes, acting on the population of plasmid molecules within the nucleus are responsible for the modification of transfected DNA that becomes eventually integrated into the genome.

This modification may not only degrade in part the plasmid DNA but is also responsible for creation of larger, that is, more extended DNA complexes. These DNA molecules seem to be created before the integration into the genome. They provide the physical basis for linking the genetic selection marker and the gene(s) of interest in those transfections that utilize separate vectors. It should be noted here that DNA transfer methods, other than the use of calcium phosphate, might deliver different quantities of DNA to the nucleus. This may have profound effects on the structure and copy number of integrated DNA molecules, as will be discussed later. In discussing the integration loci for transfected DNA in the mammalian genome, it is interesting to note that integration sites for amplified DNA sequences, detectable by fluorescence in situ hybridization (FISH) after selection in methotrexate, did not show a preference for a particular chromosome or chromosomal region. However, in light of recent data on gene amplification mechanisms, it is not possible to draw conclusions from these findings toward the identification of the initial (primary) integration site(s).

Other Methods

In addition to calcium phosphate transfection, more recently electroporation, cationic lipids, and liposome mediated transfection techniques have been developed for transfer of DNA into mammalian cells. Most of these techniques have been reported to mediate higher transfection efficiencies as compared with calcium phosphate–mediated DNA transfer. Such claims must be regarded with some caution because all DNA transfer techniques established so far suffer from high variability. Some of the variabilities observed are probably due to technical difficulties in producing reliable and consistently highly efficient carrier systems. The other factors that cause major variations in transfection efficiency are the type of cells used and the condition of the cells prior to transfection. Because carrier principles for the transfer of DNA to cells can vary widely, one suspects that very different consequences may result within the cell, events over which the person conducting the experiment has no or little control. Just one of the possible consequences should be mentioned here: The technique applied for DNA transfer may affect the copy number average of individual clones, depending on the amount of DNA transferred to the nucleus.

It has been proposed that electroporation may result in a smaller copy number of chromosomally integrated DNA than, for example, calcium phosphate–based transfection. In view of the applied selection procedures for the generation of recombinant cell "clones," very different chromosomal structures of integrated plasmid sequences may result. Another factor to be taken into account for the genomic structure of integrated plasmid sequences is the physical form of the DNA for transfection. Because linearization of plasmid DNA is a prerequisite for integration into any chromosomal site, the structure of the plasmid prior to transfection will affect the final structure when integrated. When circular plasmids are used, as predominant in the past, linearization is dependent on cellular enzyme activity. This will happen most likely at random, and the integrated DNA structure cannot be predicted. Depending on the overall size of the plasmid molecule and the size of the DHFR gene cassette relative to the vector, a significant number of molecules would be linearized within the DHFR sequence, which jeopardizes the functionality of this plasmid.

Obviously, molecules whose DHFR expression cassette are interrupted or deleted by the linearization process would be unable to transfer a functional marker gene to the host cells, and such cells would be eliminated through the selection procedure. Therefore, opening the circular plasmid molecule by restriction enzyme digestion with appropriate enzymes prior to transfection has been recommended for improved transfection efficiencies in protocols that aim at integration of plasmid DNA into chromosomal DNA. We have seen in our laboratory that linearization is, in fact, advantageous in terms of transfection efficiency with stable transfections. Linear plasmid molecules are also used routinely in gene transfer experiments that attempt to transfer the desired DNA into specific (homologous) DNA sequences of the receiving genome (gene targeting). In these experiments, linearization is executed in a way that results in cutting the homologous DNA fragment approximately in half. The molecules for DNA transfer then carry DNA sequences that are homologous at their ends. We have used such an approach in our laboratory by employing (defective) retroviral DNA sequences derived from the CHO genome in expression vector cocktails. They have quite significantly improved the frequency of high-producer cell lines upon transfection.

BIOTECHNOLOGY OF CHO CELL

Use of serum and other more or less undefined and complex media additives for growth of mammalian cells may result in problems of reproducibility in the process, and thus additional costs. To avoid a negative impact on cell growth, testing procedures have to be incorporated prior to use of these additives to identify possible variations from batch to batch. Therefore, the desire to replace some or all of these variable components in cell culture media with well-defined components having high purity and little or no variation in quality has been a priority in the industry for quite some time. Also, use of serum and animal-tissue-derived growth factors (insulin, transferrin) bears the considerable risk of transmittal of undesired adventitious agents to the production process because these additives frequently cannot be heat sterilized.

Although serum-free and growth-factor-free media may support the growth of mammalian cells, the robustness of the cell culture process, in general, and maximally achievable cell densities in particular maybe reduced when using media free of serum- and/or protein growth factors. Therefore, efforts have been undertaken to modify production host cell lines prior to the transfer of expression vector DNA. The hope was to generate hosts for production that would achieve superior growth rates, and so would be more useful for scale-up and higher in productivity. Various oncogenes, cell cycle genes, and/or hormone genes have been transferred into the genome of cells, resulting in novel and possibly superior production hosts. In view of actual discussions regarding BSE and other, so far unknown, adventitious agents, which pose the risk of entering production through process additives, the interest in such novel hosts will be very high. It needs to be seen how soon and how frequent these “super CHO” cell lines will be used in the manufacturing of novel products.

EFFECT OF METHOTREXATE ON POPULATION OF CHO CELLS

Treatment of DHFR-positive CHO cells with incremental concentrations of methotrexate has been used to select for populations and clones of populations of CHO cells with an increased (i.e., amplified) copy number of the integrated DNA within the host genome. The phenomenon of methotrexate-mediated gene amplification had been observed before the use of recombinant DNA technology, most notably in cancer patients who had been treated for extended periods of time with methotrexate. With CHO cells, cell lines containing several hundreds, and sometimes a few thousand, copies of transfected DNA fragments integrated within chromosomes have been established. Most amplified cells produce more product than unamplified host cells. However, the improvement of specific productivity is highly variable when studying individual clones and also varies from product to product.

Molecular Biology of Amplified CHO Populations

Subpopulations and clones of methotrexate-treated CHO cells may contain the transfected DNA sequences at very high copy numbers. Studies concerning genetic features of these amplified DNA sequences within the CHO genome have been ambiguous. Whether continued presence of a selective agent (i.e., methotrexate) in long-term cultures of recombinant CHO cells is required for production stability is still controversial. Whereas some studies suggest MTX is required for stable production of recombinant proteins, others indicate that continued culture of clonal cell lines in MTX may not be necessary. Cytogenetic studies using fluorescence in situ hybridization (FISH) were done in the author's laboratory with clonal and nonclonal recombinant CHO cell lines. They showed that continuous exposure to MTX at the same concentration under which the cell populations were initially established promotes genetic instability at the chromosomal level. Cell lines selected at micromolar concentrations frequently show elongated chromosomal structures that hybridize to probes representing the transfected DNA.

Some of the chromosomes, containing a large number of more or less tightly arranged bands or regions, differ dramatically from normal CHO chromosomes. This raised questions about cytogenetic stability of such aberrant chromosomes during continuous subcultivation and, in consequence, the production of recombinant proteins based on genes residing within these structures. The first of our FISH analysis studies was done with cell lines expressing the mouse c-*myc* gene in an inducible manner. The advantage of this system is that the phenomena observed in the genetic context can be attributed solely to the status (expression level, chromosomal location, degree of amplification) of the DHFR selector gene and the inhibitor MTX. The c-*myc* gene is truly silent during regular growth conditions and could not act as a metabolic burden nor result in any selective forces. When these cells are cultivated at 37°C, the inducible *Drosophila melanogaster* heat shock protein 70 promoter is in "off" status.

Even in a cell line containing 1,000–2,000 copies (selected at 320 μM methotrexate) of the transfected DNA, no trace of transcribed c-*myc* RNA and translated c-*myc* protein was observed at 37°C. However, high-level expression of c-*myc* RNA was induced upon shifting cells to a culture temperature of 43°C for 1 h. The continuous exposure of the 320-μM-MTX-selected cell line (c-*myc*) to the drug resulted in about a twofold higher copy number of plasmid DNA as compared with a population of cells that had been cultivated in the absence of any selective pressure for at least 4 weeks. A similar observation, based on a cell line with a constitutive promoter driving tPA expression, was reported by Weidle et al. These authors reported a loss of about 50% in copy number when cells were grown in the absence of methotrexate for about 50 days. This decline in gene number also resulted in a similar drop in productivity. No further copy number and productivity loss was found during the subsequent observation period of up to 100 days.

Following our initial observation of a drop in copy number in cells containing the mouse c-*myc* gene, we extended our studies with this line and also analyzed the chromosomal integration patterns of transfected sequences in two additional CHO cell lines. The three cell lines studied under this program expressed one of the three following proteins: (i) human rtPA, (ii) a truncated version of the human CD4 receptor, and (iii) the murine c-*myc* protein. The three cell lines had been established by transfecting CHO-DUKX cells using standard calcium phosphate precipitation techniques. Either individual clones or pools of clones were cultivated for extended periods in media containing stepwise increasing concentrations of MTX. In each case, a single individual clone expressing the desired level of recombinant protein was expanded for long-term growth and production.

FISH was performed using biotinylated probes complementary to the recombinant sequences. Three different sets of typical integration patterns (site, degree) of amplified sequences were found for lines under study. FISH signals in the rtPA-expressing cell lines were localized, in the majority of cells, in one of the larger submetacentric chromosomes (probably either chromosome 3 or 4, or the X chromosome). The signals (a bright band of fluorescence) covered most of the length of the short arm and showed weaker fluorescing spots over the center of the chromatids in the long arm. A number of other types of integration patterns (including an elevated percentage of quasi-tetraploid cells) were found in these cells (about 25%) when they were grown in the presence of methotrexate. When methotrexate was omitted, more than 95% of metaphases had the integration described. In cells expressing c-*myc*, continuous cultivation in media containing high concentrations of MTX was found to be associated with rearrangement and variable amplification of transfected sequences within almost half of the cells. Several types of integration patterns were found in these cells: (1) chromosomes with highly extended regions of hybridizing sequences, (2) cells containing transfected DNA integrated into multiple sites on an individual chromosome, (3) cells containing DNA integrated into multiple chromosomes, (4) chromosomes joined at amplified regions, (5) circular chromosomes, (6) cells with small derivative chromosomes or fragments, and (7) cells containing a single "normal" integration, exhibiting a bright band of hybridization in the center of the long arm of what we assume is chromosome 2.

The last type was found in about 40% of cells when grown in the presence of methotrexate. This integration type increased in frequency to 95% when cells were grown in the absence of methotrexate for 3 weeks or more. Similar results showing quite extensive heterogeneity were observed with yet another recombinant CHO cell line (the fourth studied), producing a chimeric CD4-immunoglobulin protein. When grown in the presence of MTX, CD4-IgG-producing cells showed heterogeneity in both chromosomal location and copy number of amplified sequences coding for CD4-immunoglobulin (CD4IgG). For the cell line expressing the truncated CD4 molecule, a strong signal was found over the telomeric end of a small acrocentric chromosome (possibly a derivative of chromosome 5, 6, or 7), which covered about one-third of the total length. In summary, when cultivating these cell lines in the absence of methotrexate, unique and characteristic integrations were found in 95–99% of metaphases. Subsequent to the first analysis, we continued to cultivate these cell lines for extended periods in the absence of methotrexate, and we occasionally performed additional FISH analysis. We found the chromosomal structures described here to be stable within the observation period (a minimum of 60 days and in one case of 160 days). Cytogenetically, these observations indicate a high degree of genetic stability of chromosomally amplified sequences in the absence of methotrexate.

Southern Hybridization

Southern hybridization of genomic DNA is a useful tool to acquire insights into the molecular structure of integrated plasmid sequences. Using suitable restriction enzymes it will provide information on the integrity of the expression fragments of interest. Estimates of copy numbers of the chromosomally integrated DNA can be established when, in the same experiment, an appropriate range of quantities of plasmid DNA, restricted with the same enzyme as the one used for restriction of genomic DNA, is used. Also, Southern hybridization may indicate whether one integration locus or more than one exists for the plasmid sequences. However, conclusions on multiplicity of integration must be taken with a degree of caution. Aberrations from the expected signals can be due to postintegration rearrangement in a fraction of the cell population or in the initial cointegration of a few copies of the plasmid sequences that had been subjected to nuclease attack, resulting in the deletion of the restriction enzyme site used for the analysis. All Southern hybridizations are based on DNA extracted from thousands of individual cells. Even if the cell lines are based on a "cloning" step, one must be aware that none of the so-called clonal cell lines are clonal in the true sense.

Mammalian cells, in particular immortalized cells, have an intrinsic genomic fluidity that is most evident when studying chromosome numbers in metaphase spreads. FISH can be used to obtain insights into the extent of chromosomal amplification (not to be confused with copy number estimates) and the location of recombinant DNA within the chromosomal set of cells. In order to provide some useful information, FISH studies must be supplemented by a statistical analysis of identified integration sites (and structures observed). The time point of analysis, with respect to the total period of time (since thaw) for use of these cells, should also be taken into consideration. A reasonable value (from the invested work) may be gained from a FISH study performed shortly after cells have been thawed from a bank of cells stored in liquid nitrogen. Depending on the culture conditions (with or without methotrexate, with or without serum, etc.) and the length of time during which cells have been exposed to in vitro culture, results of FISH analysis may vary considerably. In spite of variability, in the studies discussed we were able to separate without any doubt or difficulty the identity of one recombinant cell line from a second recombinant cell line by using FISH. This was possible because we were able to identify a chromosomal marker containing hybridizing DNA that was present in a large fraction of the individual cells of the populations studied. The identifying chromosomal markers containing recombinant sequences were termed *master integrations,* which we found to be the genetically stable entities in the system.

SELECTION OF PRODUCTION SYSTEM

Process scientists and their managers must eventually decide which type of production system to choose for the product in question. The choice of a production system is driven by the anticipated scale of operation for the manufacturing process, as well as by other issues. A number of very successful recombinant products from mammalian cells, such as EPO and GM-CSF, have been administered worldwide to a large number of patients. For these peptide hormones, the applied doses for treatment of patients are relatively small (about 100 μg/patient). Another successful protein therapeutic, recombinant tissue plasminogen activator (hu-rtPA, Activase) requires a dose of 100 mg/patient, a 1,000-fold difference in dose as compared with erythropoeitin. The number of patients treated per annum with the two categories of products is about the same. It is obvious that such a difference in the annual product requirement would drive very different manufacturing decisions. In the case of EPO and GM-CSF, a process based on roller bottles appears to be sufficient to supply the market with product. Using roller bottles, a "simple" and robust process can be established with minimal initial investment in hardware.

Roller Bottle to Roller Bottle Process

Cells thawed from a cell bank can be expanded by subcultivation into a fixed number of roller bottles. This approach is based on cells growing attached to the inner surface of the roller bottle. Standard 2-L roller bottles are usually filled with 300–500 ml of medium. The remaining volume (air) provides the necessary oxygen while the bottles are slowly rolled (at about 1 rpm) in an incubator at 37°C. The caps on the bottles are closed during the period of incubation. A sufficiently large number of roller bottles containing a healthy cell population represents the starting point for several production cycles. Part of the cell mass in the roller bottles can be immediately used for reproduction of cells in order to initiate the production cycle following the one already initiated. In other words, the bulk of the cell mass will serve as a seed culture to initiate the first production cycle. The remainder will provide, within one or two subcultivations, the basis for the subsequent production cycle. Media for production are usually rich in nutrient content in order to maintain viability and productivity of the cells for a minimum of 5–7 days.

Subsequently, the medium with the secreted product will be harvested, leaving the adherent cells inside the bottle. Sometimes, a refeeding with fresh medium for a second production cycle is possible, on the condition that the product of the first and the second harvest will be similar in composition and quality. Since a standard 2-L roller bottle will have about 300 ml of medium for harvest, 1,000 roller bottles can provide product from 300 L or 600 L of supernatant, from one or two harvests, respectively. If such a campaign lasts all year long, a manufacturing process based on this principle will deliver 15,000 L (30,000 L) of cell-free culture medium containing the product of interest.

Product concentrations in the 10- to 100-mg/L range are possible, resulting in 150 to 1,500 of unpurified product. Of course it must be realized that such a process is very labor intensive, requires the repeated use of trypsin for detachment of cells, and carries with it a considerable risk of contamination with adventitious agents through the handling personnel. However, because subcultivation in plastic cell-culture vessels represents a standard procedure in all laboratories, the scale-up from the lab-scale procedure to the pilot or manufacturing scale is a convenient approach for obtaining considerable (but not large) quantities of recombinant proteins.

Bioreactor to Roller Bottle Process

The ability to cultivate CHO cells both in suspension and, subsequently, in an attached form allows the roller bottle production process to be streamlined. The seed culture for the roller bottle production phase can be generated in spinners or in bioreactors. The advantage of such a process would be that fewer interactions in terms of subcultivation are needed to generate sufficient cell mass. Being less labor intensive, it also reduces the risk of contamination by adventitious agents of the feed culture.

Stirred Tank Bioreactor Process

Kilogram-scale production of recombinant proteins from CHO cells was first achieved in stirred bioreactors at very large scale (1,000 L and 10,000 L) with cells synthesizing while floating in free suspension. While such adaptation to suspension growth had been, in the early years of CHO technology, a somewhat tedious process, today multiple factors seem to have made this easier. For one, cell culture media have been developed that support the growth of cells in suspension better than those based on formulations such as DMEM and Ham's F12. Also, a number of purposeful genetic modifications (transfection of hormone genes or cell cycle genes, selection of CHO populations, or extended growth of CHO cells in media with little or no fetal calf serum and other modifications) appear to have generated CHO phenotypes that more readily support the transition from adherent growth to suspension growth. In a bioreactor-based process, the scale-up to very large volumes can occur rather fast. It is usually executed by subcultivations, that is, by diluting the entire volume of one bioreactor into 5–20 volumes of fresh medium, maintained prewarmed in the next larger reactor. Within 10–15 days, therefore, a suspension culture at the 50-L scale could inoculate a 10,000-L reactor.

It is major goal of process development work to carefully optimize media for the very last step—the production phase. Such a medium must support good growth in order to achieve the highest cell density possible, and it also must provide the nutritional basis and physiological balance needed to maintain viability and productivity for extended periods. While the termination (i.e., harvest) of such a culture is driven mainly by issues of plant capacity and volumetric productivity, the other important issue to consider is the quality of the derived product. The continuously changing composition of the culture medium during the production phase can affect the quality of earlier synthesized product through degradative activities, such as sialidase activity and proteolytic activity.

BATCH PROCESSES

The batch process is considered a very simple, robust production process for stirred bioreactors. In the step preceding production, cells in a (somewhat smaller) bioreactor are cultivated to maximal cell density and are then transferred, still suspended in the exhausted growth medium, to the production reactor. After addition of fresh medium and eventual adjustments in control of certain process parameters (O_2 supply, pH, temperature, etc.) the production process begins. It ends at a predetermined point in time when synthesis of recombinant protein seizes or slows down due to exhaustion of nutritional components in the medium and/or accumulation of toxic end products of cellular metabolism. For CHO cells the production phase will usually last, depending on the proteins' susceptibility to degradative enzymes and a number of other process- related factors, between 5 and 14 days. The advantage of such a process is obvious. Providing that the inoculating cell mass can be made reproducible and with little variation in biological quality, the resulting production process will show a high degree of reproducibility with respect to cell growth, viability, quantity, and quality of the product being synthesized. The issue of reproducibility of process parameters and of achievable product quantity and quality is of highest significance, of course. It provides confidence-generating data for the long-term future.

Another important point is the obligation to present such data to regulatory agencies that will ultimately give permission to use the process for the generation of the product. For investigational new drug applications (IND) and process licence applications (PLA), rather specific requirements must be met with respect to the minimal number of product batches analyzed. For INDs no less than three product runs are recommended to be presented. Shorter batch processes (5–7 days) have the advantage of generating more data within a given time frame. This can be a very important cost factor because the time period necessary to acquire and evaluate necessary data from the new process will eventually affect the overall time necessary for entering the market. On the other hand, there are viable arguments that could sway process-development decisions in another direction. The option to prolong the synthetic activities of cells in the production vessel would capitalize on the process investment that allows generation of the necessary cell mass for the production vessel. Therefore, more and more extended-batch or perfused-batch cultures are used. With a longer-lasting production phase in cell culture, feeding additional medium components becomes necessary.

Obviously, extending the process for a considerable period of time, (for example, from 7 days to 13 days) makes sense only when the return for this investment in labor and in occupation of the production facility results in a sufficiently high increase in product concentration within the vessel. There are various ways in which medium and medium components can be added to a culture that was initiated a few days earlier in the same tank. This could be done by feeding (batchwise) highly concentrated mixtures of essential amino acids and other medium components, thus not significantly affecting the volume in the tank. Alternatively, a culture may be started in the production vessel at half or so of the working volume, then standard concentration medium could be pumped slowly and continuously, or batchwise, into the tank until the final working volume has been reached. The choice between the modes, or combinations thereof, is in the hands of the process development scientist, who has to carefully evaluate advantages and disadvantages.

No matter what principle will be used for extending the production phase, the overall result will always be a tank that contains, in one batch, the entire protein population for subsequent recovery and purification. A very different philosophy for manufacturing is represented by continuously perfused production processes. Here, the goal is to achieve the highest cell concentrations possible within small tanks. In this process up-front investment for manufacturing equipment is reduced, and product quantities can be quite high. The much improved knowledge base in technology, and in physiology of CHO cells in culture, has made this more-complicated approach to manufacturing attractive. Perfused cultures can be maintained for many weeks and months, with product harvests occurring repeatedly throughout that period.

A protein of high interest to the pharmaceutical industry for decades, Factor VIII, is reliably being manufactured using perfusion technology with baby hamster kidney cells, a "cousin" of Chinese hamster ovary cells. While the production of Cognate Factor VIII has pioneered the use of perfusion technology for recombinant proteins with mammalian cells in a non-CHO cells, other products from CHO cells have recently been approved as well using this technology. Finally, due to the fact that many monoclonal antibodies have been produced at the laboratory scale in perfusion systems for many years by using hollow-fiber technology, the transfer of this technology to the production scale for pharmaceutical manufacturing has been an obvious conclusion for quite a while. Recently, a diagnostic monoclonal antibody for imaging in patients with prostate cancer was the first FDA-approved product for injection into humans derived from the hollow-fiber perfusion process. The application of this and other perfusion technologies, using immobilization of CHO cells on macroporous carriers to produce recombinant antibodies or other proteins, is likely to be yet another option for future manufacturing processes.

CONCLUDING REMARKS

The technology of using mammalian cells, and in particular CHO cells, for recombinant protein production is, in spite of significant success stories, still in its infancy. Much has to be done to establish production processes in a more straitforward way and to make them more productive. A data explosion is occurring with DNA sequences from microorganisms, plants, and animals, including the human genome. Subsequently, a tremendous need is arising for more detailed studies and, frequently, the urge to use the translation products of these sequences in many applications. Also, the possibility of designing new proteins, based on concepts provided by nature, will even increase the need for flexible and robust synthetic machinery.

Mammalian cells have been used as production hosts in some of the first pioneering experiments for the design of novel proteins. However, to deliver significant quantities of the desired proteins to analytical chemists, cell biologists, clinical researchers, and so on, takes far too long and is still a very demanding task. Some of the bottlenecks in the knowledge base have been addressed in this article. It is hoped that research efforts will solve some of them in the near future, resulting in more efficient, faster, and less expensive synthesis of recombinant proteins by mammalian cells. It appears that mammalian cells inherently have, due to their large size, the capacity to take up large quantities of foreign DNA without any negative effects, the responsiveness to perform a large number of synthetic tasks, and the capability to meet the challenges of the future.

10

COMMERCIAL PRODUCTION OF GLUTAMATE

Glutamic acid was isolated from wheat gluten hydralyzate by Ritthousen in 1866. Wolff carried out the first chemical synthesis of glutamic acid in 1890. In 1908, Ikeda isolated the specific taste fraction from the kelp-like seaweed *Laminariaceae*. Kelp, called konbu in Japan, is traditionally used as a soup-stock component with dried bonito, and has been identified as the salt of glutamate, used to enhance the flavor of foods. He also recognized it as "umami" taste, which does not belong to the four traditional taste categories; bitter, sour, salty, and sweet. Immediately after, Ikeda filed a patent application and began the commercial production of monosodium L-glutamate monohydrate (MSG) with his partner, Suzuki, in 1909. With the increasing demand for MSG is use as a seasoning, biochemical procedures were developed for its production. For example, α-ketoglutaric acid was first fermented from glucose by using a species of *Pseudomonas*, followed by the conversion with amination to L-glutamate in a high yield.

The vigorous study on the accumulation of L-glutamate in bacterial cultures by Kinoshita et al. opened the door to the microbial production of L-glutamate and other amino acids. The annual amount of production was estimated at more than 1,000,000 tons in 1997, an amount that increases by about 5% globally every year. Production is increased because of the demand due to changes in eating habits, and its use in commercially prepared, packaged foods; ready-made sours; and table seasonings in both Western and Eastern countries.

PHYSIOLOGICAL FUNCTION

L-glutamic acid is split into α-ketoglutaric acid and ammonia by glutamate dehydrogenase. By the reverse reaction, L-glutamic acid is synthesized from α-ketoglutaric acid, a member of the tricarboxylic acid (TCA) cycle of glycolysis. Whereas glutamate transaminase is nonspecific for the pairs of keto acids and amino acids, L-glutamate is the only amino acid in mammalian tissues that undergoes oxidative deamination at an appropriable rate. The formation of ammonia from α-amino groups requires their conversion to the α-amino nitrogen of L-glutamate. Thus, L-glutamic acid is a key substance in the nitrogen metabolism of amino acids. Glutamate dehydrogenase is widely distributed in microorganisms and higher plants as a catalyst in the synthesis of L-glutamic acid from α-ketoglutaric acid and free ammonia. Transaminase is contained in a wide variety of microorganisms.

MICROBIAL PRODUCTION

Although various kinds of microorganisms accumulate amino acids in culture medium, only bacteria have sufficient productivity to warrant the commercial production of amino acids. Since amino acids are essential components of microbial cells, and their biosyntheses are teleologically regulated to maintain an optimal level, they are normally synthesized in limited quantities and subject to negative feedback regulation. It is necessary to overcome this regulation to achieve overproduction of amino acids. Microbial amino acid overproduction can be achieved using the following procedures and has been attained by introducing mutation techniques:

1. Stimulation of cellular uptake of the starting materials
2. Hindrance of the side reaction
3. Stimulation of the formation and the activity of the enzymes for biosynthesis
4. Inhibition or reduction of the enzyme concerned with the degradation of the amino acids produced
5. Stimulation of the excretion of the product into the extracellular space

L-Glutamate production by a wild-type strain of *Corynebacterium* was explained by the release of L-glutamate regulation and efflux of L-glutamate through the cell membrane by the limitation of biotin or the addition of penicillin ester of fatty acids to the culture medium.

Fermentation Method

Excretion of L-amino acids by *Escherichia coli* when an excess of ammonium salt was added was reported in 1950. In 1957, a fermentation process was successfully commercialized that used a strain of bacteria isolated from soil, later found to be *Corynebacterium glutamicum*, which is able to excrete considerable amounts of L-glutamic acid. Numerous microorganisms have been isolated and found to be able to overproduce L-glutamic acid, including *Brevibacterium flavum*, *B. lactofermentum*, and *Microbacterium ammoniaphlum*. Because of minor differences in the character of these bacteria, which are all Gram positive, non–spore forming, nonmotile, and require biotin for growth, the name of genus *Corynebacterium* was suggested for these coryneform bacteria. For production of L-glutamic acid, the key factors in controlling the fermentation are as follows:

1. The presence of biotin in the range of 5 to 10 μg/L, which is optimal for the excretion of L-glutamic acid through cell walls
2. A sufficient supply of oxygen to reduce the accumulation of lactic acid and succinic acid as by-products

When a biotin-rich medium, such as molasses, is used as the carbon source, the addition of penicillin or cephalosporin C favors the overproduction of L-glutamic acid, supposingly due to the repression of peptide glycan synthesis on the cell membrane. The supplementation of a fatty acid or surfactant also results in an increased permeability of the cell wall, thus enhancing L-glutamic acid excretion. Kramer reports the existence of specific carrier mechanisms that are responsible for active L-glutamic acid secretion in *C. glutamicum*.

Generally, the intracellular accumulation of L-glutamic acid does not reach levels sufficient for feedback control in glutamate over-producers due to rapid excretion of glutamate. However, the regulatory mechanisms of L-glutamic acid biosynthesis have been studied intensively to obtain mutants with increased productivity. Shiio discussed two enzymes that have played key roles in the biosynthesis of L-glutamic acid.

The first, phosphoenolpyruvate carboxylase, catalyzes carboxylation of phosphoenolpyruvate to yield oxaloacetate; it is inhibited by L-aspartic acid and repressed by both L-aspartic acid and L-glutamic acid. Second, α-ketoglutaric acid dehydrogenase converts α-ketoglutaric acid to succinyl-CoA; in L-glutamic acid–overproducing strains, α-ketoglutaric acid dehydrogenase limits further oxidation of α-ketoglutaric acid to carbon dioxide and succinic acid, thus favoring the formation of L-glutamic acid.

In L-glutamic acid–overproducing strains, the K_m value of α-ketoglutaric acid dehydrogenase was nearly two magnitudes lower than that of L-glutamic acid dehydrogenase, which catalyzes the last step to L-glutamic acid. Consequently, v_{max} of L-glutamic acid dehydrogenase proved to be about 150 times higher than that of α-ketoglutaric acid dehydrogenase. Borman isolated and characterized the *C. glutamicum* glutamic acid dehydrogenase. A strain of *Microbacterium ammoniaphlum* cultured under biotin-deficient conditions produced 58% of L-glutamic acid formed from glucose via phosphoenol-pyruvate, citrate, and α-ketoglutaric acid; the other 42% was formed via the TCA cycle or glyoxylate cycle.

The mutants are reported as either having sensitivity in cell permeability, having the capability for increased carbon dioxide fixation, or having a too-low activity level of pyruvate dehydrogenase. Another approach focused on the development of thermophilic mutants. A strain of *Corynebacterium thermoaminogenes* is reported to accumulate L-glutamic acid at temperatures above 43°C.

CHEMICAL PROPERTIES

Dissociation Constant

Amino acids exit aqueous solutions as dipolar ions (zwitterions), and glutamic acid may be represented as follows. Following Bronsted, an acid maybe defined as a substance capable of combining with a proton. When an amino acid is titrated with HCl and NaOH, a curve exhibiting two points of inflection is obtained. Thus amino acid can react either as an acid or as a base, that is, as an ampholyte.

The dissociation constants for glutamic acid are pK_1 2.19 (COOH), pK_2 4.25 (COOH), and pK_3 9.67 (NH_3^+). (It is customary to designate the *pK* values of amino acids in order of decreasing acidity.) The isoelectric point (pI) of glutamic acid is the pH 3.2 calculated from the relation pI = $pK_1 + pK2/_2$, at which there is no net charge on the amino acid molecule. At this pH, no migration will occur in an electric field.

Racemization

Racemization proceeds, in general, in alkali solutions, such that α-H liberated from α-C by a base-catalyzed reaction yields carboion (C^-–C=O) on α-C, which equilibrates with α,β unsaturated from C=C–O^-. In the case of acidic solutions, racemization proceeds the same as in alkali, via carboion, C^-–C=OH^-, which equilibrates with α,β unsaturated from C=C–O^-. Racemization in neutral solution occurs at 190°C with the loss of taste characteristics after the formation of the lactam and pyroglutamic acid. This reaction is very slow, but the equilibrium shifts toward the lactam rather than glutamic acid. Under strongly acidic or alkaline conditions, the ring-opening reaction requires a very short time; therefore, neutralization of L-glutamic acid should be performed cautiously to avoid intramolecular dehydration even below 190°C. Racemization is also brought about by a microbial racemase. As in other amino acids, the racemase of glutamic acid is found in *Lactobacillus fermenti*. This enzyme acts specifically on glutamic acid but cannot act on its simplest derivatives.

```
                         OH⁻   R–C⁻=C(=O)(O)              R–C=C(O)(O)
      H                 ↗        |                ⇌          |
  R–C–C(=)                       NH₂                         NH₂
     |
    ⁺NH₃                 H⁺    R–C⁻=C(=OH⁻)(OH)           R–C=C(OH)(OH)
                        ↘        |                ⇌          |
                                ⁺NH₃⁻                       ⁺NH₃⁻
```

Figure 10.1 Mechanism of racemization of amino acid.

PROPERTIES OF L-GLUTAMATE ACID

Salt Formation

An amino acid can react with a copper ion to form a copper complex of the following type:

```
O = C--- O⁻          NH₂ ---CHR
          \          /
             Cu⁺⁺
          /          \
RCH---NH₂            O⁻ ---C = O
```

Formation of the copper complex is the basis of a quantitative procedure for the determination of the amino acid.

In alkaline solution, amino acids react with carbon dioxide to form carbamino acids, which may be precipitated as salts:

```
      O
      ||
  RCHC---O
          \
           CaH₂O
          /
   NH---O
   ||
   O
```

Some of these are quite insoluble and have been used for isolation of amino acids. Decomposition of carbamino acid occurs by boiling in aqueous solution.

L-Glu is able to react with various kind of cations, especially with metallic ions to form their corresponding specific complexes. The complex constant (log K_s) for various complexes of L-Glu is 10.3 for Ni^{2+}, 8.5 for Zn^{2+}, 8.1 for Co^{2+}, 4.6 for Fe^{2+}, and 3.0 for Mn^{2+}. Glutamic acid reacted with sodium ion to form monosodium L-glutamate, on the other hand, reacts as a base with sulfuric acid to form diglutamate sulfate crystals.

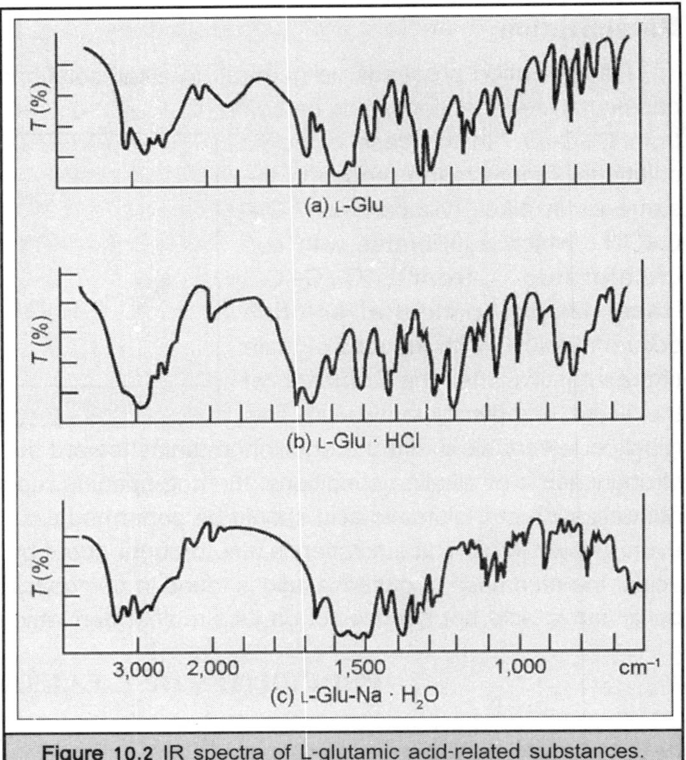

Figure 10.2 IR spectra of L-glutamic acid-related substances.

Infrared Spectra

The IR spectra of L-Glu shows strong bands at 1,620 and 1,420 cm^{-1}, which correspond to the COO; these are 1,660 cm^{-1} for COOH and 2,740 cm^{-1} for NH_3^+, respectively. The spectra of L-Glu HCl, on the other hand, does not show bands at 1,620 and 1,420 cm^{-1} for COO^- and that of L-Glu NaH_2O does not show the band at 1,670 cm^{-1} for COOH, but does show a strong band at 1,620 and 1,420 cm^{-1} for COO^-.

Ionic Forms

The four ionic forms of L-Glu exist depending on the pH of the solution. The dipolar nature of amino acids in solution has been shown by Raman spectroscopy. The Raman spectra of L-Glu shows a strong line at 1730 cm^{-1}, which corresponds to the C=O vibration. The spectrum of sodium L-Glu^-, on the other hand, does not show this frequency because the salt is completely dissociated and its structure represented is by the resonant forms

```
C=O        C–O
|          ||
O          O
```

This does not have a C=O group, but instead intermediate some between C–O and C=O. The spectra of the amino acids at the isoelectric point of pH do not show the existence of C=O, which indicates that the –COOH group is ionized. In a similar way, the ionization of the $-NH_2$ is indicated.

The optical rotation changes with changing pH in solution at 25°C; glutamic acid $[\alpha]_D$ ($G^{\pm}$; H_2O) is +12.0, $[\alpha]_D$ (G^+; 5N HCl) is +31.8, $[\alpha]_D$ (G^-; NaOH) is –4.2, and $[\alpha]_D$ ($G^{\pm}$; NaOH) is +10.9, which corresponds to the ratio of ionic form in solution.

Solubility

For the production of the umami seasoning MSG, the characteristic solubility changes in L-Glu HCl, L-Glu, and L-GluNa are used to separate it from other impurities to meet the category of food additives. The change in solubility of L-Glu for α- and β-forms, respectively, with temperature are expressed by the following:

$$\log S = 0.0174t - 0.377 \text{ (0-30°C) for } \alpha\text{-form}$$

$$\log S = 0.0153t - 0.328 \text{ (30-70°C) for } \alpha\text{-form}$$

$$\log S = 0.0159t - 0.461 \text{ for } \beta\text{-form}$$

The solubility of the β-form is lower than that of the α-form throughout the temperature range measured. It follows that the β-form is a stable form from the aspect of thermodynamic kinetics.

Table 10.1 Solubility of various salts of L-Glu (g/L)

Temperature (°C)	*L-Glu*	*DL-Glu*	*L-GluNa*	*DL-GluNa*	*L-GluHCl*	*DL-GluHCl*
0	3.41	8.55	514	158	298	471
25	8.64	20.5	627	243	479	698
50	21.9	49.3	765	372	769	1,030
75	55.3	119	933	570	1,240	1,540
100	140	285	1,140	875	1,990	2,280

The solubility changes with change in the concentration of both hydrochloric acid and sodium hydroxide, which corresponds to the existence of solid crystals, in equilibrium solution, of the hydrochloride, free, and sodium salts of L-Glu, respectively (i.e., in the range of 0 to 30% for HCl and of 0 to 20% for NaOH). In the range of 0 to 7.7% of HCl, the solubility increases linearly so that L-Glu is soluble until the equivalent amount of HCl in solution in which L-Glu exists as solid crystals. Above the concentration of 7.7% HCl, with an increase in the HCl concentration, the solubility decreases steeply.

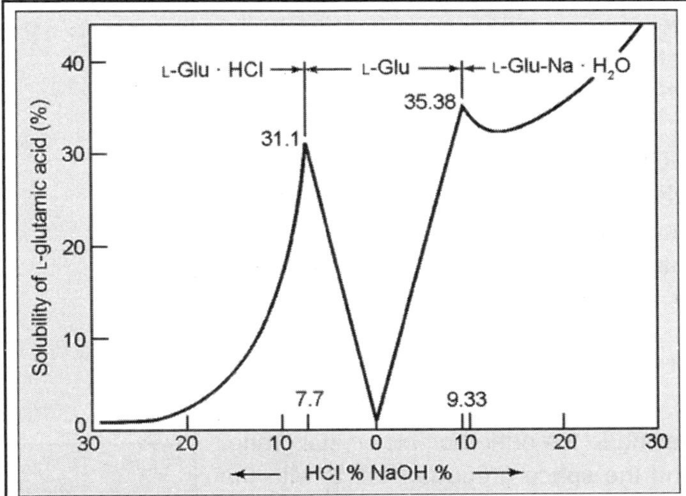

Figure 10.3 The change in the solubility of L-glutamic acid with change in the concentration of both hydrochloric acid and sodium hydroxide.

It flattens in the vicinity 20% and then becomes a constant of about 1% above 25% HCl. Above 7.7% HCl, the hydrochloride is a solid crystal in solution. At the invariant point where equilibrium is attained between L-Glu and L-Glu HCl, the maximum solubility of 31.1% is obtained. On the other hand, in the range of 0 to 9.33% of

NaOH, L-Glu solvates an equal amount of NaOH in solution and exists as a solid crystal. At the invariant point (NaOH 9.33%), the solubility of L-Glu reaches the maximum of 35.38%. Above the concentration of 9.33% NaOH, L-GluNa H_2O exists as solid crystals. A slight decrease in solubility is observed in the range from 9.33 to about 13%, and then increases with increasing concentration of NaOH. The minimum value of the solubility of L-Glu is obtained from the range of pH 2 to 4, in the vicinity of the pI.

L-Glutamic Acid (L-GLU)

$$\underset{\displaystyle NH_2}{HOOCCH_2CH_2\underset{|}{C}HCOOH};\ \text{L-Glu}$$

Chemical formula: $C_6H_9O_4N$ Molecular weight: 147.14 C = 40.82, H = 6.17, O = 43.50, N = 9.52 (%) Crystal: L-Glu is a polymorph, having two crystal forms, α and β. Both are anhydrous. Crystals of the α-form are obtained by crystallization from the saturated solution of L-Glu at 60 to 70°C by cooling rapidly with agitation or from the acidic or alkalic solution of L-Glu to which alkali or acid was added to rapidly neutralize it until the isoelectric point of L-Glu was reached (pH 3.2). The crystal shape is a column or pyramid. The α-form crystals transform gradually to β-form when kept in solution for a long time at room temperature. β-form crystals are obtained from either the saturated solution of L-Glu at 80 to 90°C, which is cooled gradually with agitation, or the relatively higher concentration of L-Glu hydrochloride or sodium salt, to which alkali or acid was added to neutralize it slowly until pH 3.2 was reached. The crystal shape is a needle or thin plate.

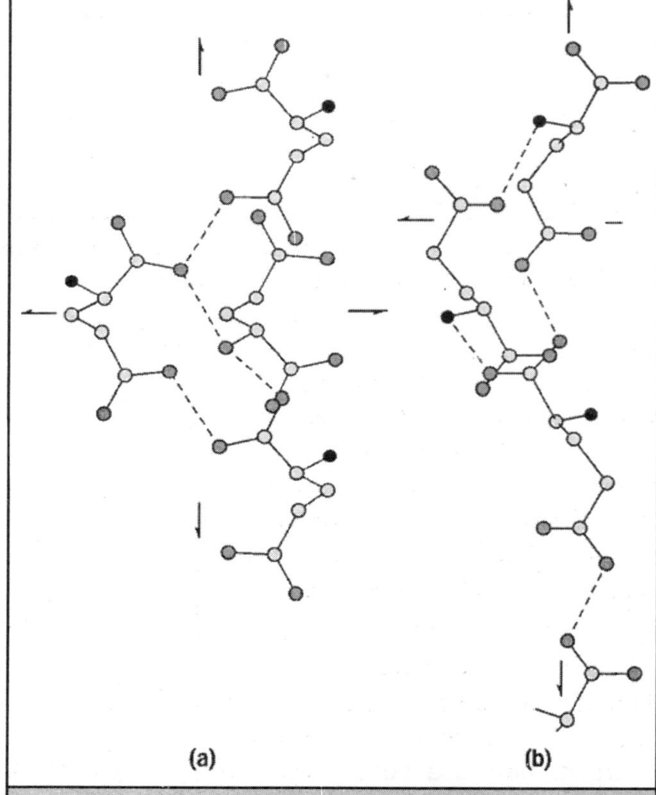

Figure 10.4 The molecular arrangement of L-glutamic acid crystals; a, α-form; b, β-form.

Crystal Structure

The α- and β-forms of the crystals belong to the orthorhombic crystal group, and the space group is $P2_12_12_1$ with the crystal-lattice constant for both crystals as follows: a = 7.06, b = 10.3, and c = 8.75 for the α-form and a = 5.17, b = 17.34, and c = 6.95 for the β-form. From the crystal-structure analysis, L-glutamic acid in both forms exists as a zwitterion;

a proton transfers from the α-carboxyl group to the amino group. The C(1)-O(1) and C(1)-O(2) distances of 1.24 and 1.27, for both crystal forms, are the values of the ionized carboxyl group.

This discrepancy of 0.03 found in both crystal forms can be attributed to differences in their intermolecular hydrogen bonds. O(2) makes a very short hydrogen bond with the OH group and a NH–O hydrogen bond with the NH_3^+ group, while O(1) makes one NH–O bond with NH_3^+. In the γ-carboxyl group, the length of C(5)-O(4) and C(5)-O(3), 1.21 and 1.37 for the α-form, and 1.19, and 1.24 for the β-form, respectively, indicate that it is unionized in both crystal forms. The bond angles around C(2), C(3), and C(4) in the α-form are significantly different from the corresponding ones in the β-form. This discrepancy may be a result of the steric repulsions that are due to the conformational differences. The OH–O bond seems to be most predominant of the four hydrogen bonds. It is reasonable to assume that considerable amounts of molecules are connected by OH–O hydrogen bonds to form chains in solution. The neighboring chains can be connected by the NH–O hydrogen bonds, and the conformation and the arrangement of the molecules are stabilized by these hydrogen bonds.

Stability

When heated as a dried crystal to 150°C, L-Glu undergoes cyclization and is converted to an interdehydrated compound of pyrrolidone carboxylic acid. Moreover, heated to 180–200°C, the optically inactive DL-pyrrolidone carboxylic acid is obtained. At greater than 200°C, glutamic acid begins to decompose. At pH 4 or 10 when the solution is heated at 100°C for about 50 h, more than 98% of L-Glu undergoes cyclization. At pH 3 and when heated at 120°C for 3 h, 100% cyclization is observed. The melting point of L-pyrrolidone carboxylic acid is 160 to 161°C, and the specific optical rotation in water is –11.35 at 20°C. This cyclization is a reversible reaction when treated with strong acid to yield glutamic acid, or when heated with 2N HCl or 0.5N NaOH at 100°C for 1 to 2 h.

L-GLUTAMIC ACID HYDROCHLORIDE

Chemical formula: $C_5H_{10}O_4NCl$ Molecular weight: 183.61

C = 32.71, H = 5.49, O = 34.86, N = 7.53, Cl = 19.33(%)

L-Glu HCl forms transparent crystals that crystallize from the high concentration of hydrochloric acid of L-Glu by cooling. The melting point is 160°C and optical rotation calculated from the ionic form (G^+) is +25.65.

The relationship between the solubility of L-GluHCl and the temperature is:

$$\log S = 1.474 + 0.00825t$$

The solubility of L-glutamic acid hydrochloride crystals at 20 and 30% HCl with changing temperature is shown in Table 10.2.

As mentioned earlier, with an increase in the concentration of HCl above 17%, the solubility of hydrochloride decreases and at more than 30% becomes constant. Conversely, when the temperature is lowered, the solubility decreases sharply. For example, at 3°C in 30% HCl, the solubility is only 0.38%. It is profitable when crystallizing L-GluHCl crystals to allow the solution to cool to as low a temperature as possible. The crystal of DL-GluHCl is racemic compound, and it is possible to resolute by the seed method. The solubility of DL-Glu HCl is log $S = 1.673 + 0.00685t$.

Table 10.2 The solubility of L-GluHCl for 20%and 30%of HCl

Temperature (°C)	Hydrochloric acid	
	20%	30%
3	1.15	0.38
7	1.31	0.44
10	1.48	0.47
15	1.38	0.58
25	2.60	1.16
35	5.29	1.19
50	8.42	2.68

SODIUM SALTS OF L-GLUTAMIC ACID

Monosodium L-glutamate (MSG), L-GluNa

$$HOOCCH_2CH_2\underset{\underset{NH_2}{|}}{C}HCOONa\ H_2O$$

Chemical formula: $C_5H_8O_4N\ Na\ H_2O$ Molecular weight: 187.13

C = 32.09, H = 4.31, O = 34.20, N = 7.49, Na = 12.29, H_2O = 9.63(%)

The monosodium salt of L-Glu crystallizes generally as a monohydrate. The melting point is 195°C and dehydration temperature is 150°C.

Table 10.3 Critical humidity of L-GluNa and NaCl

Temperature (°C)	Critical humidity (%)	
	MSG	NaCl
10	96.1	74.9
20	96.0	75.3
30	94.8	75.1
40	92.6	75.2
50	90.0	75.2

Crystal Structure

This crystal belongs to the orthorhombic crystal group $P2_12_12_1$. The crystal lattice constant is $a = 15.30$, $b = 17.86$, and $c = 5.54$, and the crystal density 1.635 g/cm^3. The packed specific volume is 1.20 to 1.25 cm^3/g. The crystal shape is an octahedral column and has bilateral planes at both termini where the atomic coordination is unstable, causing piezoelectricity. The crystal shows weak hygroscopic character with increasing temperature, which shows the relationship between the temperature and critical humidity, although the degree of hygroscopicity was relatively weak, when compared with that of NaCl. It is easily soluble in water, and the solubility *S* is designated by the following relation:

$$S = 35.30 + 0.098t + 0.012t^2$$

where *S* is g/100 g H_2O and *t* is in degrees centigrade.

L-Glutamic Acid Monosodium Salt Pentahydrate

$$\underset{\displaystyle NH_2}{\text{HOOCH}_2\text{CH}_2\overset{}{\text{CH}}\text{COONa } 5\text{H}_2\text{O}}$$

Chemical formula of pentahydrate: $C_5H_8O_4Na\ 5H_2O$ Molecular weight: 259.5 The condensed solution of L-Glu Na was cooled between –0.8 and –8.5°C, from which pentahydrate crystals were crystallized. The crystal structure belongs to the triclinic and space group P1, and lattice constant is

$$a = 6.12,\ b = 16.06, \text{ and } c = 6.01$$

$$\alpha = 99.3,\ \beta = 100.9,\ \gamma = 99.1$$

The crystal density is 1.480 g/cm³. The transition point form monohydrate to pentahydrate is – 0.8°C. Above –0.8 °C monohydrate crystals exist as the stable form, between –0.8 and –8.5°C, pentahydrate exists as the stable form, and below –8.5°C the solution itself turns into ice. The solubility of pentahydrate is changeable with temperature compared to monohydrate, and is designated as follows:

$$S = 35.72 + 0.630\ t + 0.0016\ t^2$$

CALCIUM SALT OF L-GLUTAMIC ACID

Several forms of calcium salt are reported.

1. $(Glu)_2$-CaH_2O: An equimolar amount of $Ca(OH)_2$ was added to L-Glu solution to crystallize. This is soluble in water.
2. (Glu)-CaH_2O: L-Glu HCl in solution reacts with $Ca(OH)_2$ at 80°C to form (Glu)-CaH_2O. This is sparely soluble in water.
3. $(Glu)_2$-$CaCaCl_2H_2O$: $Ca(OH)_2$ was added to L-Glu-HCl in solution, saturated with CO_2, and boiled. Concentration yielded the complex salt.

INDUSTRIAL PRODUCTION

Extraction Method

From 1909, when MSG production started, until 1965, when the extraction process ended, wheat gluten and de- fatted soybeans were used as the main raw materials, containing as much as 15 to

25% L-glutamaic acid. Crude raw materials were hydrolyzed by heating with hydrochloric acid. After concentration under reduced pressure, the hydrolyzate was cooled to crystallize L-glutamic acid hydrochloride. L-Glutamic acid can be easily separated frcm other amino acids in the form of its hydrochloride because of its low solubility in concentrated hydrochloric acid. The crude hydrochloride is dissolved in hot water and filtered to separate the humic substances formed in the hydrolyzation process. The pH was adjusted with caustic soda to 3.2, the isoelectric point of L-glutamic acid needed to crystallize L-glutamic acid crystals at a yield of more than 90%. The crude L-glutamic acid crystals are suspended in water and neutralized with sodium hydroxide. The solution is decolonized with activated carbon and crystallized MSG (monosodium L-glutamate monohydrate) under reduced pressure. Commercially preferable crystals are grown with a small amount of amino acids. The crystals are separated by centrifuge and then dried for packaging. Beet, sugarcane molasses or Stefan's waste (which contained 2-pyrrolidone carboxylic acid) has been used as raw material for MSG production in the United States and Europe. The pyrrolidone carboxylic acid was hydrolyzed at 85°C at pH 10.5 to 11.5 for about 2 h to avoid racemization.

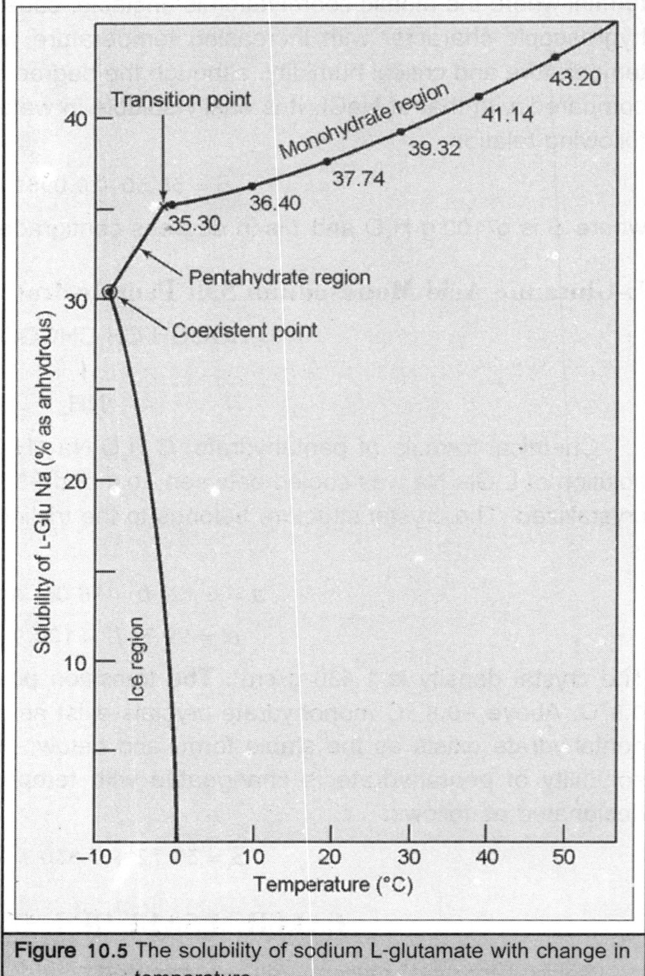

Figure 10.5 The solubility of sodium L-glutamate with change in temperature.

Industrial Fermentation Method

For industrial production of MSG, molasses or starch hydrolyzate is generally used as raw material. Sugar molasses usually contains excess amounts of biotin, so when used, it is necessary to repress the activity of biotin by adding penicillin or certain surface-active substances at an early stage of the production to retard microorganism propagation. Furthermore, sugar concentration in culture media at below about 10% is necessary so as not to retard propagation. Additional sugar was fed during fermentation to obtain a higher concentration of excreted L-glutamate. This batch-fed fermentation process made possible a concentration of more than 120 g/L of L-glutamic acid in media. Microorganisms used for L-glutamate fermentation are usually preserved under lyophilization below –80°C or, for a short period, by keeping the stock culture below 10 to 15°C. To refresh the microorganisms, stocked in either form, they are inoculated on an agar medium composed of 1% yeast extract and polypeptone, 0.5% sodium chloride, and 2% agar, at an optimum temperature of microorganisms.

The refreshed microorganisms are then cultivated in liquid medium, shaken vigorously, and transferred to a small fermenter to allow them to propagate to about several kiloliters for seed culture. Industrial-scale fermenters are pressure tight, stainless steel containers, built to hold up to several hundred kiloliters of cultivating medium. They are equipped with aeration and stirring devices, as well as other automatic controls. Fermentation takes from 30 to 45 h. Gaseous ammonia or a solution of urea and ammonium salts are convenient nitrogen sources for fermentation, not only as the initial medium, but also to maintain the pH of the medium at 7 to 7.5 for microbial growth and product formation. The culture medium becomes acidic because of assimilation of ammonium ions and the formation of L-glutamate. Gaseous ammonia can be used advantageously to maintain neutral pH and avoid dilution of culture medium, resulting in the high accumulation of L-glutamate in the fermentation broth, because it does not contain OH^- ions or water. Microorganisms required several minerals, such as ferrous and potassium ions, which play important roles in L-glutamate fermentation. Other important conditions include regulating the aeration stirring.

The biosynthesis of glutamate is performed under regulated aerobic conditions. When oxygen is not sufficiently dissolved, lactic acid and succinic acids accumulate and reduce the accumulation of glutamate. On the other hand, an excess of dissolved oxygen results in the formation of α-ketoglutaric acid. The optimal oxygen transfer rate was determined by measuring the rate of consumption of sodium sulfite, which is considered to depend on the characteristics of microorganisms used. The pressure of dissolved oxygen was usually kept above 1 kPa (0.01 atm) by aeration and agitation in the fermenter. The sterilized air is fed through a filter. Carbon dioxide is formed during fermentation and causes the medium to become heavy. Mechanical and chemical defoaming systems are provided. The optimum temperature for fermentation is dependant on the character of microorganisms used. Regulation of the temperature using heat exchangers installed inside of the fermenter is indispensable.

Fermentation is an exothermic reaction, and the temperature critically affects not only the propagation of microorganisms, but also the formation of glutamate. In large-scale production the formation of trehalose very often reduces the product yield of glutamate. Trehalose consists of two α-1,1-bond glucose molecules and is excreted by the bacteria as an osmoprotectant. A process has been developed and successfully industrialized in which trehalose formation is controlled and decreased by culturing the overproducing mutant in media with inverted sucrose from molasses. When fermentation is complete and after sterilization of the broth, the glutamate-recovery process is performed.

Separation Process

The separation process consists mainly of the following two processes:

1. Recovery of L-glutamic acid crystals from the fermentation broth
2. The purification of L-monosodium glutamate monohydrate from crude L-glutamic acid crystals

Hydrochloric acid at 35%% is added to the fermentation broth with agitation to crystallize the L-glutamic acid crystals at pH 3.2, the isoelectric point of L-glutamic acid. The crude L-glutamic acid crystals, retaining some mother liquor when removed from the slurry, are dissolved in water with equimolar sodium hydroxide; then the solution is applied to the activated carbon column to obtain the decolonized L-MSG solution, on which most of the color-related and hydrophobic substances are adsorbed. This purified L-MSG solution is fed continuously to a vacuum crystallizer to crystallize L-MSG crystals. L-MSG crystals separated by centrifugation are dried, sieved, and finally packaged to deliver to the consumers.

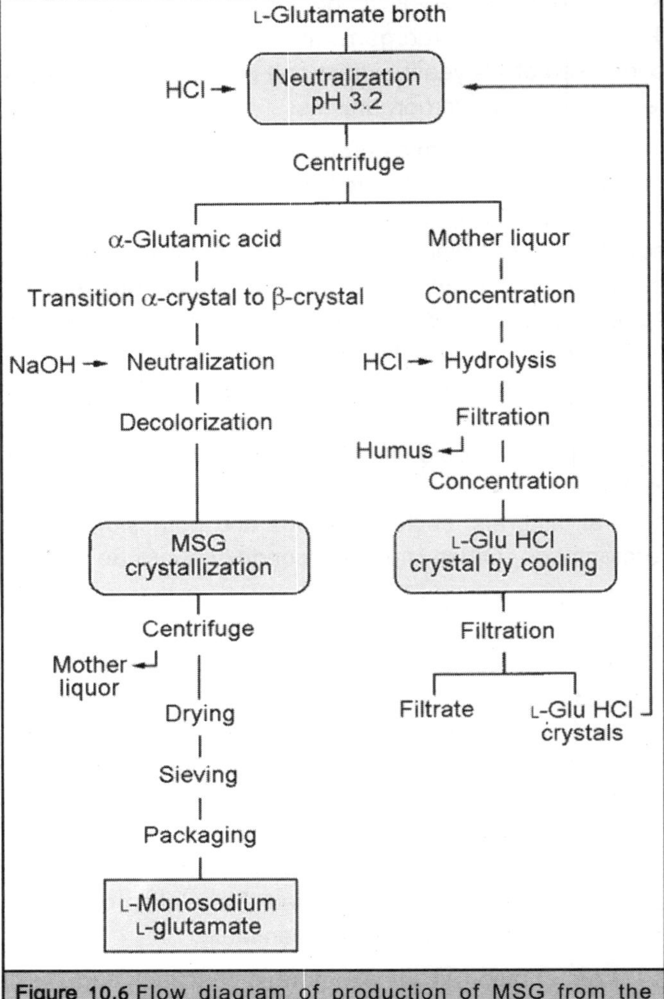

Figure 10.6 Flow diagram of production of MSG from the fermentation broth.

Crystallization of L-Glutamic Acid

In the fermentation broth, L-glutamic acid exists as an ammonium salt in the vicinity of pH 7. For obtaining L-MSG directly from the fermentation broth, it is necessary to exchange ammonium ion for sodium ions. This process is rather complicated, taking into account an environmental approach. To avoid this complexity, as a first step, crystallization of the *α*-form of L-glutamic acid crystals, which show lower solubility as compared to L-MSG and fairly good sedimentation as compared to *β*-form crystals, is employed in actual industrial production. The presence of salts can affect solubility of L-glutamic acid in solution. For example, the chloride-related inorganic substances NaCl, $CaCl_2$, NH_4Cl, and KCl increase the solubility of L-glutamic acid. When crystallized under ordinary conditions, pyramid- type *α*-form crystals are obtained by the effect of the impurities contained in the fermentation broth.

Pyramid Type

In the growing of crystals, only those faces having the lowest translation velocities survive. The face of [001] of the *α*-form are fast-growing (high translation velocities) and this face tends to disappear, as overlapped by slower faces, and finally results in the pyramid type. If Ostwald's step theory is obeyed, as α-form crystals are unstable from the thermodynamic point of view, α-form crystals first appear and then gradually transform to β-form. But this theory is not based on the thermodynamics, but on the kinetics, so exceptions frequently occur. The conditions to obtain α-form crystals are summarized as follows:

1. When neutralized, 35% hydrochloric acid is added as quickly as possible to the fermentation broth and then cooled to below 10°C instantly with vigorous agitation.
2. A small amount of a third substance is added to affect the stability of α-form, such as a peptide, active surfactant, protein hydrolyzate, or amino acid, especially L-phenylalanine and L-cysteine.

Centrifugation

The α-form of L-glutamic acid crystals are separated by a centrifuge. Residual mother liquor after centrifugation amounts to 5 to 15% of the weight of the crystals, depending on the viscosity of the mother liquor and the degree of uniformity of crystal size. It is more effective to wash the crystals on the centrifuge with fresh water to reduce impurities than recrystallization. In this process, the bacterial cells remain effectively in the mother liquor, and without another bacterial cell separation process, L-glutamic acid crystals remain effectively separated from the bacterial cells.

Transformation of α-Crystals to β-Crystals

When the α-form crystals are poured into the water and the temperature is elevated to 60 to 80°C, the transformation of α-form to β-form occurs within about 1 h. During this solvent-mediated transformation, a large amount of adhering impurities are rejected out of the crystals, and higher purified β-form miceous plate-shape crystals are obtained. This transformation process has so marked an efficiency for removing impurities, especially color-related substances, that when higher purity is required, a small amount of activated carbon is used to obtain a transparent MSG solution. For another purification of L-glutamic acid solution, the application of UF membrane or NF membrane is investigated: When these membranes are applied to permeate the fermentation broth, the permeate solution obtained is neutralized followed by cooling. Purified β-form crystals are obtained directly, without α-form crystallization.

Crystallization of L-MSG

Neutralized L-glutamic acid solution (L-MSG solution) is applied to the activated carbon column to obtain decolonized solution, which is fed to the continuous evaporator to concentrate until the saturation of L-MSG (50% at 60°C). To meet the requirements for L-MSG monohydrate crystals—crystal size, crystal shape (aspect ratio), and size distribution—it is necessary to strictly control the operational conditions of the crystallization process, in which nucleation and crystal growth occur simultaneously. The result of the combined processes on the mentioned requirement for crystals depends on the relative rates of nucleation and growth. To obtain a more uniform product, nucleation is controlled by adding the desired number of crystals, usually crushed, to the crystallizer when the solution is either saturated or supersaturated.

In a continuous crystallizer, the number of nuclei formed per unit time will be continuous and will equal the number of crystals withdrawn from the crystallizer, so that all nuclei can receive the same time of growth. Otherwise, there must be a classifying action in the crystallizer that will retain the small crystals under treatment until they have grown to the proper size, when they can be removed from the crystallizer. For practical control, it is necessary to reduce as much as possible the local variation in the temperature or concentration of the solution that causes excess nucleation. Usually, the objective of the crystallizer operator is to achieve the maximum growth rate, consistent with low nucleation rate. More importantly, impurities in solution may inhibit the formation of new nuclei or retard crystal growth. The degree of the effect of a given impurity cannot be predicted; in general, higher molecular weight materials seem to be more effective inhibitors.

Crystal Shape (Aspect Ratio)

When no impurity is contained in the solution the aspect ratio is 7 to 8 which is not convenient for umami seasoning for cooking. To overcome this problem, the appropriate amount of amino acids is added. The crystal shape is determined by the change in the difference in the relative crystal growth of constituted crystal faces. For L-MSG crystals, a small amount of amino acids changes the crystal shape, the amino acids such as L-alanine, L-lysine, and L-arginine are easily adsorbed on the face of the L-MSG crystal and inhibit the growth of this face, those having the lowest translation velocities survive, and the aspect ratio is 4 to 5, which satisfies the demand of users.

Recovery of L-Glutamic Acid from Mother Liquor

In mother liquor–separated L-glutamic acid crystals other than L-glutamic acid, many impurities derived from the raw material used as the carbon source, as well as ammonium hydrochloride and bacterial cells, are found. For recovery of L-glutamic acid from this mother liquor, the following process is proposed: The mother liquor is condensed and ammonium chloride crystallized by cooling for a day and night followed by filtration to obtain the ammonium chloride crystals that are used as fertilizer. The separated mother liquor is hydrolyzed with excess hydrochloric acid to convert the L-glutamic acid to hydrochloride and the bacterial cells and other debris to insoluble matters (so-called humic acid). The humic acid is separated by filtration. The filtrate is condensed, then cooled to crystallize the hydrochloride crystals at lower than 10°C for more than several days. This is recycled in the process of L-glutamic acid crystallization from the fermentation broth to curtail the consumption of hydrochloric acid. The residual mother liquor combined with humic acid is used as raw material for the organic fertilizer.

USES

Seasoning

MSG is widely used as a long-established seasoning or flavor enhancer to improve the palatability of foods. The effect of MSG is due to its characteristic taste, umami. The development of food science, including electrophysiology, psychophysics, and nutrition, have provided evidence that umami should be regarded as a basic taste, independent from the traditional four basic tastes—sweet, salty, sour, and bitter. Glutamate is ubiquitous in foods such as tomato, cheese, human milk, and seaweed. Other umami substances, that is, the 5′-ribonucleotides disodium 5′-inosinic acid and disodium 5′-guanylic acid, are found in meats, fish, vegetables, and mushrooms. When glutamate and nucleotides coexist, a very strong synergistic effect occurs, and even a small amount of glutamate added to a food containing nucleotide causes umami to be dramatically enhanced—seven to eight times as much as the original umami of the food. The taste threshold of MSG is about 0.03% in aqueous solution. The intensity of umami increases linearly with a logarithmic increase in the concentration of MSG. The synergistic effect of MSG with 5′-ribonucleotides is expressed by the following relation:

$$y = u + Auv$$

where y is the intensity equivalent to the concentration of MSG alone, and u and v are the concentrations of MSG and 5′-ribonucleotides, respectively. A is a constant that is 1,200 for disodium 5′-inosinate (IMP) and 2,800 for disodium 5′-guanylate (GMP). In some commercial umami seasonings, 5′-ribonucleotides are mixed in to take advantage of the synergistic effect with MSG.

Although MSG increases the palatability of food, it is not always palatable by itself. When MSG is tested in crystal form or in simple solutions, it does not cause a pleasant sensation. It gives a pleasant taste only in flavored solution or in actual foods. Another important property of MSG is the concentration dependence of its pleasantness rating. An excess of MSG causes an unpleasant sensation, and thus the amount of MSG that is added to foods should be limited. The desirable usage level of MSG depends on the properties of the food, including the amount of original umami substances. Results of extensive taste studies indicate that the optimum MSG level in food ranges from 0.1 to 0.8 wt %, varying from food to food.

In most cases, MSG is used in combination with salt. A relationship between the levels of MSG and salt, and palatability of foods, has been clarified. The lower the salt concentration, the higher the MSG concentration required to maintain the palatability, and vice versa. When umami substances are added to meals, the amount of salt can be reduced without a decrease in the palatability or degree of satisfaction. Umami taste is considered to be a universal taste sensation for humankind. An ethnic comparison of taste sensitivity has shown that there is no difference in the thresholds of MSG and IMP between Japanese and European-American Caucasians.

Industrial Use

Depending on the pH, glutamic acid can be used to neutralize acidic or basic compounds, that is, arginine glutamate is used as a pharmaceutical and as a raw material for cosmetics. The reaction of glutamic acid with hydrochloric acid affords glutamic acid hydrochloride, which is used as a gastric acidifier. The property of glutamic acid as an amphoteric electrolyte is exploited in chelating agents, buffers, and builders for detergents. The sodium salt of pyroglutamic acid, which is obtained by dehydration of glutamic acid followed by neutralization with sodium hydroxide, is very hydroscopic and used as a component of a natural moisturizing factor for human skin and as a humectant for cosmetics. The *N*-acylated compound produced by reaction with a long-chain fatty acid is an anionic surfactant and is popularly used for skin and hair cleansers due to its mildness and high safety. The dibutylamine of the *N*-acyl-glutamate gelatinizes nonpolar oils and can be used as a recovering agent for marine-oil spills.

SAFETY

Monosodium L-glutamate is metabolized in the same way as glutamic acid from any digested protein. After oral ingestion, glutamate is absorbed from the intestine by an active-transport system. During this process, a large portion is metabolized to alanine and α-ketoglutarate, which enters the tricarboxylic acid cycle. Glutamate is further metabolized in the liver to give glucose, lactate, glutamine, and other amino acids. Consequently, blood glutamate levels do not rise significantly unless very large doses are administered. Blood glutamate levels transiently rise when large doses are ingested on their own, but the coingestion of foods that contain metabolizable carbohydrate increases the metabolism of monosodium L-glutamate and eliminates or greatly attenuates this rise. Human infants, including premature infants, metabolize glutamate similar to adults.

The acute toxicity of MSG is low, the oral LD_{50} for humans, calculated on the basis of doses administered in different ways to various animals, would represent a single dose greater than 1 kg for a person weighing 70 kg. In contrast, the oral LD_{50} for sodium chloride in rats is 3.75 kg/kg body weight. The main use of MSG is as a umami seasoning or a food ingredient, and so its safety when used in the diet is the most important aspect of its use. Both short-term and chronic toxicity studies on MSG in the diet of several species at doses of up to 4% (approximately 6 to 8 g/kg body weight/day) in the diet showed no specific toxic effects and no evidence for carcinogenicity or mutagenecity. Reproduction studies of up to three generations did not reveal any adverse effects of dietary MSG ingestion.

Fertility, gestation, viability, and lactation indexes, or the pre- and post-weaning performance of offspring, were unaffected. In primates, it was found that the placenta serves as an effective barrier to the transfer of glutamate from the oral ingestion of large doses of MSG. The concentration of glutamate is higher in the brain than in the blood. The demonstration that injected or force-fed neonatal rodents given extremely high doses of MSG showed evidence of brain leisons has led to much additional research to determine any possible link between neurotoxicity and human use of MSG. However, from animal tests on monkeys, guinea pigs, rat, and mice, glutamate levels in the brain remain unchanged after the administration of large oral doses of MSG. Numerous experiments on rodents, as well as dogs and monkeys, using dosage levels up to 43 g MSG/kg body weight have failed to show any link between dietary use of MSG and brain damage. In the case of dogs and monkeys, even experiments involving injection of MSG have not shown any effects on the brain. An anecdotal report in 1968 triggered interest in Chinese restaurant syndrome (CRS), a reaction associated with transient subjective symptoms of burning, numbness, and a tight sensation in the upper part of the body.

Possible association with food ingredients such as MSG was suggested. No objective changes in skin temperature, heart rate, ECG, or tone were observed, and no correlation was seen between plasma glutamate levels and symptoms. In 1979 a questionnaire survey of more than 3,000 people in the United States was conducted for CRS. Only 1.8% of those surveyed acknowledged that they experienced possible CRS feelings, and 0.19% of those with symptoms attributed them to eating in Chinese restaurants. These feelings were also found to occur after consumption of spicy tomato juice, orange juice, coffee, and tea. In 1986 self-identified MSG responders were challenged in a properly controlled double-blind study, and it was concluded that the link between MSG and CRS was not supportable.

In 1987, the Joint FAO/WHO Expert Committee on Food Additives (JECFA), a scientific advisory body to the World Health Organization and Food and Agriculture Organization of the United Nations, concluded that properly conducted double-blind studies among individuals who claimed to suffer from the syndrome did not confirm MSG as the causal agent. In 1981 Chinese restaurant asthma was reported following capsule administration of MSG to several asthmatics, but the researchers failed to account for other allergens to which the subjects could have been exposed and did not utilize the scientific practice of a "control" substance that would have helped to determine if glutamate triggered this response. In a double-blind crossover study in which chronic asthmatics were challenged with MSG or a placebo, no decrease in pulmonary function was observed. JECFA reviewed the safety studies of glutamate and endorsed its safety by allocation of an acceptable daily intake (ADI) for L-glutamic acid and its monosodium salt, potassium salt, ammonium salt, calcium salt, and magnesium salt as being "not specified." This result is ascertained by the scientific committee for food of EC.

11

ANALYZING CSTR

A continuous stirred-tank reactor (CSTR) is defined as an agitated vessel with continuous addition and removal of material and energy. The CSTR is one of the basic continuous reactor types widely used in the chemical process industries because of its amenability to process control and scale-up, although in biotechnology applications the CSTR is used more often as a research tool than as a production technology. An idealized, well-mixed CSTR can be modeled as having no spatial variations in temperature, concentration, fluid properties, or reaction rates. Thus, the properties of the exit stream maybe considered the same as those throughout the vessel. Although such ideal mixing is never observed, the vessel is designed to provide good mixing through selection of operating conditions and vessel, baffle, and impeller geometries.

The stirred tank used in continuous bioprocesses is similar to that used in batch bioprocesses, with the exception that the CSTR likely has an overflow or other level control device. Oxygen can be introduced into the vessel by sparging through inlets at the base of the vessel, where impellers then disperse the bubbles. Vessel jacketing or internal cooling coils provide a means for heat transfer. Continuous systems that are not agitated vessels often are modeled as CSTRs when their behavior approximates that of the ideal CSTR. CSTRs are also known as backmix reactors, continuous-flow stirred-tank reactors (CFSTRs), or chemostats, when used for cell growth.

BASIS FOR CSTR ANALYSIS

The performance and analysis of a CSTR is based on the material and energy conservation balances and the underlying processes governing the reaction kinetics. Because of the large variety of CSTR applications and limited space for discussion in this article, a brief outline of the steps in systems analysis will be beneficial in understanding any CSTR-based process. The strategy for analyzing CSTR performance first requires defining the problem statement and goals. The second step is system identification, which includes defining the system boundaries and the interactions between the system and its environment across the system boundaries. The system could be a cell, the fluid in the reactor, or an entire bioprocessing plant. The third step is to identify the state variables that characterize the system. During the course of the analysis, new state variables may be identified and added to the original list. The fourth step is to characterize the state of the system using material and energy balances that account for the accumulation of mass and energy. In a general balance for a particular quantity, the rate of accumulation of that quantity in the system is equal to the net influx of the quantity across the system boundaries plus the rate at which the quantity is generated. Separate material balances are written for the reactants, products, and the catalyst (e.g., cells or enzymes). The final step is to calculate performance metrics and revisit the assumptions to determine the conditions under which they are valid.

THEORIES FOR IDEAL CSTR

This section discusses the theory governing ideal CSTR performance in two important bioprocess applications (cell growth and enzyme reactions), the underlying assumptions of the theory, and some practical aspects associated with CSTR operation.

Cell Growth

Cell growth occurs in response to the environment. It is useful to classify growth in four mechanistic categories and identify those features relevant to the analysis of CSTRs. These classes are fission, budding, mycelial, and viral growth. The modes of growth serve to highlight the impact of morphology on bioprocessing considerations. Bacteria grow by a process of binary fission yielding two identical daughter cells with doubling times typically between 0.5 and 3.0 h. The high specific growth rates of bacteria make them especially suitable for many CSTR applications; however, high O_2 requirements and metabolic heat generation become important considerations in bioprocess scale-up. Animal cells (5 to 20 μm) are much larger than bacteria (–1 μm) and yeast (5 to 10 μm) cells but grow more slowly, with typical doubling times between 18 and 48 h. Because of their larger size and lack of a protective cell wall, mammalian cells are particularly sensitive to the fluid shear in the vessel as well as to the osmotic pressure of the medium.

Mammalian cells grow over a narrow range of osmotic/pressures and pH and typically have more complex nutritional requirements than bacteria and yeasts do. Cell division in yeast occurs primarily by budding, with typical doubling times between 1 and 3 h. The budding process leads to a mother and a daughter cell, each having different growth rates and cell surface characteristics. Unlike bacterial cultures, yeast cell populations have a broad and time-varying distribution of ages and properties that may influence the formation of a desired product. Mycelial growth occurs in molds, actinomycetes, and some yeasts by a process of hypha chain elongation and branching. Mycelial cultures also are characterized by a distribution of ages, with younger cells located at the hyphal tips. The hyphae form intertwined cellular strands, or mycelia, that increase the broth viscosity and lead to nonideal fluid mixing.

High broth viscosity can be problematic for process monitoring and control, cell separation and recycle, and oxygen and heat transfer. Furthermore, fluid shear in the vessel can cause hyphal breakage and the formation of denser, more highly branched pellets or flocs. Cells in these pellets may be exposed to different micro-environments because of mass transfer limitations, which can vary with culture conditions and influence important cell properties. Viral growth initially requires the infection of a host cell, which occurs by attachment of the virus to the cell surface and injection of viral nucleic acids into the cell interior. New viruses are constructed from biological molecules synthesized by the host cell under the direction of the viral genome. Viral nucleic acid is replicated many times (e.g., >500) and encapsulated in coat proteins to form a large number of new viral particles.

Viral growth can proceed to one of two phases. In the lytic cycle, the host cell will lyse or break open and release infectious viral particles, whereas in the lysogenic cycle, the viral DNA will be integrated into the host cell DNA and the host cell will continue to reproduce normally. CSTRs used for cell growth are commonly referred to as chemostats or turbidostats, depending on the strategy used to control the vessel environment. The most common arrangement is the chemostat, in which the medium fed to the vessel is designed so that all but a single nutrient essential for growth are present in excess of the cells' requirements. Any nutrient necessary for growth can be used to control the size of the cell population in the vessel, making the chemostat a flexible tool to study cellular behavior under different nutrient limitations.

In a turbidostat, the cell concentration in the vessel is maintained constant by monitoring the optical density of the culture and modulating the medium feed rate to achieve a set point optical density. When the optical density rises above the set point, the feed rate is increased and, because the fluid volume is maintained constant by an overflow device, the well-mixed culture is diluted and the optical density approaches the set point value. The turbidostat is less commonly used because of difficulties in continuously monitoring the cell concentration. Its main utility is to control the growth rate near the maximum growth rate, an operating region in which the chemostat is less stable.

Materials for balanced growth

Cell mass is most often used to quantify microbial growth and usually is proportional to cell number under conditions of balanced growth, in which cellular chemical and physical properties are preserved in subsequent generations. Material balances based on cell number may have particular utility for some applications, such as mammalian cell culture, where number rather than mass is the conventional method of analysis. One may need to be wary of variations in cell size and morphology that may not be apparent from measurements of cell mass or number. The material balance for a uniform cell population in a CSTR can be written as shown in equation 1, in which $\mu[h^{-1}]$ represents the specific growth rate and $\alpha[h^{-1}]$ represents the specific rate of cell lysis and/or endogenous metabolism (i.e., resulting in a decrease in cell mass). The specific growth and death rates differ among organisms and are functions of the cell environment (e.g., pH, temperature, nutrients).

$$\underbrace{F \cdot X_0}_{\text{Cells in}} - \underbrace{F \cdot X}_{\text{Cells out}} + \underbrace{V \cdot \mu \cdot X}_{\text{Cell growth}} - \underbrace{V \cdot \alpha X}_{\text{Cell lysis}} \quad \underbrace{\frac{d(V \cdot X)}{dt}}_{\text{Accumulation}} \qquad \ldots(1)$$

It may be necessary to reformulate the cell balance if the cell population is significantly differentiated; examples include mixed cultures, mammalian cell culture, and recombinant fermentations with plasmid instability. If we assume the system is operating at steady state and there is no accumulation of fluid or cells in the vessel, then the time derivative can be set to zero. Under normal bioprocessing conditions, cell death is assumed to be negligible (i.e., $\mu \gg \alpha$). Bacterial and yeast cells maintain approximately complete viability except in suboptimal environments and at very low dilution rates, whereas in mammalian cell culture, viability and cell lysis may vary significantly with time.

Assuming that the feed to the reactor is sterile ($X_0 = O$), the material balance reduces to equation 2, in which the specific growth rate is equal to the liquid flow rate from the vessel divided by the liquid volume. This quantity is called the dilution rate D (the inverse residence time for the CSTR) and has units of h^{-1}.

$$\mu = \frac{F}{V} = D \qquad \text{...(2)}$$

Equation 2 illustrates one of the most important attributes of the chemostat: that the specific growth rate can be controlled by manipulating the dilution rate. Control of the specific growth rate, combined with the ability to maintain a constant, defined cell environment, makes the chemostat a powerful experimental tool with which to investigate the many factors that influence cell growth, metabolism, and product formation. It is as an investigative tool, rather than as a production technology, that continuous culture has found its widest and most successful application.

A second material balance can be written for the growth-limiting substrate (S) using an *allocation model* for substrate utilization, in which substrate uptake is divided into cell growth, cell maintenance, and product formation components.

$$\underbrace{F \cdot S_0}_{\text{Substrate in}} - \underbrace{F \cdot S}_{\text{Substrate out}} - V \cdot \left(\underbrace{\frac{\mu \cdot X}{Y_{X/S}}}_{\text{Growth}} + \underbrace{m \cdot X}_{\text{Maintenance}} + \underbrace{\frac{q_P \cdot X}{Y_{P/S}}}_{\text{Product formation}} \right) = \underbrace{\frac{d(V \cdot S)}{dt}}_{\text{Accumulation}} \qquad \text{...(3)}$$

In this equation, $Y_{X/S}$ is the cell yield (or dry cell weight, DCW) on substrate (g DCW/g substrate), m is the cell maintenance coefficient (g substrate/g DCW $\times$ h), q_P is the specific product formation rate (g product/g DCW x h), and $Y_{P/S}$ is the mass yield of product on substrate (g product/g substrate).

Each specific yield coefficient describes the allocation of substrate to cells, product, or maintenance. Depending on whether the fermentation goal is to produce cells (biomass) or a metabolic product, it may be possible to simplify the substrate balance by assuming that some uptake terms are dominant. The following sections describe the analysis of biomass production and product formation.

Ecological production

In the case of biomass production, in which the goal of the fermentation is to produce cells, the large majority of nutrient uptake goes toward cell growth. The rate of substrate uptake for growth is assumed to be much greater than that for maintenance (i.e., $\mu/Y_{X/S} \gg m$) and product formation (i.e., $\mu/Y_{X/S} \gg q_P/Y_{P/S}$). Assuming the system is at steady state, the substrate balance (equation 3) can be rewritten and solved for the cell concentration using equation 2.

$$X = Y_{X/S} \cdot (S_0 - S) \qquad \ldots(4)$$

To determine the relationship between the specific growth rate and the cell environment, a suitable growth model must be adopted. The simplest and most common relationship used is the unstructured Monod growth model, in which cell growth is a function of a single limiting substrate, usually the carbon source. These unstructured growth models are empirically derived from observations of chemostat behavior, and their applicability to dynamic-batch or fed-batch processes should not be assumed. Some unstructured models (e.g., substrate inhibition) can have more than one solution, making different steady states possible depending on the starting conditions. Structured models, which typically account for either changes in cell composition, intracellular concentrations, or cell morphology, have been proposed in varying degrees of complexity. Structured models can have utility when such properties significantly influence the kinetics and are required to accurately describe behavior (such as in dynamic process modeling).

$$\mu = \frac{\mu_{max} \cdot S}{K_S + S} = D \qquad \ldots(5)$$

In the Monod model, μ_{max} is the maximum specific growth rate of the organism, and K_S, called the saturation constant, is inversely proportional to the cell's affinity for the substrate. The value of K_S is typically quite low (1 to 5 mg/L for *Escherichia coli* on glucose), which means that $\mu \approx \mu_{max}$ when $S > 10\ K_S$ and that μ only becomes a strong function of the substrate concentration when $S < 10\ K_S$.

Note that the maximum specific growth rate of the organism limits the extent to which the dilution rate can be increased.

$$\mu_{max} = D_C \qquad \ldots(6)$$

This threshold dilution rate is called the critical dilution rate (D_C), and increasing the dilution rate beyond D_C results in "washout"; cells are removed from the vessel at a rate faster than their growth rate. This can limit the productivity of the simple CSTR and motivates cell retention or recycle strategies that enable operation at higher throughputs.

Rearranging the growth model to solve for the substrate concentration in the vessel gives

$$S = \frac{D \cdot K_S}{\mu_{max} - D} \qquad \ldots(7)$$

At dilution rates above D_C, cells have been washed out of the vessel, and the substrate concentration equals the inlet concentration (S_O). By substituting equation 7 into equation 4, the cell concentration becomes an explicit function of D and S_O.

$$X = X_{X/S} \cdot \left(S_0 - \frac{D \cdot K_S}{\mu_{max}} \right) \qquad \ldots(8)$$

The biomass productivity of the CSTR (R_{CSTR}), defined as the cell output per reactor volume, is calculated as

$$R_{CSTR} = \frac{F \cdot X}{V} = (D \cdot X) = D \cdot Y_{X/S} \cdot \left(S_0 - \frac{D \cdot K_S}{\mu_{max}} \right) \qquad \ldots(9)$$

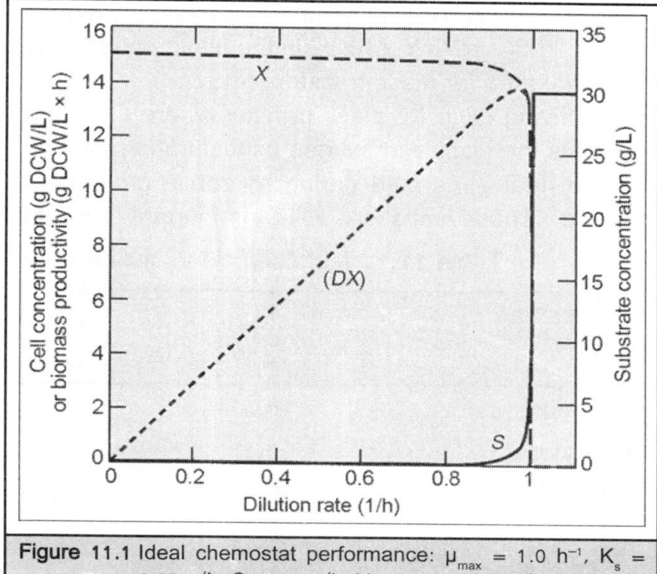

Figure 11.1 Ideal chemostat performance: μ_{max} = 1.0 h^{-1}, K_s = 0.05 g/L, S_0 = 30 g/L, $Y_{X/S}$ = 0.5 g DCW/g substrate.

Figure 11.1 shows the steady-state cell and substrate concentrations and biomass productivity for the ideal chemostat. Note that the substrate is almost completely utilized over most of the operating dilution rates. This high conversion of substrate is a key attribute of the ideal CSTR system, improving the economics of processes that depend on efficient substrate utilization and minimizing the effects of substrate inhibition. Also shown in Figure 11.1 is the high concentration of cells in the vessel until washout at the critical dilution rate. The maximum biomass productivity is located close to the washout point, making operation at the maximum productivity point very sensitive to deviations in dilution rate. The dilution rate associated with maximum biomass productivity for a fixed limiting nutrient concentration in the feed (D_M) can be calculated by setting the derivative of (DX) with respect to D to zero and solving for the dilution rate. D_M is then given as

$$D_M = \mu_{max}\left(1-\sqrt{\frac{K_S}{K_S+S_0}}\right) \qquad \ldots(10)$$

By substituting D_M into equation 8, we can solve for X_M, the cell concentration corresponding to D_M.

$$X_M = Y_{X/S}\left[S_0 + K_S - \sqrt{K_S(S_0+K_S)}\right] \qquad \ldots(11)$$

Considering the limiting nutrient concentration (S_0) as an independent design variable, equation 11 suggests that the substrate concentration in the feed can be increased arbitrarily to achieve extraordinary cell densities and productivities. In reality, the productivity of an aerobic reactor system is ultimately limited by the rates of heat and/ or mass transfer when the reaction kinetics are fast (i.e., high X and D). Oxygen transfer, and not the carbon substrate, is often growth limiting because oxygen is an essential nutrient for aerobic metabolism. It is poorly soluble in the medium (typically around 7 mg/L for air at 1 atm), and its transfer rate is restricted by the physical capabilities of the oxygenation system. Oxygen-limited growth may be expressed as a steady-state balance between the oxygen uptake rate (OUR) and the oxygen transfer rate (OTR). Oxygen transfer is usually limited by transfer from the gas to liquid phases, leading to the steady-state balance:

$$\left(\underbrace{\frac{\mu \cdot X}{Y_{X/O_2}} + m_{O_2} \cdot X}_{\text{OUR}}\right) = \underbrace{k_L a(C^* - C_L)}_{\text{OTR}} \qquad \ldots(12)$$

in which Y_{X/O_2} is the cell yield on oxygen (g DCW/g O_2), m_{O_2} is the maintenance coefficient for oxygen (g O_2/g DCW x h), k_L is the liquid phase mass transfer coefficient (cm/h) a is the specific interfacial area for mass transfer (cm^2/cm^3), k_La is the mass transfer coefficient (h^{-1}), and ($C^* - C_L$) is the driving force for mass transfer where C^* is the equilibrium oxygen concentration (mmol/L) and C_L is the dissolved oxygen concentration (mmol/L). The ability of the heat transfer system to remove heat generated during microbial growth can also limit $R_{CSTR,}$ as discussed in "Energy Balance". These limitations will be important in evaluating and comparing CSTR performance.

Table 11.1 Yield coefficients for bacteria on different carbon substrates

Substrate *(i)*	$Y_{X/S}$ *[g DCW/g substrate]*	Y_{X/O_2} *[g DCW/g O_2]*	Y_{HI} *[g DCW/kcal]*
Acetate	0.36	0.70	0.21
Glucose	0.51	1.5	0.42
Methanol	0.40	0.44	0.12
Ethanol	0.68	0.61	0.18
n-Paraffins	1.0	0.50	0.16
Methane	0.62	0.20	0.061

The biomass productivity of the CSTR can be compared to the productivity of a batch fermentor (R_{BATCH}) by defining a relevant batch productivity. The batch fermentation cycle consists of a lag phase, an exponential growth phase, cell harvest, and a batch turnaround time associated with cleaning, sterilizing, and filling the vessel. The lag, harvest, and turnaround activities can be grouped into a term $t_{TURNAROUND}$ in order to determine the batch cycle time (t_{CYCLE}) using equation 13, in which X_i is the concentration of cells in the vessel following inoculation (typically $X_i \sim 0.1 \times X$).

$$t_{CYCLE} = \frac{1}{\mu_{max}} \ln\left(\frac{X}{X_i}\right) + t_{TURNAROUND} \quad \text{...(13)}$$

Cell growth can be calculated from the cell yield on the growth-limiting substrate and the initial concentration of substrate, again assuming cell maintenance and product formation are negligible.

$$X - X_i = Y_{Y/S} S_0 \quad \text{...(14)}$$

Subsequently, the ratio of biomass productivities in the CSTR at D_M, $(R_{CSTR})_M$, and in the batch fermentor is given by equation 15, in which it is assumed that $S_0 \gg K_S$ (as it often is).

$$\frac{(R_{CSTR})_M}{R_{BATCH}} = \frac{(D_M \cdot X_M)}{\left(\dfrac{X - X_i}{t_{CYCLE}}\right)} = \ln\left(\frac{X}{X_i}\right) + \mu_{max} \cdot t_{TURNAROUND} \quad \text{...(15)}$$

Equation 15 often appears as a measure of relative biomass productivity, in which the CSTR is favored over batch operation at high growth rates and long turnaround times. However, equations 10 and 11, associated with maximum productivity, do not reflect the ultimate limitation posed by heat and mass transfer in industrial aerobic processes. Given that productivity has a limit dictated by the system, the independent parameters, D and S_0, can be adjusted so that the maximum productivity is attained during operation. Recognizing that (R_{CSTR}) is ultimately limited by the maximum oxygen transfer rate, equation 12 can be rearranged to give the biomass productivity.

$$(D \cdot X) = \frac{D \cdot Y_{X/O_2} \cdot OTR_{max}}{(D + m_{O_2} \cdot Y_{X/O_2})} \quad \ldots(16)$$

Notice that the cell concentration is fixed once an operating dilution rate is specified. Solving equation 16 for X and substituting into equation 4 yields

$$S_0 = \frac{Y_{X/O_2} \cdot OTR_{max}}{Y_{X/S} \cdot \delta_S \cdot (D + m_{O_2} \cdot Y_{X/O_2})} \quad \ldots(17)$$

where δ_S is the fractional substrate conversion. High substrate conversion leads to lower raw materials costs and reduces the burden of residual substrate on downstream purification and waste treatment operations. When operating near the maximum reactor productivity (i.e., maximum OTR), the dilution rate and the inlet substrate concentration can be adjusted independently to achieve target levels of cell concentration and/or substrate conversion. For biomass production when O_2 transfer is not limiting productivity, the CSTR is favored over batch operation when the specific growth rate is high and the batch turnaround time is long. However, most industrial processes will be operating at or near the oxygen or heat transfer limitation.

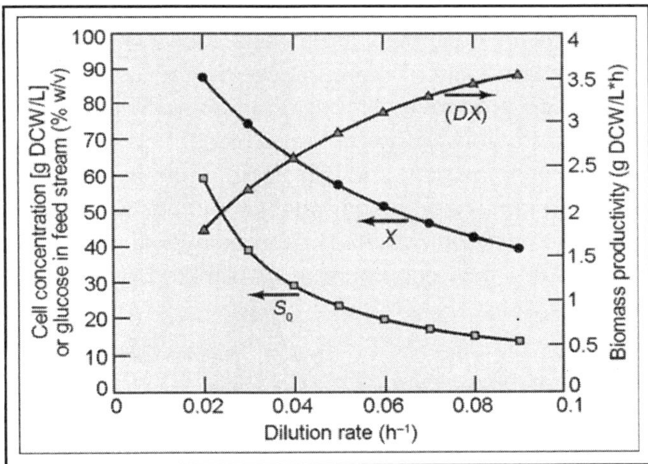

Figure 11.2 Chemostat operating at maximum oxygen transfer rate: OTR_{max} = 100 mmol O_2/L x h, μ_{max} = 0.09 h^{-1}, $Y_{X/O2}$ = 1.56 g DCW/g O_2, m_{O2} = 0.024 g O_2/g DCW x h, $Y_{X/S}$ = 0.45 g DCW/g substrate, 95% substrate conversion.

The desirability of enhanced O_2 transfer has motivated the development of novel bioreactor designs (e.g., bubble columns, loop or airlift reactors). For therapeutic products, however, the overwhelming majority of fermentations are batch or fed-batch processes. In these applications, the choice of operating mode is not based on biomass productivity. Performance metrics such as volumetric productivity of the product, product yield, and product concentration become more important in evaluating potential operating strategies than simply the biomass concentration.

Fermented products

When the goal of the fermentation is to produce a product other than biomass, the criteria used to evaluate alternative operational modes are less straightforward than biomass productivity. In order to evaluate process alternatives, a proper set of performance metrics must be identified that relate to overall process economics. When process economics are dominated by fermentation costs (e.g., fuel alcohol or gluconic acid production), volumetric productivity and conversion yields are especially relevant criteria as they relate to the size and cost of the reactor system and the cost of raw materials. In processes where recovery costs dominate (e.g., antibiotics), however, the size and operational costs of the recovery system are proportional to the fluid volume processed and inversely proportional to the product concentration. As a result, the final product concentration, or titer, is more important than biomass productivity.

The material balance on the product can be written as shown below, in which product formation is expressed using a specific product formation rate q_P (g product /g DCW × h), and product degradation by a specific rate constant $k_p(h^{-1})$.

$$\underbrace{q_p \cdot X}_{\text{Product formation}} - \underbrace{k_p \cdot P \cdot D}_{\text{Product degradation}} - \underbrace{D \cdot P}_{\text{Product out}} = \underbrace{\frac{dP}{dt}}_{\text{Product accumulation}} \quad \text{...(18)}$$

Product formation can be characterized in relation to growth, being growth (primary metabolites) or nongrowth (secondary metabolites) associated. Examples of growth-associated products are direct catabolic products of the carbon substrate, such as ethanol and citric or acetic acid. Nongrowth-associated products, comprising many antibiotics, are metabolites that are not necessary for cell growth and typically are only produced during slow or stationary growth phases. Some products, such as xanthan gum and lactic acid, are mixed growth associated in that they are produced during slow and stationary growth phases. Numerous models of product formation have been proposed, taking into account variables such as hyphal morphology, cell age, surface area, metabolic carbon flux, and plasmid copy number. A simple model expresses the growth dependence of q_P as

$$q_p = \underbrace{\alpha \cdot \mu}_{\text{Growth associated}} + \underbrace{\beta}_{\text{Nongrowth associated}} \quad \text{...(19)}$$

The steady-state product balance can be rewritten and solved for P.

$$P = \frac{q_p \cdot X}{D} = \frac{(\alpha \cdot \mu + \beta)}{D} \cdot X = \left(\alpha + \frac{\beta}{D}\right) \cdot X \quad \text{...(20)}$$

For growth-associated products (i.e., $\alpha \gg \beta$) product concentration is proportional to biomass and is independent of the dilution rate when X is approximately constant. Increasing the dilution rate results in increased product formation up to the region near D_C. The growth dependence of product formation and intracellular metabolic fluxes can be determined using a chemostat. An example of growth-associated product formation is shown in Figure 11.3, in which product concentration is proportional to biomass and the growth dependence of q_P is evident. In the case of secondary metabolites (i.e., $\alpha \ll \beta$), product concentration is inversely proportional to the dilution rate, and the productivity ($D \times P$) is independent of dilution rate.

The low dilution rates favorable for secondary metabolite production approach batch operation, which is generally favored over the CSTR in such instances. Other issues can impact the decision between batch and continuous culture. The ability of the CSTR to maintain an ideal environment for product formation may offer a competitive advantage over the batch fermentation, with its time-varying environment and prolonged lag and stationary growth phases. Regulatory and market factors also play an important role in deciding on the operating mode. The CSTR is a dedicated manufacturing system used to produce a single product. Such a system may not be well suited for the production of specialty chemicals and pharmaceuticals because it can neither adapt to variable market demand nor satisfy demand for multiple products.

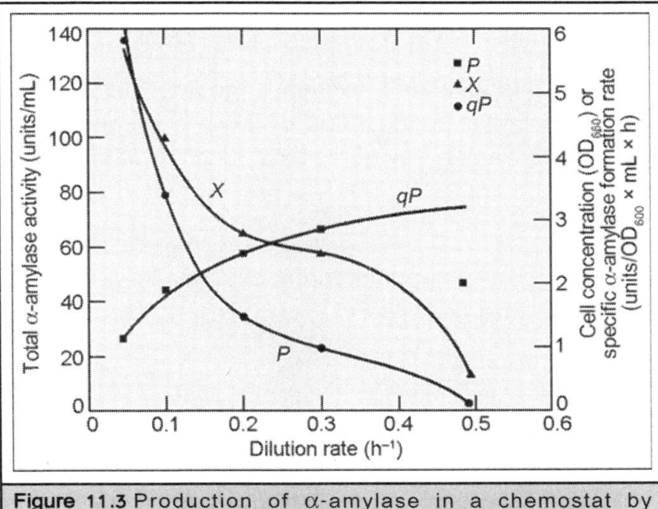

Figure 11.3 Production of α-amylase in a chemostat by recombinant *Escherichia coli.*

Energy cycle

Heat transfer is an important consideration in fermentor design, scale-up, operation, and sterilization. The energy balance is used to determine the time–temperature profile of the fermentation broth by accounting for the transfer and accumulation of energy. The heat transfer rate limits the ability to reduce the cycle time for sterilization. More importantly, the rate of heat removal from the broth during cell growth can constrain volumetric productivity, an issue in very large reactors with reduced area-to-volume ratios. Because considerable heat generation accompanies rapid cell growth, the high specific growth rates favored in industrial CSTR applications will exacerbate the problem of heat transfer in large reactors. The steady-state energy balance for the fluid in the CSTR is written as

$$Q_{AGIT} + Q_{MET} + \Delta Q_{SENS} - Q_{LOSS} - Q_{EVAP} - Q_{EXCH} = 0 \quad \text{...(21)}$$

in which Q_{AGIT} is the mechanical energy imparted to the fluid through impeller agitation (equal to the gassed power input), Q_{MET} is the metabolic heat generated by cell growth, ΔQ_{SENS} is the net sensible heat added to the system by streams entering and leaving the system, Q_{LOSS} is the sum of the heat losses from the system to the surroundings, Q_{EVAP} is the latent heat removed by evaporation, and Q_{EXCH} is the heat removed from the system by an appropriate heat exchanger system. In some cases, the heats of solution and mixing must be accounted for, but in most cases they are negligible. The terms ΔQ_{SENS}, Q_{LOSS}, and Q_{EVAP} are comparatively small, leading to the simplified energy balance

$$Q_{EXCH} = Q_{MET} + Q_{AGIT} \quad \text{...(22)}$$

For fast-growing microorganisms, the heat exchanger duty can be as high as 7.7 to 23.2 kW/m^3, of which Q_{MET} and Q_{AGIT} typically represent about 75% and 25% of the total, respectively.

Approximately 40 to 50% of the energy contained by a substrate is converted into useful chemical energy, whereas the balance is released as heat. If this metabolic heat is not removed from the fermentation broth, the temperature will rise and possibly hinder performance. Metabolic heat generation is a function of the growth rate of the organism, the cell concentration, the fluid volume, and the efficiency of cell growth on a particular substrate (i), which can be expressed as the metabolic heat released per gram of cell produced ($1/Y_{Hi}$) (kcal/g DCW).

$$Q_{MET} = V \frac{\mu \cdot X}{Y_{Hi}} \quad ...(23)$$

In aerobic fermentations, oxygen is the final electron acceptor in substrate metabolism, enabling a correlation between the rates of oxygen uptake and heat generation. The following empirical correlation gives Q_{MET} (kcal/h) as a function of specific oxygen consumption, q_{O_2} (mmol O_2/L x h), and V(L):

$$Q_{MET} = 0.12 \cdot V \cdot q_{O_2} \quad ...(24)$$

where the oxygen demand during exponential growth can be expressed using the cellular yield on oxygen, Y_{X/O_2} (g DCW/g O_2).

$$Q_{O_2} = \frac{\mu \cdot X}{Y_{X/O_2}} \quad ...(25)$$

Table 11.1 gives values of Y_{Hi} and Y_{X/O_2} for bacterial growth that can be used to estimate oxygen demand and metabolic heat generation. Notice that the more reduced substrates result in greater O_2 demand and heat generation and subsequently a larger burden on the O_2 transfer and heat exchange system.

Heat exchange systems are chosen based on the expected heat exchanger duty, influence on fluid mixing, impact on cleaning and sterilization, utility economy, and maintenance, operating, and capital costs. Heat exchangers for fermentors commonly consist of a jacket or shell around the vessel, internal coolant coils, or occasionally an external heat exchanger. The rate of heat removal by a heat exchanger system can be described using a Fourier's law expression:

$$Q_{EXCH} = U \cdot A \cdot (T - Tc) \quad ...(26)$$

in which A is the surface area available for heat transfer, U is an overall heat transfer coefficient accounting for all heat transfer resistances, and $(T - Tc)$ is the driving force for heat transfer, where T is the bulk temperature of the fermentation broth and T_C is the temperature of the cooling fluid used in the heat exchanger. Typical values of the heat transfer coefficient are 50 to 150 BTU/ ft^2 x h x °F (280 to 850 W/m^2 x K). An important consequence of fermentor scale-up is a decreasing surface-area-to-volume ratio (A/V): at increasing scale the capacity for heat removal relative to heat generation diminishes and often becomes limiting at larger volumes.

The rate of heat removal can be increased by increasing the temperature driving force, increasing the heat transfer surface area, or reducing resistance to heat transfer (increasing U). Water is primarily used as the coolant fluid because of its availability and low cost relative to a refrigerated coolant system. The temperature of the cooling water increases as it passes through the heat exchanger, so an arithmetic or logarithmic mean Tc may give a more representative measure of the coolant temperature. Increasing the coolant flow rate can decrease the mean coolant temperature but with a subsequent increase in utility consumption. Furthermore, the pressure drop of the exchanger and pump capacity limit the extent to which coolant flow rate can be increased. The heat exchanger area is dictated by the size and type of heat exchange system chosen during process scale-up. The overall heat transfer coefficient and constituent resistances are thoroughly discussed elsewhere. In general, poor mixing contributes to decreased heat transfer. It is sufficient to focus on the convective heat transfer coefficient on the fermentation broth side, h [W/m^2 * K], which is often the dominant heat transfer resistance. The convective heat transfer coefficient is a function of the Reynolds number (*Re*), Prandtl number (*Pr*), viscosity ratio (*Vi*), and the system geometry ($F_{GEOMETRY}$). In many cases $Vi \approx 1$, and the exponents *a* and *b* have typical values of 0.8 and 0.3, respectively.

$$Nu = C_1 \cdot Re^a \cdot Pr^b \cdot Vi^c \bullet F_{GEOMETRY} \quad \frac{h \cdot D_r}{k} = C_1 \cdot \left(\frac{D_1^2 N \rho}{\mu}\right)^a \left(\frac{Cp\mu}{k}\right)^b \left(\frac{\mu}{\mu_w}\right)^c F_{GEOMETRY} \quad \ldots(27)$$

Commercial values

Operating costs are those costs associated with maintaining a given production level dictated by the scale and scheduling of the process, typically grouped as raw materials, direct expenses, and indirect expenses. Although the focus here will be on raw materials and direct expenses (e.g., utilities, maintenance, operating supplies, and labor) associated with the CSTR, it is important to view the reactor system in the context of the entire process. The process flow diagram will form the basis for determining not only the reactor operating costs but also the capital and operating costs of air compressors, pumps, holding tanks, and various downstream units. It is important to remember that the relative costs associated with the reactor system, batch or continuous, may not have a significant impact on the total process economics if product recovery costs are dominant. Raw materials costs include the material and handling costs of components added to the system to satisfy metabolic (e.g., C, N, O sources) or process (e.g., acids or bases and antifoam agents) requirements. Direct expenses include the cost of utilities, maintenance, operating supplies, operating labor, direct supervision, laboratory charges, and patent royalties. Continuous systems are easier to automate and offer the potential of lower labor costs than batch production systems, with their labor intensive startup and shut-down operations. Indirect expenses include taxes and depreciation, usually expressed as a percentage of the plant cost.

Bioprocess media

The selection of bioprocess media typically involves a trade-off between media cost and the product yield and titer. It is an important stage of development that can influence the design and performance of the entire process. Selecting an apparently cheap media, for example, may result in more expensive downstream recovery and waste treatment operations. In general, media selection includes a number of technical and economic considerations: yield or titer of desired and undesired products; cost; variability in composition and price; availability; effect on downstream processes; need for pretreatment or supplements; shipment, storage, and handling, and need for testing and validation. Media can be classified as defined or undefined with respect to chemical composition. Laboratory-scale chemostat investigations commonly use defined media to allow precise control over the growth-limiting nutrient and medium composition.

Small bioreactors (1 to 4 L) are preferred for continuous operation in the laboratory because media preparation and storage are less of a burden. Industrial microbial fermentations predominantly use undefined media because they generally are less costly and perform better than defined media. Undefined microbial media often contain agricultural by-products (e.g., molasses or corn steep liquor) and thus are subject to source market fluctuations in quality and price. Processes that have been validated with a variety of media may change the production media to take advantage of market changes. The costs of raw materials as a percentage of operating costs for primary metabolites can range from 40% for citric acid to 70% for ethanol from sugar cane, whereas for secondary metabolites media costs can be around 10 to 20%. At small scale (less than 10 m^3), mammalian cells are often grown in undefined serum, an expensive and sometimes scarce media derived from mammalian plasma. Viral contamination and the presence of serum proteins can complicate cultivation, product recovery, and quality control and quality assurance (QA–QC) in industrial processes. These complications, combined with regulatory pressure to safeguard against viral contamination, are a driving force toward the development of defined, serum-free media in order to avoid use of animal-derived media components.

Requirements

Steam, cooling water, and power requirements comprise the majority of utility expenses. Process demand for water-for-injection (WFI) must also be determined for clean-in-place (CIP) systems. Steam usage occurs predominantly during media and equipment sterilization and can be calculated using knowledge of the sterilization cycle. Continuous reactors readily lend themselves to the use of continuous, as opposed to batch, media sterilization systems, offering advantages in reduced thermal degradation of heat-sensitive media, reduced sterilization time, more efficient fermentor use, and greater steam economy (about 20 to 25% of the steam used by batch sterilization). The cooling water requirement, W_{H_2O} (kg/h), can be calculated knowing the heat exchanger duty from the energy balance (equation 21):

$$W_{H_2O} = \frac{Q_{EXCH}}{C_p \cdot (T_{out} - T_{in})} \qquad \ldots(28)$$

where C_p is the heat capacity of water [~1 kcal/kg * K] and T_{out} and T_{in} are the outlet and inlet cooling water temperatures, respectively. The availability of abundant low-temperature water can reduce cooling water requirements and possibly allow the use of larger reactors because of the improved ability to remove metabolic heat.

Oxygen demand and mixing requirements drive power consumption by agitators and compressors, whereas larger volumes, higher cell densities, higher specific O_2 uptake rates, and higher broth viscosities result in increased power requirements. Total power consumption for agitated vessels is typically in the range of 2 to 10 kW/m^3. Centrifugal pumps are predominantly used in bioprocesses for their relatively low cost and ability to handle suspended solids. Although pumps represent only a small fraction of power consumption, they can be a significant percentage of overall maintenance costs. In addition, the performance of continuous processes is particularly susceptible to pump failure and may warrant additional capital investment toward the installation of backup pumps in parallel.

Behavioral studies

In this section, the assumptions used in the development of the ideal CSTR theory are revisited in order to determine when they are invalid and to gauge the impact of nonidealities on performance.

Nonideal mixing

The major assumption of the ideal CSTR is that there are no spatial variations of properties inside the vessel. Such ideal mixing is never observed in actual systems and even deteriorates on scale-up, although it can be a good approximation of behavior. The challenge is to determine when nonidealities can be expected and how they will influence performance. Equipment design, operating conditions, and broth properties all influence the quality of fluid mixing in the vessel. Agitated tanks are designed to provide good mixing through selection of tank geometry, baffle placement, and impeller design, although mixing quality decreases with increasing scale. Agitator power input and gas sparging rates, although typically associated with oxygen transfer, are critical to mixing. High broth viscosities, typical of mycelial fermentations, contribute to poor mixing.

Poor mixing can affect oxygen transfer, product formation, heat transfer, process monitoring and control, and the distribution of components added to the system. Nonideal mixing in continuous systems is typically characterized by either the residence time distribution, the distribution of fluid residence times around the ideal residence time (τ), or the mixing time (t_M), the time required for a system to respond to a feed disturbance. The ideal mixing time is equal to zero (instantaneous), and the mean value of the residence time distribution is τ (or D^{-1}). Residence time distributions are typically measured by pulse or step addition of a tracer into the reactor feed and tracking the appearance of the tracer in the exit stream as a function of time. Correlations obtained from dimensional analysis can be used to predict mixing times, which increase with scale and broth viscosity. Small vessels (<500 L) are generally well mixed ($t_M \sim$ s), but large fermentors (>5,000 L) typically have poor mixing (t_M – min).

Compartmental mixing models, in which the bulk fluid is modeled as discrete CSTRs and plug flow reactors (PFRs) with fluid interchange, have been used to describe heterogeneity in large vessels. Wall growth is a special case of nonideal mixing in which cells adhere and proliferate on vessel surfaces, where they are hidden from cell mass measurements obtained from samples of the bulk fluid. The system, in addition to being heterogeneous, is no longer at steady state because cells are accumulating in the vessel. This can pose serious problems if the accumulating organism is a contaminant or otherwise undesirable organism. The metabolism and growth of wall-bound cells can be quite different from the suspended population because of mass transfer limitations. Wall growth can reduce heat transfer, create sterilization and cleaning problems, and corrupt measurements in experimental systems. It can be a significant factor when the surface-area-to-volume ratio (A/V) is high (e.g., laboratory-scale systems and vessels with internal cooling coils) and may require modifications to chemostat design or operation. For example, the glass walls of laboratory vessels are sometimes treated with organosilane compounds to minimize wall growth. Yeast and mycelial cells with a propensity to form pellets are prone to wall growth under the same conditions that favor flocculation. A discrete washout point does not exist with wall growth, because cells are effectively immobilized in the vessel even past D_C.

Importance of substrates

Deviations from ideal chemostat behavior may arise when assumptions regarding the magnitudes of nutrient uptake, the consistency of cell composition, or the identity of the growth-limiting nutrient become invalidated. Substrate uptake for growth in equation 3 was assumed to be much larger than that for maintenance and product formation. At low dilution rates (i.e., low growth rates) the maintenance term becomes significant ($m \sim \mu / Y_{X/S}$) and less substrate goes toward cell growth, causing the actual X to be less than predicted by the ideal theory at low D. Similarly, at high dilution rates, the production and accumulation of growth-related products and intermediates can become significant, leading to a reduced cell yield at higher D. The ideal chemostat derivation assumes that cell composition does not change over the operating region. In actuality, cell composition varies with pH, temperature, growth rate, and medium composition. As cell composition changes, the demand for essential nutrients will change in ways that were not accounted for in the derived equations.

Proper evaluation of experimental data from a chemostat may require consideration of the variation of nutrient uptake and cell composition with environment and growth rate. When using complex and undefined media or an organism with complex nutritional requirements (e.g., mammalian cells), it is often difficult to identify the growth- limiting nutrient. In addition, the limiting nutrient itself may change because the nutrient demand, and subsequently the media composition, may change with operating conditions. The cell concentration profile in these situations likely would be constantly decreasing with dilution rate, unlike the ideal chemostat where Xis approximately constant over the majority of $0 < D < D_C$. More complicated, structured growth models would be required to account for such behavior.

Potential sources

The potential sources of process variability and non-steady-state behavior are perhaps too numerous to mention. However, typical instances include chemostat start-up, execution of control actions, induced disturbances (e.g., pulse and shift methods), variations in feed composition, culture degeneration (e.g., plasmid loss or apoptosis), wall growth, or equipment failures. Steady state often is declared when the measurable process states are maintained constant for 3 to 5 residence times. Sustained oscillations are sometimes observed in continuous culture and often are the result of growth inhibition resulting from either an accumulated product or the burden of product formation.

Selection/mutation and contamination

By controlling the culture conditions in the CSTR, a highly selective environment for the selection and proliferation of certain microorganisms can be created. Cells in this selective environment with growth rates less than the dilution rate will be washed out of the reactor, leaving only those cells with the properties that have been selected for. In this way, continuous culture can be used as a strain improvement tool to select organisms that possess a desirable trait, such as yeast with higher ethanol tolerance. Because of the metabolic burden imposed by high levels of product formation, the production strain has a growth disadvantage relative to unproductive strains that are present in the reactor. Without selection pressure in favor of the production strain, a gradual decline in productivity will be observed over time as nonproductive cells dominate the culture. Examples include the reversion of specially selected antibiotic strains to low productivity mutants or the domination of recombinant protein processes by plasmid-free cells.

The configuration of CSTRs in series can be used to circumvent this problem by providing separate environments for growth and product formation. Selective pressure in favor of the production strain (and against contaminants) may be exerted through the application of elevated temperatures, extremes of pH, the use of narrowly defined or modified media, and the use of specially selected cultures (e.g., antibiotic-resistant strains). The prolonged operating periods of continuous culture increase the probability of contamination by a foreign organism. The threat posed by contamination depends on the ability of the undesirable microorganism to complete and thrive in the CSTR environment. Consider the case of two types of microorganisms with concentrations X (the desired strain) and Z (the contaminant) competing for the same limiting substrate in a CSTR. The material balances on cell mass, neglecting cell lysis, can be written as

$$\frac{dX}{dt} = \mu \cdot X - D \cdot X \qquad \text{...(29)}$$

$$\frac{dZ}{dt} = \mu_2 \cdot Z - D \cdot Z \qquad \text{...(30)}$$

Subtracting equation 30 from equation 29 and rearranging leads to equation 31:

$$\frac{d\ \ln[X/Z]}{dt} = \mu - \mu_Z \qquad \text{...(31)}$$

which shows that the growth rates and their dependence on the limiting substrate will determine the fate of the culture. The contaminant (Z) could be washed out ($\mu > \mu_z$), remain at a stable level ($\mu = \mu_z$), or dominate ($\mu < \mu_z$) the culture. This simple analysis of selection can be complicated if the contaminant has properties that prevent it from being washed out (e.g., adhesion to reactor surfaces), if the contaminant competes for a different substrate than the production strain, or if there are interactions from inhibitory cellular products. For example, lactobacilli are often a persistent contaminant of continuous ethanol production processes because of their intimate association with flocculant yeast aggregates and ability to adapt to high ethanol concentrations. Selective recycle of desirable organisms back to the vessel has been used to prevent domination of the culture by undesirable strains.

Biological Catalysts

Biological catalysts or enzymes are biological catalysts with high selectivity toward reactants and products, making them attractive for use in a number of industrial applications. Enzyme activity is strongly influenced by the environment (e.g., pH, temperature, metal ions). Loss of activity or denaturation can be reversible or irreversible, depending on the type, strength, and duration of an unfavorable interaction. A benefit of using the CSTR for enzyme reactions is that the constant, controlled reactor environment can be designed for maximum enzyme activity and life.

Material for balanced growth

Assuming that the inlet and outlet flow rates are approximately equal (i.e., solutions are dilute), the steady-state material balance on the substrate can be written as

$$F \cdot (S_0 - S) = v \cdot V \qquad \text{...(32)}$$

where F is the volumetric flow rate, v is the rate of substrate consumption by reaction, V is the fluid volume in the reactor, and S_0 and S are the substrate concentrations in the feed and vessel, respectively. Rewriting equation 32 in terms of the fluid residence time (τ) and the fractional substrate conversion (δ_S) yields the CSTR design equation

$$\tau = \frac{V}{F} = \frac{S_0 \cdot \delta_S}{\nu} \qquad \ldots(33)$$

Using a valid rate expression for v, the design equation can be used to determine the reactor volume required to yield a given conversion rate ($S_0 \times \delta_S \times F$). Fast reaction kinetics are obviously favorable because reactor cost scales with reactor size. Enzyme loading in the reactor can be increased beyond the solubility limit by immobilization on inert support particles, which increases v, reduces the necessary reactor volume, facilitates enzyme retention and recycle, and may improve enzyme stability.

Enzyme reaction kinetics

Numerous mechanistic models have been developed to describe enzyme reaction rates as a function of enzyme and substrate concentrations. Some of the more common models accompanied by the corresponding solution to the design equation.

Unlike a plug flow reactor (PFR), in which the substrate enters at a high concentration and leaves at a lower concentration, the substrate concentration in a CSTR is at a uniform, low concentration. The reduced substrate concentration leads to a slower reaction rate, so that the CSTR requires more of the active enzyme than the PFR to attain the same substrate conversion rate. Substrate inhibition is less problematic in a CSTR than in a PFR because of the lower substrate concentration in the bulk fluid, whereas product inhibition is generally more of a problem in CSTRs than PFRs. Arranging CSTRs in series can reduce the effects of product inhibition (approaching PFR behavior) while taking advantage of the good mixing characteristics of the CSTR to provide optimal pH control.

Thermal denaturation

Like many chemical reactions, increasing the temperature enhances the rate of enzyme reactions. Higher temperatures also result in increased rates of thermal denaturation and loss of the active biocatalyst. Process economics often depend on optimal temperature control to maintain high substrate conversion and long catalyst life. The effects of temperature on the catalytic rate constant (k_2) can be described using an Arrhenius expression

$$k_2 = A \cdot e^{-Ea/RT} \qquad \ldots(34)$$

where A is the Arrhenius constant, E_a is the activation energy, R is the gas constant, and T is the absolute temperature. The activation energy of enzyme-catalyzed reactions ranges from 4 to 20 kcal/mol, with most reactions near 11 kcal/mol.

Thermal denaturation usually can be described as a first-order decay reaction:

$$\frac{dE}{dt} = -k_d \cdot E \quad \text{or} \quad E = E_0 e^{-k_d t} \qquad \ldots(35)$$

where k_d is the thermal denaturation constant, which also follows an Arrhenius temperature dependence. For thermal denaturation, E_a varies from 40 to 130 kcal/mol, with most in the vicinity of 70 kcal/mol. Increasing temperature has a greater effect on the rate of denaturation than catalysis. For a typical enzyme (i.e., 11 and 70 kcal/mol), an increase in temperature from 30 to 40°C results in a 1.8-fold increase in the rate of catalysis, but a 41-fold increase in the denaturation rate.

Energy cycle

An optimal temperature control strategy requires good heat removal because most industrial enzyme reactions are exothermic. Although heat transfer is generally good for soluble enzymes in agitated tanks, the high enzyme concentrations attained with immobilization can result in fast reaction rates and appreciable heat generation. Heat transfer resistance within the catalyst pellet can reduce heat removal rates, resulting in higher pellet temperatures and shorter catalyst life. Catalyst degradation from insufficient heat removal is more of a concern in packed beds with high enzyme loading, where heat transfer resistances can be significant. The steady-state energy balance for an exothermic ($\Delta H_{RXN} < 0$) enzyme reaction in a well-mixed CSTR is

$$F\rho c_P(T_F - T) + v' \cdot E \cdot V(-\Delta H_{RXN}) - Q_{EXCH} = 0 \qquad \text{...(36)}$$

where F is the volumetric flow rate (m³/h), ρ is the fluid density (kg/m³), c_P is the fluid heat capacity (kJ/kg x K), v' is the specific reaction rate (kmol substrate/kg enzyme x h), E is the enzyme concentration (kg enzyme/m³), ΔH_{RXN} is the heat of reaction (kJ /kmol substrate), Q_{EXCH} is the heat removed by the heat exchanger (kJ/h) from equation 26, and T_F and T are the feed and bulk fluid temperatures (K), respectively.

Commercial values

For enzyme CSTRs, the primary operating costs are associated with enzyme replacement. Prolonged catalyst activity and marked reductions in raw materials costs can be achieved by maintaining an optimal environment for the enzyme during operation. Preserving enzyme activity reduces the number and frequency of labor-intensive cleaning and changeovers, facilitating downstream operations by consistently providing a constant-quality product stream. With multiple reactors installed in parallel, changeovers can be scheduled to minimize production variations and downtime. Industrial enzymes are often sold as a crude mixture containing only a fraction of active enzyme. Selecting enzymes among different vendors may involve a trade-off between cost and purity (percent active enzyme) and a consideration of how the impurities may affect the process.

REACTOR PERFORMANCE

The rate of mass transfer ultimately will limit the maximum aerobic reactor performance. Oxygen transfer to the fermentor broth, for example, can limit both the extent and rate of cell growth. Mass transfer limitations to microbial flocculants and immobilized catalyst pellets can result in reduced reaction rates and inefficient conversion. Liquid–liquid mass transfer rates from hydrocarbon substrates to suspended cells may limit productivity in two-phase systems. To determine the rate-controlling regime, it is useful to characterize the relative rates of mass transfer and reaction using the dimensionless Damkohler number (Da):

$$\text{Da} = \frac{\text{Maximum rate of reaction}}{\text{Maximum rate of diffusion}} \qquad \ldots(37)$$

The observed reaction rate may be limited by the rate of diffusion depending on the value of the Damkohler number: if $Da \gg 1$ the diffusion rate is limiting, if $Da \ll 1$ the reaction rate is limiting, and if $Da \sim 1$ then the reaction and diffusion rates are comparable. As with all dimensionless numbers, the Damkohler number is only meaningful if it is calculated using the proper time and length scales for a given system. Consider spherical pellets ($r = 400\ \mu$m) of *Penicillium chrysogenum*, assuming a pellet cell density of 0.1 g/cm³, an effective oxygen diffusivity ($D_{O_2}^{eff}$) of 1 x 10⁻⁶ cm²/s, a particle size of 400 μm, an oxygen concentration (c_{O_2}) of 7 mg/L, and the following cell parameters: $\mu = 0.075\ h^{-1}$, $Y_{X/O_2} = 1.56$ DCM/g O_2, and $m_{O_2} = 0.024$ g O_2/(g DCW x h). The Damkholer number shows that internal mass resistance is considerable and that cell growth in the pellet is likely to be limited by oxygen transfer.

$$\text{Da} = \frac{\overbrace{\left(\frac{\mu}{Y_{X/O_2}} + m_{O_2}\right)}^{O_2\ \text{Demand}} \cdot \overbrace{X}^{\text{Pellet cell density}} \cdot \overbrace{\left(\frac{4\pi r^3}{3}\right)}^{\text{Pellet volume}}}{\underbrace{\left(D_{O_2} \cdot \frac{c_{O_2}}{r}\right)}_{\text{Diffusive flux}} \cdot \underbrace{\left(4 \cdot \pi \cdot r^3\right)}_{\text{Pellet surface area}}}$$

$$= \frac{\left(\frac{0.075\ h^{-1}}{1.56\ \text{g DCW/g}} + \frac{0.024\ \text{g}\ O_2}{\text{g DCW} \cdot h}\right) \cdot \frac{0.1\ \text{g DCW}}{cm^3} \cdot (4 \times 10^{-2}\ cm)^2}{3 \cdot (1 \times 10^{-6}\ cm^2/s) \cdot (7 \times 10^{-6}\ g/cm^3)} \cong 5.5 \times 10^5 \qquad \ldots(38)$$

Gas–liquid mass transfer is often rate-limiting for gases that are sparingly soluble in the broth, such as oxygen and methane. Although highly soluble, carbon dioxide exhibits pH-dependent partitioning between gaseous and dissolved forms (CO_2, H_2CO_3, HCO_3^-, CO_3^{2-}) that is influenced by the rates of both reaction and mass transfer. Proper interpretation of the respiratory coefficient (RQ) in fermentations operated at neutral pH requires consideration of CO_2 dynamics. The OTR has already been used to determine the productivity limit of a CSTR used for biomass production in equation 12. In general, the rate of mass transfer from the gas to the liquid phase is given as

$$N_A = k_L a(C^* - C_L) \qquad \ldots(39)$$

where N_A is the rate of gas transfer (mmol/l x h) and the remaining terms have the same definitions as in equation 12. For sparged, agitated tanks, $k_L a$ has typical values in the range from 50 to 1,400 h⁻¹. Several correlations have been developed for $k_L a$ as a function of the gassed power input per unit volume and the superficial gas velocity for Newtonian broths in a variety of fermentors. The correlations can offer wide variability in mass transfer estimates and should be used in conjunction with knowledge from past experience or empirical measurements of $k_L a$ (e.g., dynamic or sulfite oxidation methods). Oxygen transfer to shear-sensitive mammalian cells requires gentle agitation combined with surface or membrane aeration, or light sparging, as opposed to the large power inputs and high rates of gas sparging in microbial fermentations. This limitation is somewhat offset by the fact that mammalian cells have lower O_2 requirements (0.05 to 0.5 mmol/10^9 cells x h) and grow to lower cell densities (10^6 to 10^7 cells/mL) than microbial cultures.

VOLUMETRIC PRODUCTIVITY

Recycling

Volumetric productivity is related to the concentration of active catalyst. Cell or enzyme concentrations greater than the steady state obtained from the simple CSTR can be achieved by separating cells from the effluent stream and recycling them to the vessel or by retaining them within reactor. Higher catalyst concentrations enhance substrate conversion and reduce the reactor size necessary to attain a given conversion.

Recycle operation improves system stability in the face of feed disturbances by retaining cells in the vessel even under conditions that would cause washout in the simple CSTR. Recycle systems can be operated at dilution rates, or throughputs, greater than the specific growth rate of the organism. Productivity improvements achieved with cell recycle are demonstrated in Table 11.2 for *Saccharomyces cerevisiae* ATCC 4126 and *Zymomonas mobilis* ATCC 10988 at 100 g/L glucose feed.

Table 11.2 Ethanol productivity enhancements for *S. cerevisiae* and *Z.mobilis*

		Dilution rate *(h^1)*	*Ethanol productivity* *g(L x h)*	*Cell density* *(g DCW/L)*
S. cerevisiae	CSTR	0.17	7	12
	CSTR with recycle	0.68	29	50
Z. mobilis	CSTR	0.175	8	2.5
	CSTR with recycle	2.7	120	38

Implementation of cell recycle

Cell recycle is implemented through a cell separation step, often by a unit operation commonly used in the initial stages of downstream processing. Typical methods of continuous cell separation include centrifugation, filtration, and sedimentation. Cell separation can be viewed as having two often equally important purposes: (i) the recovery or retention of cells for reuse and (ii) the removal of potentially inhibitory by-products or products from the culture environment. The separation step often has to satisfy additional performance requirements such as handling of shear- or temperature-sensitive materials, selectivity in rejection or recovery, containment, maintenance of asepsis, corrosion resistance, brief retention time, and ease of cleaning, sterilization, maintenance, and validation.

Recycle operation is standard for reactors using stable enzymes, because discarding expensive active catalyst is economically unfeasible. Cell separation operations are discussed elsewhere in the context of downstream processing, although a brief description is presented here in relation to cell recycle.

Sedimentation

Sedimentation is the settling of particles in a gravitational field. With low energy requirements and simple equipment, sedimentation is a relatively inexpensive way of separating a dilute cell phase. Waste treatment is by far the largest application of sedimentation- based cell recycle, in which cells are typically separated in large sedimentation tanks using lime or clay to enhance flocculant formation. The settling velocity (u_0) for an isolated spherical particle can be described using Stoke's law:

$$u_0 = \frac{d_P^2(\rho_P - \rho_F)g}{18\eta} \qquad \text{...(40)}$$

in which d_P is the particle diameter, ρ_P is the particle density (the specific gravity of a typical cell is 1.05), ρ_F is the fluid density, g is the gravitational acceleration constant, and η is the fluid viscosity. The functional dependence of Stoke's law suggests ways of increasing the settling velocity. The easiest and most common method is to increase the effective cell size by promoting flocculation (cell aggregation) through physiological, chemical, and physical factors: selection of flocculant strains; modification of cell wall structure or surface charge; changing the pH, temperature, or shear stress; addition of inorganic salts (e.g., Ca^{2+} and Mg^{2+} salts) or clays; controlling the concentration of certain nutrients or products (e.g., extracellular polysaccharides); and controlling the cell age or growth phase. Selective cell recycle has been implemented using the differential sedimentation properties of a desired and unwanted microorganism. The properties of a particular broth are generally unchangeable and will probably only impede particle settling. Although equation 40 holds for dilute suspensions of cells, the interactions among settling particles in concentrated slurries results in hindered settling. The hindered particle velocity (u_h) is influenced by the particle concentration and can be expressed with the following correlation:

$$\frac{u_h}{u_0} = \frac{1}{1 + \lambda \cdot \varepsilon_P^{1/3}} \qquad \text{...(41)}$$

in which ε_P is the volume fraction of particles and λ is an empirical function of ε_P. For dilute suspensions, $\varepsilon_P < 0.15$, whereas in slurries $0.15 < \varepsilon_P < 0.50$.

The limiting settling velocity for a system has a strong influence on equipment design and operation. Consider the case of a continuous sedimentation tank with volumetric throughput (F) and constant cross-sectional area (A). The sedimentation tank performance can then be described by equation 42, in which throughput is directly proportional to A and independent of tank depth:

$$F = u_{\lim} \cdot A \qquad \text{...(42)}$$

The throughput and limiting settling time will thus dictate equipment size and costs. Another design consideration is the residence time of cells in the settling tank, which must be considered in the context of nutrient depletion (particularly for oxygen) and its potential effects on performance.

Sedimentation at laboratory scale may be implemented with an external settling column. Similar devices may be used at bioprocessing scales, whereas large open-air tanks must be used in high-volume wastewater treatment. Internal sedimentation has been implemented in tower fermenters, in which immobilized cells and enzymes or microbial flocculants are retained in the vessel by a sedimentation zone within the vessel. Unlike the ideal CSTR, tower fermenters may exhibit spatial variations in nutrient concentrations and broth properties along the height of the tower that can significantly influence reactor performance. In addition, the productivity in these reactors may be limited by the need to maintain low upward velocities (e.g., low aeration or CO_2 evolution) to allow adequate cell sedimentation.

Mechanical separation

The operating principle behind centrifugation is the same as that of sedimentation; however, much higher settling velocities than in sedimentation may be obtained in the centrifugal field. Centrifugal separators enable high-volume continuous processing of fluids containing many particles, with short retention times and small space requirements. To determine the unhindered particle velocity in a centrifugal field (u_{0C}), equation 40 is multiplied by the centrifugal coefficient (C), also known as the G-value, which describes the increase in sedimentation rate due to centrifugation relative to gravitational settling:

$$u_{0C} = \frac{d_P^2(\rho_P - \rho_F)}{18\eta}\underbrace{\left(\frac{r\omega^2}{g}\right)}_{C} \quad ...(43)$$

where r is the radial distance from the axis of rotation and ω is the angular velocity.

Industrial centrifuges are most often classified by internal structure (e.g., disk stack, tubular bowl) and mode of operation (e.g., solids retaining, continuous or intermittent solids ejecting). The selection of sturdier construction and materials will enable higher rotation speeds for separation of smaller particles. The equation describing throughput in a centrifuge is analogous to equation 42, except that the centrifuge area is expressed using the Σ value, which is the area equivalent for a given centrifuge and rotation speed. Centrifuge manufacturers will often provide machine-specific Σ values, although the Σ value for simple disk-stack and tubular-bowl centrifuges can be calculated directly.

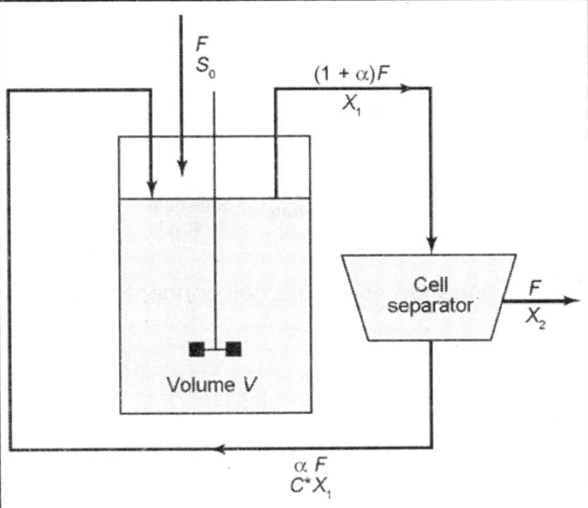

Figure 11.4 CSTR with recycle. Cell separation can be achieved through centrifugation, sedimentation, or filtration.

Filtration

Filtration is separation based on size, allowing retention of molecules larger than the pore size of the filter and passage of smaller molecules. Membrane filtration thus offers the twin benefits of cell retention and inhibitory by-product removal. In cell recycle systems, the most common arrangements are internal filters for cell and enzyme retention or external membrane filters in plate and frame, spiral cartridge, and hollow fiber configurations. In all these configurations, flow patterns tangential to the membrane surface can reduce fouling and improve the filtrate flux across the membrane.

Compared to internal filters, external filters have higher surface-area-to-volume ratios and may be easier to maintain; however, they may be less easily sterilized (particularly for some polymer membranes) and could introduce problems of nutrient depletion in the external recycle loop. Membrane selection depends primarily on the critical particle size, with other criteria being cost, mechanical stability, and susceptibility to plugging and fouling. Because membranes have the potential for complete cell recycle, a purge or bleed stream is typically split from the recycle stream to prevent accumulation of inert particles and debris in the vessel.

Materials for balanced growth

A material balance on cell mass for the CSTR with recycle system, neglecting cell death, may be written

$$\underbrace{F \cdot X_0}_{\text{Cells in feed}} + \underbrace{\alpha \cdot F \cdot C \cdot X_1}_{\text{Cells in recycle stream}} - \underbrace{(1+\alpha) \cdot F \cdot X_1}_{\text{Cell out}} + \underbrace{V \cdot \mu \cdot X_1}_{\text{Cell growth}} = \underbrace{\frac{d(V \cdot X_1)}{dt}}_{\text{Cell accumulation}} \quad \ldots(44)$$

where α is the recycle ratio equal to the recycle volume divided by the feed volume, C is the concentration factor (cell concentration in the recycle divided by the effluent concentration) related to the efficiency of the separation step, and X_0, X_1, and X_2 are the cell concentrations in the feed, recycle, and separator effluent streams, respectively. Note that the low substrate concentrations in waste treatment create suboptimal growth environments in which cell death cannot be neglected.

Assuming the system is at steady state ($dX_1/dt = 0$) and that the feed is sterile ($X_0 = 0$), equation 44 yields

$$\mu = (1 + \alpha - \alpha \cdot C) \cdot D \quad \ldots(45)$$

The dilution rate is no longer equal to the specific growth rate; in fact, because $C > 1$ and $\alpha < 1$, the dilution rate is greater than the specific growth rate. A material balance on the limiting substrate, again neglecting maintenance and product formation, may be written

$$\underbrace{F \cdot S_0}_{\text{Substrate in feed}} + \underbrace{\alpha \cdot F \cdot S}_{\text{Substrate in recycle}} - \underbrace{(1+\alpha) \cdot F \cdot S}_{\text{Substrate out}} - \underbrace{\frac{\mu \cdot X}{Y_{X/S}} \cdot V}_{\text{Substrate consumed}} = \underbrace{V \cdot \frac{dS}{dt}}_{\text{Substrate accumulation}} \quad \ldots(46)$$

Solving equation 46 for the cell concentration, assuming steady-state ($dS/dt = 0$) operation, yields

$$X_1 = \frac{Y_{X/S} \cdot (S_0 - S)}{[1 + \alpha(1 - C)]} \quad \ldots(47)$$

in which the steady-state cell concentration with recycle is greater than that in the simple CSTR by a factor of $1/[1 + \alpha(1 - C)]$. By adopting a suitable expression for cell growth, the substrate concentration can be determined. Using the Monod expression, as before, and solving for the substrate concentration gives

$$S = K_S \frac{\mu}{\mu_{max} - \mu} = K_S \frac{D(1 + \alpha - \alpha C)}{\mu_{max} - D(1 + \alpha - \alpha C)} \quad \ldots(48)$$

Substituting equation 48 into equation 47 yields

$$X_1 = \frac{Y_{X/S}}{(1+\alpha-\alpha C)}\left[S_0 - K_S\frac{D(1+\alpha-\alpha C)}{\mu_{max} - D(1+\alpha-\alpha C)}\right] \quad \ldots(49)$$

A material balance on cell mass around the separator gives the cell concentration in the outlet:

$$X_2 = (1+\alpha-\alpha\cdot C)\cdot X_1 \quad \ldots(50)$$

The higher biomass productivity of the recycle system results from a dilution rate higher than the specific growth rate and the increased cell concentration in the vessel. Implementation of a recycle system is often critical to the economic viability of processes using expensive biocatalysts (e.g., enzymes). Typically, this is accomplished by immobilizing the enzymes on inert support particles to facilitate either internal or external recycle. The potential use of cell recycle in an industrial process involves weighing the effectiveness and economics associated with the cell separation step against the marginal improvement in process performance. It should also be noted that higher cell densities exacerbate the oxygen transfer and heat removal burden of the system.

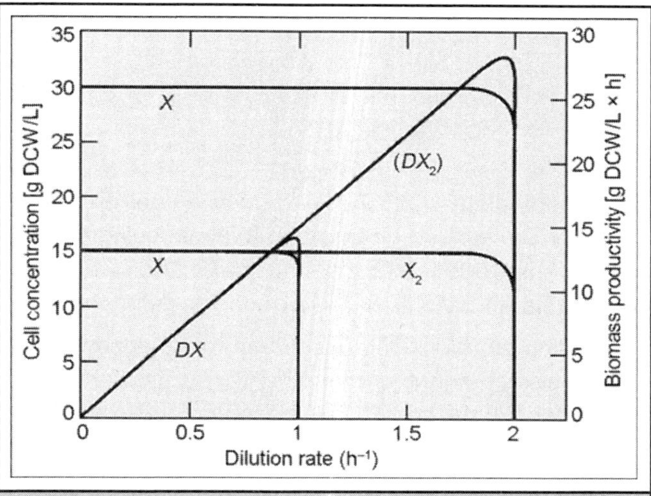

Figure 11.5 Comparison of steady-state behavior of a chemostat and a chemostat with recycle.

CSTRs in Series

In the single CSTR, the constant, controlled environment gives the advantage of being able to control the cellular–enzyme environment for maximum utility. Sometimes, however, a particular cell system will exhibit multiple properties of interest that can only be realized in different environments. The optimal environments for cell growth and product formation, for example, may be characterized by different temperatures, pH, and limiting nutrients. The configuration of CSTRs in series lends itself to those applications in which multiple environments are required.

Study of cellular growth

Bacterial growth in the presence of multiple carbon substrates often results in diauxic growth, in which cells preferentially metabolize a single substrate over all others. In a waste treatment application, the preferred substrate would be consumed by the microorganisms and the remaining substrates would pass through the system untreated. Configuring CSTRs in series provides a partitioning of cell metabolism so that less-favored substrates are consumed in subsequent stages. CSTRs in series have been used to improve recombinant protein fermentations in which performance is threatened by plasmid instability and lethal protein overproduction. Cells are grown to high density in the first stage without inducer so that plasmid-free cells have little growth advantage over plasmid-containing cells. Induction in the second stage results in higher productivity than the simple CSTR because the continuous introduction of plasmid-containing cells from the first stage reduces the ability of nonproductive cells to dominate the culture. Consider the two-stage system for biomass production, in which a separate feedstream can be added to the second stage. The steady-state material balances for cells and substrate in the first reactor are identical to the single CSTR case (equations 1 and 3), with steady-state solutions as equations 5 and 8 for Monod growth. Considering the case without the second feedstream, the material balances on cell mass and growth-limiting substrate in the second stage can be written as

$$\underbrace{\frac{F \cdot X_1}{V_2}}_{\text{Cells in}} - \underbrace{\frac{F \cdot X_2}{V_2}}_{\text{Cells out}} + \underbrace{\mu_2 \cdot X_2}_{\text{Cell growth}} = \underbrace{\frac{dX_2}{dt}}_{\text{Accumulation}} \qquad \ldots(51)$$

and

$$\underbrace{\frac{F \cdot S_1}{V_2}}_{\text{Substrate in}} - \underbrace{\frac{F \cdot S_2}{V_2}}_{\text{Substrate out}} - \underbrace{\frac{\mu_2 \cdot X_2}{Y_{X/S}}}_{\text{Consumption for growth}} = \underbrace{\frac{dS_2}{dt}}_{\text{Accumulation}} \qquad \ldots(52)$$

with steady-state solutions. A growth model on the limiting substrate must be adopted to further complete the system description. Minimal cell growth will occur in the second stage if no additional substrate is added, because the majority of substrate is consumed in the first stage. Thus, non-growth-related cellular behavior and product formation could be studied in the second CSTR.

Adding an additional feedstream to the second stage provides the opportunity to introduce more of the limiting nutrient, other nutrients required for growth or product formation, inducers, or inhibitors. The material balances on cell mass and substrate on the second stage can be written as equations 53 and 54, respectively. In equation 53, F' is the volumetric flow of the second feedstream, which is assumed to be sterile ($X' = 0$).

$$\underbrace{\frac{F_1}{V_2} X_1}_{\text{Cells in from stage 1}} - \underbrace{\frac{F_1 - F'}{V_2} X_2}_{\text{Cells out}} + \underbrace{\mu_2 \cdot X_2}_{\text{Cell growth}} = \underbrace{\frac{dX^2}{dt}}_{\text{Accumulation}} \qquad \ldots(53)$$

The dilution rate for the second stage is given by $D_2 = (F_1 + F')/V_2$, and the concentration of the limiting nutrient in the second feed is S_0' .

$$\underbrace{\frac{F_1}{V_2} S_1}_{\text{Substrate in from stage 1}} + \underbrace{\frac{F'}{V_2} S_0'}_{\text{Substrate in from second feed}} - \underbrace{\frac{F_1 + F'}{V_2} S_2}_{\text{Substrate out}} - \underbrace{\frac{\mu_2 \cdot X_2}{Y_{X/S}}}_{\text{Consumption for growth}} = \underbrace{\frac{dS_2}{dt}}_{\text{Accumulation}} \qquad \ldots(54)$$

Feeding additional substrate to the second stage allows for more growth to occur. In addition, the dilution rate in the second stage is larger than the maximum specific growth rate of the organism because the second stage has a continuous feed of cells.

Role of enzymes

Enzyme reactions may also be carried out in multiple CSTRs. The performance of CSTRs in series approaches that of a single PFR while maintaining the good mixing characteristics of the stirred-tank reactor. Considering a first-order enzymatic reaction for substrate conversion ($v = k \times S$), the reactor design equation for a single CSTR can be written as

$$\tau = \frac{S_0 - S}{k \cdot S} \quad \text{...(55)}$$

Assuming that there is no volume change upon reaction, the conversion in the single CSTR is given by

$$X = \frac{\tau \cdot k}{1 + \tau \cdot k} \quad \text{...(56)}$$

Then for a system of *n* CSTRs in series with equal volumes and reactor conditions (constant *k*) the conversion in the *n*th CSTR is given by equation 57

$$X_n = 1 - \frac{1}{(1 + \tau \cdot k)^n} . \quad \text{...(57)}$$

12

OXIDATIVE ENZYME

Cholesterol oxidase is the enzyme that catalyzes the oxidation of the 3β-hydroxy group of various kinds of steroids. The oxidation is the important step in the catabolism of the sterols. This catalytic property plays an important role in various bioprocesses The catalytic and structural properties of cholesterol oxidases from microbial origins are described. Further, some applications are described, including cholesterol determination in medical diagnosis, reduction of cholesterol content in the food industry, and pest control (based on its insecticidal activity) in the agriculture industry.

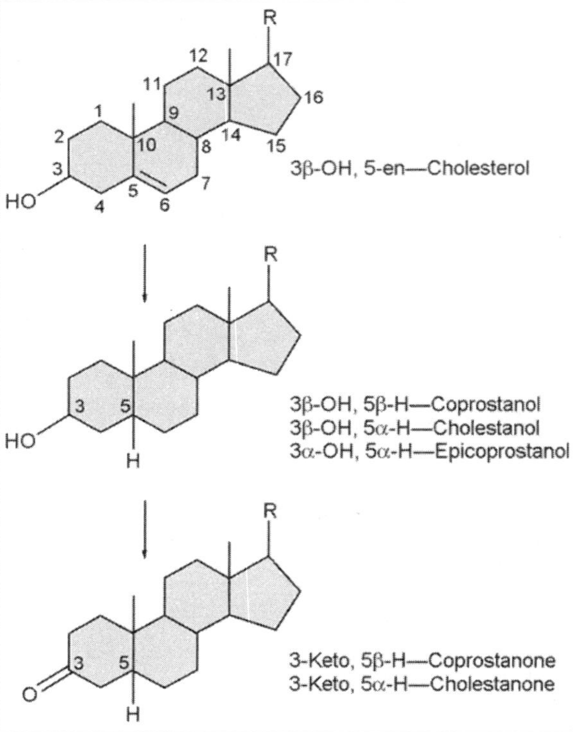

CHOLESTEROL

Cholesterol (cholest-5-en-3β-ol) is principal sterol of higher animals. Cholesterol is practically insoluble in water, but it is soluble in aqueous solutions of bile salts. It is found in all body tissues, especially the brain and spinal cord, and in animal fats or oils. Cholesterol is considered to be a key intermediate in the biosyntheses of fecal sterols, bile acids, and steroid hormones in animals. Turfitt first demonstrated the formation of 4-cholesten-3-one from cholesterol by *Proactiomyces erythropolis* (*Nocardia erythropolis*) and showed

Figure 12.1 Anaerobic metabolism of cholesterol.

that this oxidation of the 3β-hydroxy group to the keto group of cholesterol was the initial step in the aerobic catabolism of the sterol. On the other hand, in the intestinal tract of humans, cholesterol can be anaerobically metabolized by bacterial microflora to coprostanol (65%) and coprostanone (10%) and to a lesser extent to cholestanone and epicoprostanol.

ENZYME CHOLESTEROL OXIDASE

Cholesterol oxidase catalyzes the oxidation by dioxygen of the 3β-hydroxyl group of cholesterol and of related steroids (other 3β-hydroxy steroids) to a keto group and catalyzes the successive isomerization of the $\delta 5$ double bond to the conjugated position: $\delta 4$. Thus, in many cholesterol oxidases, both the oxidase and isomerase activities are catalyzed by the single enzyme protein, although the two kinds of enzymes, cholesterol dehydrogenase and $\delta 5$-3-ketosteroid isomerase are present in some microorganisms containing *Pseudomonas testosteroni* and *P. putida*. The oxidation is estimated by the following methods.

1. Conjugated enone formation by the increase in UV absorption
2. Monitoring oxygen uptake
3. Hydrogen peroxide production by the oxidative coupling of 4-aminoantipyrine with phenol by peroxidase to give a quinoneimine dye absorbing at 500 nm

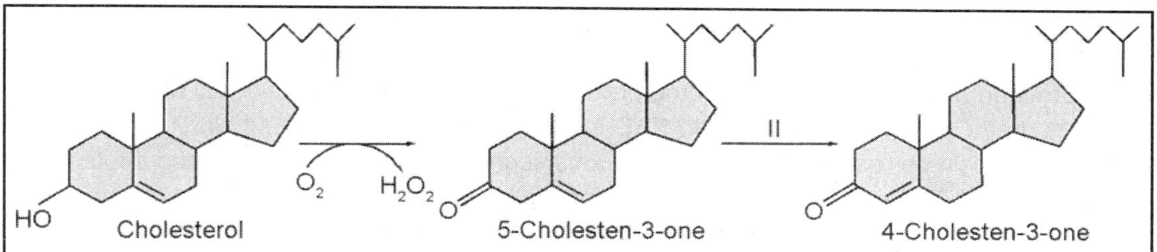

HO Cholesterol → (O_2 → H_2O_2) → 5-Cholesten-3-one → II → 4-Cholesten-3-one

Figure 12.2 Scheme for the cholesterol oxidase reaction. I, oxidation step; II, isomerization step.

Generally, the enzyme is a flavin-dependent oxidase containing one mole of tightly bound flavin adenine dinucleotide (FAD) per mole of protein. However, cholesterol oxidases from *Nocardia rhodochrous, Streptomyces griseocarneus,* and *Streptomyces violascens* lack FAD or any other cofactor. The enzyme has been isolated from a number of different soil bacteria that are capable of using cholesterol as the sole source of carbon and energy. They include Gram- positive bacteria such as *Arthrobacter simplex, Brevibacterium sterolicum, Corynebacterium sterolicum,* and *Rhodococcus equi*; Gram-negative bacteria such as *Pseudomonas* spp., actinomycetes such as *Actinomyces lavendulae, N. erythropolis, N. rhodochrous, Nocardia* spp., *Streptomyces griseocarneus,* and *Streptomyces violascens*; and fungi such as *Schizophyllum commune*.

There are also two types of enzymes, secreted and membrane bound, according to their localization. Commercial preparations of cholesterol oxidase can be obtained from the following suppliers: *Streptomyces* spp. (15 units/mg) from Toyobo Co.; *B. sterolicum* ATCC21387 (13 units/mg) from Kyowa Hakko Kogyo Co.; *Pseudomonas* spp. (3 units/mg) from Wako Pure Chemical Industries; *R. erythropolis* (25 units/mg) from Boehringer Mannheim; and *Cellulomonas* spp. (20–60 units/mg) and *Schizophyllum commune* (5 units/mg) from Sigma Chemicals. The enzymes from

B. sterolicum, Schizophyllum commune, and *Arthrobacter simplex* are purified, and their enzymatic properties have been examined. This enzyme can oxidize 3β-hydroxy groups in cholesterols, the biosynthetic precursors such as desmosterol, 17 ketosteroids such as dehydroepiandrosterone, and plant sterols such as stigmasterol, β-sitosterol, and campesterol. Of these, the length of the C17 side chain is especially an important factor in determining the reaction rate. On the other hand, 3α-hydroxysteroids, estrogens, and cholecalciferol do not serve as substrates.

Further, the relative activities are also dependent on the origin of the enzymes. Murooka et al. cloned the cholesterol oxidase gene from a *Streptomyces* sp. SA-COO and demonstrated high productivity of the extracellular cholesterol oxidase by using the *Streptomyces lividans* host-vector system. Ohta et al. cloned the cholesterol oxidase gene from *B. sterolicum*. The gene was also overexpressed into *Streptomyces lividans*. Both genes show identities of 64% in nucleotide sequence level and 58% in amino acid sequence level, respectively. The gene analysis shows that the mature enzyme consists of 503 or 507 amino acids and is preceded by a 42 or 45 amino acid signal sequence for *Streptomyces* sp. or *B. sterolicum*. There are at least six highly conserved regions between the two enzymes, which may play structurally or functionally important roles such as substrate binding or catalysis.

The FAD-binding domain in a number of FAD-dependent enzymes contains a pattern of three glycine residues, Gly-X-Gly-X-X-Gly, common to NAD and FAD-binding proteins. This pattern of residues is also found near the N-terminal of the sequences of cholesterol oxidases from *Streptomyces* sp. and *B. sterolicum*. The crystal structure of the enzyme from *B. sterolicum* has been determined using the method of isomorphous replacement and refined to 1.8-□ resolution. The structure is made up of two domains: an FAD-binding domain and a steroid-binding domain. The FAD-binding domain consists of three noncontinuous segments of sequence, including the N-terminal and the C-terminal. The overall topology of this domain is very similar to other FAD-binding proteins. The steroid-binding domain consists of two non- continuous segments of sequence and contains the roof of the active-site cavity. The active site lies at the interface of the two domains. The FAD lies on the floor of the cavity.

The steroid substrate is buried within the internal cavity. The hydroxyl group of the steroid is hydrogen bonded to the flavin ring system of the FAD. The active-site cavity contains one charged residue, Glu361. This residue is appropriately positioned to act as the proton acceptor in the proposed mechanism for the isomerization step. FAD-dependent cholesterol oxidase shares significant structural homology with another flavoenzyme, glucose oxidase, suggesting that it might also be a member of the glucose-methanol-choline (GMC) oxidoreductase family. A histidine residue is completely conserved in the known sequences of GMC oxidoreductases and may act as the general base in the oxidation reaction.

APPLICATION OF OXIDATIVE ENZYME

Clinical Approach

Cholesterol oxidase is used in the determination of serum cholesterol concentration for the assessment of arteriosclerosis and other lipid disorders and of the risk of thrombosis. Richmond and Flegg first showed the possibility of using cholesterol oxidase from microorganisms as a basis for a specific assay for serum cholesterol. In the procedure, 4-cholesten-3-one produced was assayed by its UV absorption after solvent extraction of the reaction mixture. However, their procedures were tedious and time consuming.

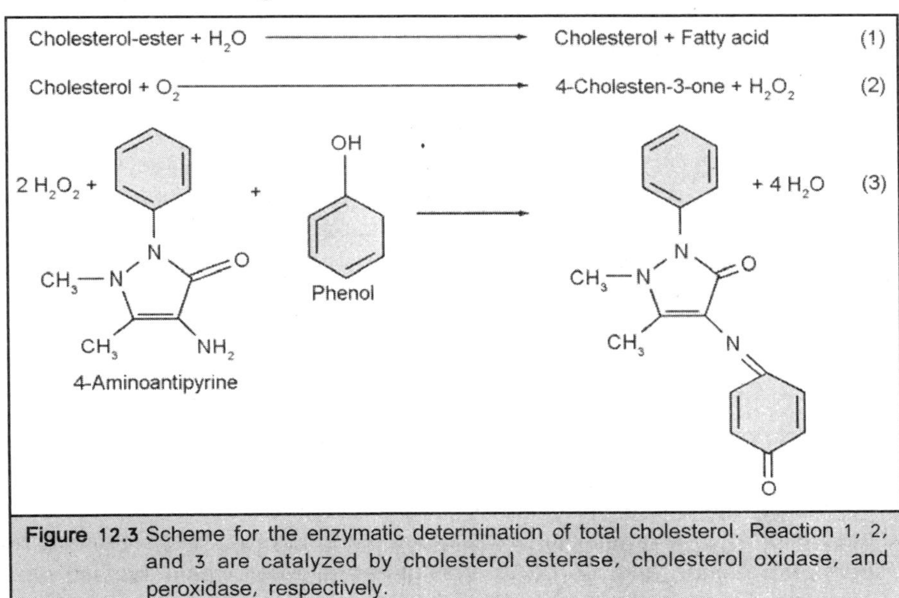

Figure 12.3 Scheme for the enzymatic determination of total cholesterol. Reaction 1, 2, and 3 are catalyzed by cholesterol esterase, cholesterol oxidase, and peroxidase, respectively.

Allain et al. developed the excellent method in which the hydrogen peroxide produced during the cholesterol oxidase reaction was oxidatively coupled with 4-aminoantipyrine and phenol by peroxidase to form a quinoneimine dye. First, cholesterol esters are able to be converted to free cholesterol by the action of cholesterol esterase. Then, the liberated cholesterol can be determined by cholesterol oxidase and peroxidase. This enzymatic method is simple and specific under extremely mild conditions. The results with serum samples agreed very well with those obtained by chemical procedures based on the Liebermann-Burchard reaction. This assay system is marketed as a serum cholesterol determination kit; worldwide annual sales amount to $100 to 200 million.

Dietary Value

Dietary cholesterol has been implicated as one of the causes of coronary artery disease. Because of this, reducing dietary cholesterol is said to cause a gradual reduction in serum cholesterol and regression of atherosclerotic plaque. The National Cholesterol Education Program coordinating board recommends a daily cholesterol intake in the American diet of 300 mg or less. As a result, reduction of cholesterol in food items relatively rich in cholesterol (dairy foods, eggs) has become a challenging problem. Some organisms have been shown to degrade cholesterol, as described earlier, but only a few do so without accumulating any steroid intermediates.

Watanabe et al. described the degradation of cholesterol in egg yolk and lard by crude cholesterol oxidase in *R. equi* no. 23 and the bacterial cell suspension. *R. equi* no. 23 degraded cholesterol with little or no accumulation of steroid intermediates such as 4-cholesten-3-one and 1,4-androstadiene3,17-dione, which are recognized as key intermediates in a comprehensive pathway of cholesterol degradation. A biotechnological approach to reducing the cholesterol content of milk involves the genetic engineering of starter culture bacteria (lactococci, lactobacilli, streptococci), leading to the expression of phenotypic traits that are related to cholesterol metabolism. These cultures are then used in the production of fermented dairy foods (yogurt and cheeses).

Somkuti et al. have investigated the possibility of introducing the cholesterol oxidase gene from *Streptomyces* sp. SA-COO into a common dairy starter culture bacteria such as *Streptococcus thermophilus* and *Lactobacillus casai.* The biosynthesis of a functional cholesterol oxidase in *Streptococcus thermophilus* was confirmed by TLC analysis of reaction products after incubation of sonically disrupted cells with cholesterol. However, unlike *Streptomyces lividans, Streptococcus thermophilus* did not secrete cholesterol oxidase into the culture medium. At present, the use of cholesterol-degrading enzymes in the food industry remains unrealistic because of the lack of toxicological data on enzyme extracts and the intermediate products of cholesterol degradation.

Role in Transgenic Crops

The discovery of novel insecticidal proteins is vital for pest control strategies using transgenic crops. Filtrates from several thousand microbial fermentations were assayed for insecticidal activity against crop-damaging insects. Two *Streptomyces*-culture filtrates killed boll weevil larvae in feeding studies, and the bioactive component was identified as cholesterol oxidase. The enzyme was active in a dose-dependent manner, and the LC50 against neonate boll weevil larvae was 20.9 µg/mL after 6 days in diet-incorporation bioassay. *Bacillus thuringensis* subsp. Kurstaki protein has an LC50 of 1 µg/mL against tobacco budworm and 37 µg/mL against European corn borer in similar diet-incorporation bioassay. Thus, cholesterol oxidase is active against the boll weevil in the same concentration range that *B. thuringiensis* protein is active against lepidopterans. Morphological changes induced by ingestion of cholesterol oxidase suggest that the enzyme has a direct effect on the midgut tissue of boll weevil larvae. Thus, cholesterol oxidase disrupted the midgut epithelium at low doses and lysed the midgut cells at higher doses. The cholesterol oxidase gene from *Streptomyces* spp. was expressed in tobacco and tomato, and successful control of boll weevil was achieved by the Monsanto company.

13

BIOPROCESS SEPARATION

The design of a process for the large-scale purification of biological molecules is one of the several complex tasks that must be successfully completed to bring a product to market. Most downstream processes include some chromatography steps as well as other conventional separation methods. Chromatography is generally recognized as the technique that enables the highest degree of purification of biomolecules. For example, the U.S. Food and Drug Administration (FDA) recommends that at least two chromatography steps be used in the purification of therapeutic-grade monoclonal antibodies. Chromatography separations are sophisticated and can be expensive processes, which explains the continuous development of effective approaches by scientists and engineers to improve the productivity of chromatography technologies.

Fluidized bed adsorption, various column technologies, computerized optimization strategies, and continuous and semicontinuous processes are some of the recent approaches that have been proposed. However, one of the most important aspects of chromatography technology is the choice of matrix. In this article, we examine the use of inorganic rigid matrices in chromatography separation processes of biological molecules, with emphasis on the preparation, properties, and applications of controlled pore glass (CPG).

METHODOLOGY

The subject of matrices for process chromatography and the requirements of an ideal matrix have been reviewed by many authors. In process-scale purification, the important characteristics of a matrix are chemical and physical stability, rigidity of the matrix to withstand high flow rates and differing hydrodynamic pressure, uniform porosity, and high surface area. In addition, for most chromatography applications, surface modification of the matrix will be necessary to facilitate ligand immobilization, which requires the presence of suitable functional groups on the matrix. Finally, the cost and commercial availability of the matrix are also important considerations in selecting a matrix. The majority of matrices used in chromatography processes can be divided into two categories: natural and synthetic. Natural matrices include glass, silica, ceramic, alumina, agarose, chitosan, and cellulose. Synthetic matrices are prepared by polymerization of functional monomers. The incorporation of monomers with suitable functional groups can provide activation sites on these supports for ligand attachments. There are also the composite matrices, which are mixed networks constituted of at least two components. Generally, one component acts as a rigid skeleton, and the other component has active groups that interact with the biological molecule to be separated. Dextran-coated silica and polyacrylamide-agarose are examples of composite matrices that have been applied to the separation of biological molecules.

CHROMATOGRAPHIC SEPARATION

A large selection of inorganic matrices are commercially available for incorporating homogenous catalysts, and some have been applied to chromatography separations. These matrices include silica, alumina, glass, zeolite, clay, celite, zerconia, titania, magnesium oxide, and carbon. Only a limited number of these matrices with suitable pore structure, particle size, and mechanical stability can be used for chromatography separation of biomolecules. Silica, alumina, and glass are available with a large variety of surface areas, pore structure, and particle sizes, and some have been used extensively in both analytical and preparative chromatography of biological molecules.

Controlled Pore Glass

Controlled pore glass (CPG) has excellent mechanical stability and can be manufactured with a wide range of pore diameters, a high surface area, and a range of particle sizes. The material is almost pure silica and has a very narrow pore size distribution. Surface-modified CPG with the trade name PROSEP manufactured by Bioprocessing Limited has been extensively used in large-scale purification of biomolecules.

Silica Matrices

Silica is extensively used in the analytical field (HPLC) and to a much lesser extent for preparative chromatography of biomolecules. Silica matrices for chromatography applications are generally produced by a wet process, that is, by acidification of sodium silicate in water, and are available in particle sizes of up to 40 μm and pore sizes of up to 30 nm. In most cases, these materials are unsuitable for large-scale purification processes. The small particle size causes excessive back-pressure in preparative chromatography columns, and the pore size of 30 nm is not large enough for the separation of most proteins, especially in the field of affinity chromatography, where both the ligand and the ligate could be large proteins.

In our laboratories, we have used Matrex silica (Amicon) for the large-scale purification of biomolecules. This material is available in a range of pore diameters and particle sizes and is manufactured by first forming a hydrosol from organosilicic acid. Hydrosol is then allowed to set to a jelly-like mass called hydrogel. After setting, the hydrogel is thoroughly washed to remove salts resulting from the reaction. Temperature and pH of the washing medium and the washing time influence the specific surface area, pore volume, and purity of the hydrogel. The hydrogel is subsequently dried to give a structurally stable xerogel. This material is 99.5% silica, with impurities including Na^+, Ca^{2+}, Fe^{3+}, SO_4^{2-}, and Cl^- ranging from 600 to 60 ppm.

Table 13.1 Properties of matrex silica

Nominal pore diameter (nm)	*Surface area (m^2/g)*	*Pore volume (ml/g)*	*Particle size range (μm)*
6	540	0.93	20–500
10	320	1.20	45–300
25	160	1.10	45–70
50	90	1.10	50
100	45	1.10	50–130
150	25	1.10	50

Alumina

Alumina-based materials have a much wider pH stability than silica matrices and therefore are of great interest in chemical separations and industrial-scale chromatography. Commercially available alumina matrices are all aluminium oxides prepared by thermal dehydration of aluminum hydroxide. The origin of the starting material together with other factors, such as the heating rate, drying time, and temperature, influence the properties of the final product. Alumina matrices can be obtained in a wide range of particles and pore sizes. The physical stability of alumina particles in chromatography columns, even under moderate flow rates and pressure, has been a major problem in our laboratories. Generally, the particles break up and cause a substantial increase in the back-pressure generated by the column. Furthermore, any risk of contamination of the purified product with aluminum may not be acceptable to the pharmaceutical and biotechnology industries, particularly when the reports relating to Alzheimer's disease to aluminum are considered.

Table 13.2 Properties of alumina matrix

Nominal pore diameter (nm)	*Surface area (m^2/g)*	*Particle size range (μm)*
6	120–150	60–120
10	100	60–120
15	75	60–120
30	80	40
50	50	40
100	8	40
180	6–8	20

Zirconia and Titania Based Matrices

Recently, there has been an increasing interest in zirconia and titania-based matrices in chromatography applications, mainly because of their chemical stability over a relatively wide pH range and their physical characteristics. Zirconia or titania supports with properties suitable for preparative chromatography or HPLC are not commercially available, although zirconia and titania powders can be obtained commercially from a number of sources and have been used by researchers to prepare matrices with a small particle size (1-10 μm) and pore sizes of up to 100 nm, suitable for HPLC applications. More recently, larger porous zirconia particles (30-150 μm) have been prepared and used in packed and fluidized beds for protein purification. Zirconia and titania as base stable matrices not only enable separation of biomolecules at high pH values, but they can also be sanitized with alkaline media, which is the standard sanitization procedure in downstream processing of biological molecules. Various adsorbents such as affinity, ion-exchange, and reversed phase can be prepared from these matrices for protein purification by either polymer coating or reaction with organosilanes.

Calcium Phosphate

Hydroxyapatite is a special crystalline form of calcium phosphate that has been used in the separation of biological molecules. The adsorbent consists of a mosaic of crystals that result from a combination of phosphate and calcium salts after a number of steps of wet alkaline treatment and heat aging. The application of hydroxyapatite to the field of protein separation was first introduced by Tiselius et al. The standard procedures for the operation of hydroxyapatite columns were developed by Bernardi and Gorbunoff using a number of different proteins in several solvent systems, Bernardi also proposed a mechanism for protein adsorption to, and desorption from, hydroxyapatite. Proteins bind reversibly to hydroxyapatite by a complex ionic mechanism different from the principle of ion- exchange chromatography. Both the amino and carboxyl groups on the protein molecules act in the adsorption mechanism.

The adsorption of proteins to hydroxyapatite via the amino groups is the result of their positive charges and the negative charge on the adsorbent when the column is equilibrated in phosphate buffer. The surface of hydroxyapatite crystals consists of sites with both positive (calcium) and negative (phosphate) charges. When the hydroxyapatite columns are equilibrated in phosphate buffers, normally at pH 6.8, the surface of the adsorbent can be regarded as negative because of partial neutralization of the positive calcium loci by phosphate ions. The carboxyl groups act in two ways. First, they are repelled electrostatically from the negative charge of the adsorbent; second, they bind specifically by complex formation with the calcium sites. The elution of basic proteins from hydroxyapatite is generally carried out by specific displacement with Ca^{2+} and Mg^{2+} ions that neutralize the negative charges of the hydroxyapatite phosphate sites by complex formation.

Acidic proteins are eluted by displacement of their carboxyl groups from hydroxyapatite calcium sites by ions, such as fluoride or phosphate, which form stronger complexes. The difficulty with using hydroxyapatite in medium- and large-scale purification of biological molecules is the physical nature of the material. The crystals are highly fragile, generating small particles resulting in column clogging and precluding any long-term usage of the column. An improved form of hydroxyapatite crystals has recently been prepared and is commercially available from a number of sources, including Bio-Rad. The new ceramic material from Bio-Rad is spherical and has been reported to be stable both chemically and physically.

PREPARATION OF CPG

The method used for the preparation of CPG was originally invented by Hood and Nordberg. They discovered that by heat treating a certain sodium borosilicate glass composition, the glass separated into two intermingled and continuous glassy phases. One phase, rich in boric oxide, was soluble in acids; the other phase was high in silica and stable toward acid solutions. This process was used by the inventors to prepare nonporous silica particles by first removing the boron phase with an acid solution to obtain porous particles and then revitrifying this material by heating to 900°C or above to yield transparent homogenous nonporous particles having a composition of greater than 95% silica. The porous glass material obtained by this method typically had an average pore diameter of about 4 to 5 nm.

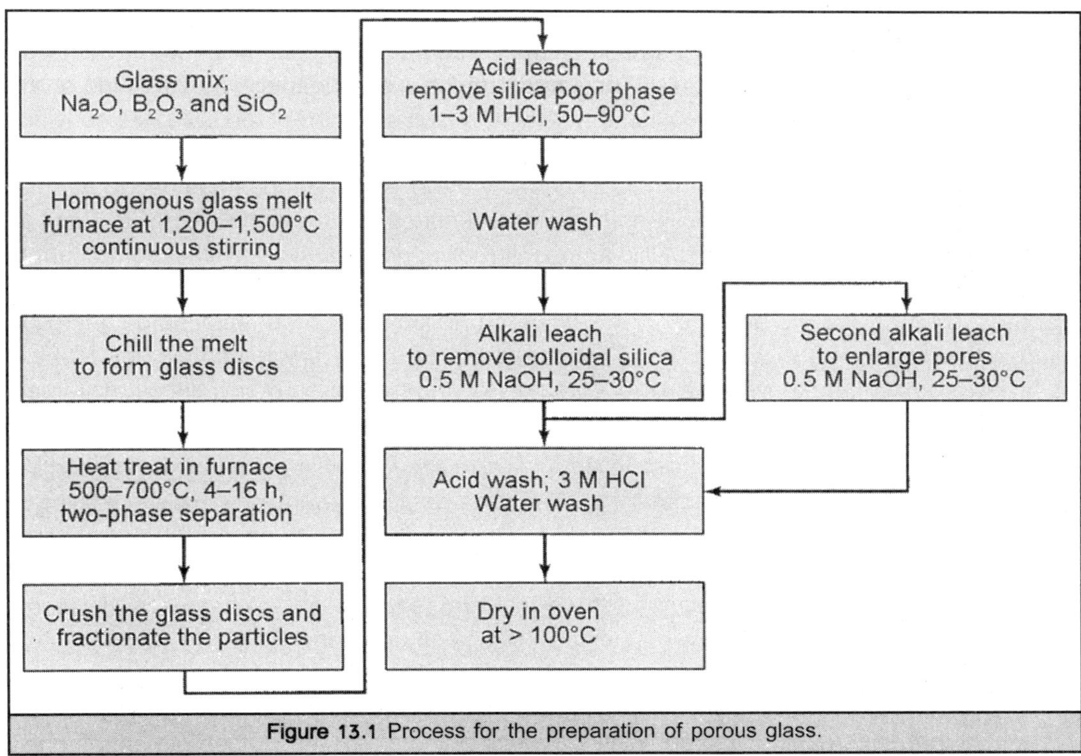

Figure 13.1 Process for the preparation of porous glass.

A procedure for enlarging the pore size to approximately 20 nm was invented by Chapman and Elmer. This was carried out by impregnating a porous glass body with an aqueous solution of a weakly reactive fluorine-containing compound, reacting the compound in situ with a mineral acid to release hydrofluoric acid at a temperature sufficient to dissolve a portion of the glass body and washing the body to remove the soluble constituents.

Ammonium bifluoride and nitric acid at 95°C were applied to a porous glass disc to enlarge the pore size from 4 to 24 nm. The process for making porous glass was modified by Haller to produce the material with a range of controllable pore sizes. This was achieved by varying the temperature and duration of the heat treatment to obtain porous glass with pore sizes from 12 to 250 nm. Larger pore sizes were also obtained by the addition of a sodium hydroxide leaching process.

Industrial Production

CPG can be commercially obtained from a number of sources, and the manufacturing processes are all based on the process invented by Haller. The composition of the starting metal oxides mixture and the exact conditions used for the heat treatment and leaching of the glass, which influence the quality and physical properties of the porous glass, are the propriety technology of the manufacturing companies.

CPG is manufactured in a range of particle sizes, 40 to 200 μm, and different pore sizes, 7 to 300 nm. The manufacturing batch sizes are generally much smaller than silica particles, possibly because of the heat treatment process. The largest batch size commercially available is in the region of 50 L (20 kg).

Table 13.3 Properties of controlled pore glass

Nominal pore diameter (nm)	*Surface area (m^2/g); pore volume = 1.0 cm^3/g*	*Surface area (m^2/g); pore volume = 0.9 cm^3/g*
7.5	340	240
12	210	150
17	150	100
24	110	75
35	75	50
50	50	35
70	36	25
100	25	18
140	18	13
200	13	9
300	9	7

Properties

Most commercially available CPG granules are irregularly shaped and highly porous. CPG has a narrow pore size distribution, with 80 to 90% of the pores showing a deviation less than ±10% from the nominal pore diameter. CPG has a very high surface area that varies according to the pore diameter and pore volume. The surface area increases with increasing pore volume and decreases with increasing pore diameter. CPG is a rigid insoluble matrix, compatible with most reagents used in the purification of biological molecules. Because CPG is essentially pure silica, it generally exhibits a high degree of solubility in alkali solutions. The solubility of CPG is a function of temperature, time, and pH of the solution.

The surface of native CPG is slightly hydrophobic and contains acidic hydroxyl groups that could denature biological molecules and cause strong, often irreversible binding of proteins. For most separation applications, surface modification of CPG with organosilanes is necessary, although native CPG has been used for the purification of a number of biological molecules, such as factor VIII. Surface modification also enhances the stability and durability of CPG.

Effect of Organosilane Compounds

Siliceous materials such as CPG can be reacted with the organosilane compounds in a number of different ways to obtain the desired bonding through hydrolyzable groups of the organosilane. Depending on the type of organosilane, the reaction can be performed in organic, aqueous, or vapour phase, resulting in matrices suitable for different applications. The majority of the work cited on this subject relates to the surface modification of silica material, which is not always applicable to CPG because of the difference in the surface characteristics of the two matrices. Surface modification of CPG is carried out by treating the material with an organosilane containing an organic functional group at one end and a silylalkoxy group at the other.

Attachment of the silane compound to the matrix is through the surface silanol or oxide groups to the silylakoxy groups with polymerization between the adjacent silanes. The result is an inorganic matrix with available organic functional groups that can be used in the preparation of chromatography adsorbents. A large number of different organosilanes are commercially available with functional groups such as epoxy, amine, and sulfhydryl. The first step in the surface modification of porous glass is to clean the surface by washing the matrix in nitric acid at elevated temperature. The matrix is then washed with water and dried in the oven at temperatures not exceeding 180°C.

Table 13.4 Surface modification of CPG

Organosilane	*Reactant*	*Activated CPG*
3-Aminopropyltriethoxysilane	—	CPG-amine
3-Aminopropyltriethoxysilane	Glutaraldehyde	CPG-aldehyde
3-Aminopropyltriethoxysilane	Succinic anhydride	CPG-carboxyl
3-Glycidoxypropyltrimethoxysilane	—	CPG-epoxy
3-Glycidoxypropyltrimethoxysilane	Acid solution	CPG-diol
3-Glycidoxypropyltrimethoxysilane	Sodium periodate	CPG-aldehyde
Mercaptopropyltrimethoxysilane	—	CPG-sulfyhyryl

The dry matrix is reacted with the silane compound dissolved in water or an organic solvent at temperatures ranging from 40°C to 120°C, depending on the solvent. After removing the excess silane by washing the matrix with suitable solvents or water, the functional groups on the matrix can be further modified to obtain various adsorbents. The exact conditions used in the surface modification of CPG will depend on the nature of the silane compound and the application of the resulting adsorbent. Extensive optimization of the surface modification procedure is usually required to obtain a suitable adsorbent for a specific application. At Bioprocessing, special techniques for preparing surface modified CPG (PROSEP) have been developed for applications in affinity and other types of separation processes. PROSEP has been designed for effective immobilization of both large and small ligands, resulting in affinity adsorbents with a high capacity for the target molecules, minimum nonspecific binding of contaminating proteins, and low ligand leakage.

Functions

CPG has been effective in the separation of a range of biological molecules, using a variety of chromatography techniques that include adsorption, size exclusion, affinity, and ion exchange. It has also been used extensively in the field of enzyme immobilization for the modification of biological molecules. The rigidity and incompressibility of the matrix allows it to be used in large columns at very high flow rates generating low back-pressure, which makes this matrix especially useful for processing large volumes of feedstock in short cycle times. Application of native CPG in adsorption chromatography has been generally limited because of its denaturing surface characteristics and irreversible binding of proteins. A number of proteins, such as factor VIII, have been purified from plasma by first blocking the surface of CPG with albumin.

Table 13.5 Application of surface-modified CPG (PROSEP) in separation processes

Product	*Trade name*	*Application*
Protein A adsorbent	PROSEP-A High Capacity	Purification of IgG from all species and subclasses except rat IgG
Protein G adsorbent	PROSEP-G	Purification of IgG from all species and subclasses
Heparin adsorbent	PROSEP-Heparin	Purification of antithrombin III and blood clothing factors
Lysine adsorbent	PROSEP-Lysine	Purification ofplasminogen and tPA
Metal-chelating adsorbent	PROSEP-Chelating	A metal-binding adsorbent used in purification of metal-binding proteins and metal removals from dilute solutions
Thiophilic adsorbent	PROSEP-Thiosorb	Purification of antibodies
Immobilized antibodies	—	Purification of specific antigens and second antibodies

The use of CPG in ion exchange chromatography applications has also been very limited because of the instability of the matrix in sodium hydroxide solution, which is the standard cleaning and sanitization solution for ion exchange adsorbents. Furthermore, ion exchange adsorbents prepared with CPG as the matrix have generally shown low functional capacity when compared to the polymetric matrices. Nevertheless, some applications of surface-modified CPG in ion exchange chromatography have been reported. The use of CPG in size-exclusion chromatography has generally been for the separation of macromolecules such as viruses, cell components, and polymers. Separation of these species from low molecular weight contaminants has been achieved by taking advantage of the very sharp exclusion limits and narrow pore size distribution of CPG. Generally, surface-modified CPG has been used for size-exclusion chromatography to eliminate the problem of nonspecific and irreversible binding of biological molecules to native CPG. The surface of the matrix is modified with glycidoxypropyl-trimethoxysilane, converting the solid phase to diol-CPG. This is a nonionic, hydrophilic coating compatible with aqueous and most solvent systems.

Surface-modified CPG has been widely applied in the area of affinity chromatography for the purification of biological molecules. Biologically active molecules will often bind to other molecules in a reversible and highly specific manner. The formation of these stable, specific, and dissociable complexes forms the basis of this powerful separation technique. Affinity adsorbents prepared using PROSEP and other surface-modified CPG have been used in many applications for the purification of biological molecules, such as antibodies and enzymes. PROSEP adsorbents have been used in packed-bed chromatography columns and in fluidized or expanded bed mode for the direct capture of biological molecules from feedstocks containing particulate or colloidal solids.

CONCLUDING REMARKS

The progress of inorganic matrices such as silica and CPG as adsorbents in the downstream processing of biological molecules is hampered by their instability toward alkaline solutions. Sodium hydroxide solution up to 0.5 M concentration is the standard cleaning and sanitizing reagent for chromatography adsorbents used in the separation processes of biomolecules. Furthermore, certain chemistries used in the surface modification of matrices and some purification protocols require alkaline conditions. Alkaline-resistant inorganic matrices are under development using different approaches. Development of zirconia and titania matrices by sol-gel techniques have already been mentioned. Surface coating of silica and CPG with an inert metal oxide such as zirconium oxide and preparation of glass–ceramic matrices may result in suitable products that satisfy all the requirements of an ideal matrix for the purification of biomolecules. Methods for coating silica matrices with zirconium oxide were studied by Meijers et al. and applied to CPG at Bioprocessing.

Great improvement in the stability of the coated CPG to alkaline solutions were observed; however, total stability has not yet been achieved. A preferred approach for making alkaline-stable inorganic matrices is the glass–ceramic route. Porous glass– ceramic with a skeleton of $CaTi_4(PO_4)_6$ crystals has been prepared by controlled crystallization of glasses in the $CaO\text{-}TiO_2\text{-}P_2O_5\text{-}Na_2O$ system and subsequent acid leaching of the resulting dense glass–ceramic. The medium pore diameter has been reported to be controllable in the range of 40 to 150 nm by changing the temperature of the heat treatment. Such an approach is being examined at Bioprocessing in collaboration with a glass–ceramic research center for developing alkaline stable inorganic matrices suitable for downstream processing of biological molecules.

14

TECHNOLOGY OF BIOREMEDIATION

As a result of the manufacture, use, spillage, and improper disposal of anthropogenic chemicals and their waste products during past decades, toxic chemicals have become ubiquitous contaminants of soils and waters around the globe. These compounds include everything from petroleum and synthetic organic chemicals to heavy metals and radionuclides. Governments throughout the world now require that the worst of these contamination problems be cleaned up to eliminate, insofar as possible, their hazards to human health and the environment. Traditional methods such as excavation and incineration of contaminated soil or pumping and aboveground treatment of groundwater are proving to be either too expensive, politically unpopular, or both. Thus, innovative technologies are being sought to address the needs of the environmental restoration industry. *Bioremediation* is such a technology.

Bioremediation involves the use of microorganisms, plants, or products yielded by living organisms to degrade, detoxify, or permanently sequester toxic chemicals present in environmental systems. Bioremediation can be used either ex situ or in situ to treat soil, sediment, sludge, water, or air. In ex situ bioremediation procedures, contaminated materials are first removed from the polluted site and then treated, whereas in situ techniques are used to treat contaminants in place without relocating them to a treatment facility.

Bioremediation can be promoted by augmenting natural systems with exogenous biological materials or by stimulating indigenous contaminant-transforming microbial populations. Natural systems can be augmented by the introduction of bacteria that degrade the targeted contaminants or plants that sequester pollutants. Native microbial populations are stimulated by adding nutrients or modifying environmental conditions to favor the growth and activity of appropriate organisms. Although intrinsic bioremediation, or microbial processes that occur without human intervention, may ultimately be able to clean up a contaminated site, a great variety of techniques are used by the bioremediation industry to enhance and accelerate degradative processes.

MECHANISM

Treatment of Contaminated Materials

Bioreactors, or vessels of various configurations containing degradative microbes, are a common means of ex situ treatment of contaminated materials, typically soil that has been excavated from a polluted site or groundwater that has been pumped from a polluted aquifer. A bioreactor may contain, for example, a slurry of soil and water to which nutrients have been added, such as sources of nitrogen, phosphorus, and/or carbon that support the growth and metabolism of degradative microorganisms. Sometimes the microorganisms themselves may have to be added. Slurry reactors can be run either aerobically, by stirring and mixing in air or pure oxygen, or anoxically, with little or no mixing. Slurry phase reactor treatment cells can be designed to hold a few thousand to millions of gallons. Aeration systems for aerobic slurry reactors must be able to maintain dissolved oxygen concentrations at 2.0 mg/L or higher. Aeration is a major expense for such systems. To increase oxygen transfer rates, such systems may be sparged with pure oxygen. Often reactor designs include a series of eductors to oxygenate the mixed liquor while minimizing air emissions.

Anaerobic slurry systems usually require some mixing devices to evenly distribute soil or sediment materials throughout the aqueous phase, but they spare the need and cost of transferring large amounts of oxygen to the solution phase. Aerobic slurry reactors have been used to treat materials contaminated with chemicals such as polynuclear aromatic hydrocarbons (PAHs) and chlorinated phenols. Anaerobic slurry reactors have been used to treat materials polluted by compounds such as chlorinated solvents, polychlorinated biphenyls (PCBs), nitrated munitions like 2,4,6-trinitrotoluene (TNT), and nitrated herbicides such as dinoseb. Water may be treated in simple batch reactors that hold a few hundred to many thousands of liters; nutrients and microorganisms may be added as required.

Fluidized beds that contain biomass-bearing carriers are popular for water treatment and can be run aerobically or anaerobically, depending on the contaminants being treated. Even normally recalcitrant compounds such as aniline and nitrobenzene can be treated in such systems. Alternatively, biomass can be grown on a supporting matrix that is packed into a vessel such as a column, and the water to be treated can be passed through the packed bed to remove contaminants by degradation or absorption. Packed beds can also be used to treat vapor-phase contaminants such as chlorinated solvents, which are sparged through the reactor as a separate gaseous stream. A novel, mechanically mixed packed-bed reactor that works as a biotrickling filter is a variation on packed-bed technology for treatment of vapor-phase contaminants. Another variation on reactor design for water treatment is the up-flow sludge blanket reactor using granules containing anaerobic bacteria acclimated to the degradation of a particular waste stream. Bioreactors can be used in treatment trains that vary conditions, for example, by alternating anaerobic and aerobic conditions, so that mixtures of contaminants or highly chlorinated compounds can be treated successfully.

Treatment for Vapor-phase Contamination

Biofiltration is an ex situ treatment for vapor-phase contaminants. Typically, it involves passing contaminated air through a bed of soil, peat, or compost that has been preacclimated to develop a microbial population that completely degrades the targeted pollutants. This type of technology works well for volatile components of petroleum, and for many years it has been used in Europe for odor control. Biofilters have been developed for toxic airborne industrial contaminants as diverse as styrene, pentane and isobutane mixtures, toluene, chlorinated benzenes, dimethylsulfide, and volatile organic compounds (VOCs). For example, a biofiltration process was developed for styrene-containing industrial off gases. This process used peat as filter material. Styrene reduction after 190 days of operation was 70% to 98%.

The styrene elimination capacity was 12 g m^{-3} h^{-1} (maximum, 30 g m^{-3} h^{-1}). Degradation required addition of nutrients to the peat, adjustment of the pH, and control of humidity. The optimum operation temperature was about 23°C, but the styrene removal was still satisfactory at 12°C. Composting of leaves and alfalfa was used for the treatment of toluene-contaminated air in a biofilter. During the thermophilic phase (45–55°C), toluene biodegradation rates reached 110 g toluene m^{-3} h^{-1} at an inlet concentration of about 5 g/m^3 and a gas residence time of 90 s. Removal rates by biofilters for chlorinated benzenes reached 300 g solvent vapor h^{-1} m^{-3} of filter volume.

Role of Oxygen

The biodegradation of many contaminants in groundwater is limited by the shortage of electron acceptors for oxidative processes. The most effective electron acceptor for the bioremediation of many contaminants is oxygen, which is present in groundwater at only 8–10 mg/L even under optimal conditions. Air sparging is an in situ technology for introducing oxygen into contaminated aquifers. Air or pure oxygen is injected into the aquifer through sparging wells or infiltration galleries. Potential problems associated with the introduction of oxygen into groundwater include biofouling, in which injection wells become plugged with biomass, and obstruction of well screens or subsurface geologic formations by iron complexes that precipitate as a result of the oxidation of Fe(II) to Fe(III). Hydrogen peroxide, which decomposes in the presence of iron or biomass to produce oxygen, is sometimes substituted for air or pure oxygen.

Even when oxygen is supplied by sparging, it can be used so rapidly that its supply can still be rate limiting for bioremediation processes. Thus, the bioremediation industry has developed oxygen-releasing compounds (ORCs) composed of materials such as magnesium peroxide that allow controlled, long-term release of oxygen when introduced into boreholes as a slurry. Most successes with air sparging, as with many other bioremediation systems, have been reported in petroleum-contaminated aquifers.

The reintroduction of indigenous microorganisms isolated from a contaminated site after culturing is one approach to in situ hydrocarbon bioremediation. It is effective especially when microorganism growth is supplemented by addition of oxygen and nitrogen. At the Moffett Field (Sunnyvale, Calif.) groundwater test site, in situ removal of *cis*-dichloroethene (*c*-DCE) and trichloroethene (TCE) coincided with biostimulation through phenol and oxygen injection and utilization, with c-DCE removed more rapidly than TCE. Greater TCE and c-DCE removal was observed when the phenol concentration was increased. More than 90% removal of *c*-DCE and TCE was observed in a 2-m biostimulated zone. Thus, introduction of oxygen into contaminated aquifers can be of use even for recalcitrant compounds such as VOCs.

Bioventing System

Bioventing is related to air sparging in that the objective of this in situ technology is to provide oxygen to the microbial populations in the subsurface. Bioventing systems use equipment similar to that used in soil vapor extraction (SVE) processes. Air injection or extraction wells or trenches are connected by pipes to blowers or vacuum pumps which either push or pull air through the vadose zone (unsaturated soil horizons) to stimulate microbial populations capable of degrading contaminants. The objective is to provide sufficient oxygen to the subsurface to promote in situ contaminant degradation, but not so much that volatile contaminants are released undegraded to the atmosphere. Thus, bioventing is best used with contaminants that are moderately volatile, such as diesel or jet fuels or residues of aged petroleum. For bioventing to succeed, the vadose zone must be sufficiently permeable to allow air exchange at least every 24 h, so this process is not suitable for all sites.

It may be possible in some instances to open up impermeable soils by fracturing them with hydraulic or air pressure injections, allowing bioventing strategies to be employed. Bioventing has produced excellent results at hundreds of sites worldwide in recent years. A typical bioventing system involves soil vapor extraction followed by bioventing via vent wells located within an aquifer in areas of high pollutant contamination. Monitoring wells in both the vadose zone (unsaturated soils above the aquifer) and within the aquifer are used to monitor bioventing progress. Neutron probes may sometimes be used for soil moisture monitoring, to assure that venting does not dry out the soil so that microbial populations are inhibited. A plastic liner is usually installed over the surface of part or all of a contaminated area to prevent infiltration of air that might bypass the desired path of air to the vent wells.

A catalytic incinerator to destroy extracted vapor that was not degraded by microbes in the vadose zone is often installed as a safety device, especially at sites near populated areas. If bioventing proceeds effectively, there may be no need for treatment of extracted vapors because the concentration of contaminants in the vented gases may be very low. In bioventing, as with all biological processes in soil, soil moisture must be maintained within a range of 6% to 12%, and supplemental phosphorus or nitrogen may be needed (e.g., N as ammonia provided in the gas phase) to maintain a C:N:P ratio of about 100:10:1. Not all sites are amenable to bioventing. The Battelle Corporation has used a special site evaluation protocol to determine the suitability of a site for remediation by this technique. Four steps are performed sequentially for characterization of a potential bioventing site:

1. A soil-gas survey is conducted to determine if the subsurface is oxygen limited. Soil gas is analyzed for concentrations of oxygen and carbon dioxide. Oxygen is, of course, required for successful bioventing of many contaminants, such as petroleum hydrocarbons.
2. A soil-gas permeability test is conducted to see if air can be injected at rates sufficient to aerate the vadose zone. This test requires injection and withdrawal of air for measuring changes in subsurface pressures at discrete distances from the point of injection. Soils must be sufficiently permeable to be candidates for bioventing. "Tight" soils (e.g., clay- rich soils) can usually be eliminated at this stage.

3. Oxygen-limited but sufficiently permeable soils are then subjected to an in situ respiration test to measure biological contaminant degradation rates for comparison to respiration rates in similar, but uncontaminated, soils. This involves aerating the soils for 24 h and then monitoring the oxygen and carbon dioxide concentrations in the soil-gas phase after the final air injection. An oxygen utilization rate is calculated and used to estimate a contaminant degradation rate, which can in turn be used to estimate the time that will be required to biovent the site.
4. Finally, a pilot- and/or full-scale field bioventing re- mediation demonstration is conducted at the site. There may or may not be a need for pilot-scale testing before a full-scale process is implemented. This is determined on a site-specific basis according to the prior testing.

Biodegradation of Contamination

Landfarming is an aboveground ex situ process in which contaminated soils are excavated, supplemented with nutrients (primarily sources of nitrogen or phosphorus), and spread in relatively thin layers on a prepared site so that biodegradation of contaminants is promoted. These soils are aerated by periodic tilling and kept moist by irrigation, thus explaining the name of this procedure. Sometimes soils are spread directly on the ground at a graded site, if there is no danger that the contaminants will leach through the soil to groundwater or vaporize to dangerous levels in the atmosphere. More commonly, soils are spread on sites with preinstalled plastic liners or clay barriers and leachate collection systems to prevent any possible leaching of contaminants off site or to groundwater, as well as vapor containment enclosures, especially if the site is located near populated areas. Petroleum is commonly treated in this manner.

Highly toxic or biologically recalcitrant compounds generally should not be land-farmed, and the presence of heavy metals such as lead usually indicates that landfarming is not an appropriate technology for a particular soil. The presence of vegetation can enhance the degradation of contaminants during landfarming of contaminated soil. Greenhouse experiments have shown the degradation of PAHs to be greater in the presence of plants than in their absence. PAHs were detectable in the plant tissue, but overall uptake was not significant. Enhanced microbial activity appeared to be responsible for the increased removal of PAHs. Thus, planting of vegetation is promising as an inexpensive means of speeding landfarm-based remediation of sites contaminated with petroleum.

Operating Biopiles

Biopiles are basically soil-heaping operations, a variation on landfarming. The primary advantages of biopiles over landfarms are that biopiles save land space, and air emissions can be more inexpensively captured and treated. In this technique soil is excavated and piled into a mound within a lined treatment cell. The pile is typically covered by a plastic sheet, and an internal system of perforated pipes is usually installed to draw air in by means of vacuum pumps. Alternatively, air may be blown into the pile through the internal pumping system, or supplied from oxygen-releasing compounds incorporated into the piles. Water, nutrients, and microbial inoculants are added as needed, and leachates and gaseous emissions are collected for separate treatment. In theory, biopiles can be used for most biodegradable contaminants, but the limitations of landfarming generally apply. In practice, most biopile operations have treated petroleum-contaminated soils. "Biomounding" is a low-technology, low-cost variation of biopiles. It has been used to treat petroleum-contaminated soil, with the addition of bulking materials like animal manure and wood chips to promote good conditions (sufficient moisture retention, good aeration, source of nitrogen, etc.) for building up microbial populations capable of hydrocarbon degradation. This simple process usually is well accepted by local governments and the general public.

Biomounding can reach fairly stringent soil cleanup standards in the range of 10 ppm as total petroleum hydrocarbons (TPH). Generally, indigenous bacteria present in petroleum-contaminated soil proliferate under biomounding conditions, so inoculation with exogenous microbes is not normally needed. By engineering a favorable environment for microorganisms in a contaminated soil, it is possible to construct biomounds that are reasonably inexpensive to manage, yet allow fairly fast breakdown of petroleum compounds in soil, without the need for continually turning of the soil as in composting. Simple passive aeration systems normally are used in biomounds. For example, a grid of horizontal piping constructed of flexible perforated drain tile, along with vertical piping (risers) of PVC, can be used in the system. Multiple layers of piping can be employed, depending on the volume of soil to be treated. Moisture and heat loss from biomounds are usually controlled by covering the mounds with plastic sheeting.

Role of Agricultural Residues

In *composting*, excavated soil is mixed with bulking agents such as straw, horse manure, or other agricultural residues to form compost piles, which are then managed for good aeration by periodic turning with specially designed machines. Vigorous microbial activity generates considerable heat, so the piles are termed thermophilic, operating at temperatures around 50–60°C. Composting has been examined as a treatment for a variety of contaminants, including petroleum sludges, chlorophenols, and munitions residues. This approach is still experimental, since some questions remain as to its overall efficacy in treating some contaminants. For example, nitroaromatic compounds such as 2,4,6-trinitrotoluene (TNT) tend to be polymerized or incorporated into soil humus fractions, with little or no conversion to carbon dioxide or products that might themselves be degradable to carbon dioxide.

The long-term stability and potential for release of toxic compounds from such uncharacterized materials in soil is not known. In one study, a reduction in direct lethality of TNT to earthworms reflected the removal of explosive compounds by composting, as indicated by chemical analysis. However, assays by the Mutatox procedure indicated increased mutagenicity, a result that would not have been shown by chemical analysis alone, which implied that mutagenic metabolites of explosives were formed during composting and were incompletely degraded. In another study, [^{14}C]TNT was subjected to composting during which much of the ^{14}C becomes incorporated into the "humic" fraction of the compost where it supposedly becomes unavailable. The composted TNT resulting from this process was then fed to rats. Although only 2.3% of the total ^{14}C dose appeared in the rats' urine during the first 3 days after feeding, low levels of ^{14}C continued to be excreted in urine from compost-treated rats for more than 6 months. The total amount of ^{14}C excreted in the urine was comparable to that excreted by animals treated with pure TNT. Thus, it is clear that composted TNT is not irreversibly immobilized, but that TNT or its biotransformation products in compost are bioavailable. Thus, composting may not be the treatment of choice for some wastes where there is little mineralization of the target pollutant.

Despite the experimental nature of composting, it is likely to become a useful tool for treatment of some chemically polluted soils. Variations on composting appear to be possible to overcome some of these problems. For example, TNT is degraded more effectively when the compost is allowed to first undergo a fairly long anaerobic treatment phase before traditional aerobic composting. The critical stage of the traditional composting process is the thermophilic (hot) phase. The fate and mechanism of removal of chemicals such as pesticides and other xenobiotic compounds in composting processes are largely unknown and in need of comprehensive analysis. The hazards of composting stem not only from chemical agents but also from microbiological agents, whose presence depends heavily on composting methods. Further research clearly is needed on the risks related to human exposure to chemical residues and microorganisms in composted wastes. The fate of toxic heavy metals in compost applied to land is of particular concern, as is the danger to the public health from pathogenic microorganisms such as *Aspergillus fumigatus*. Spores of *A. fumigatus* (the causal agent of the lung disease aspergillosis) are abundant in the outdoor air at some composting sites, indicating the importance of site location among process and design criteria for composting operations. In summary, although composting is basically a simple, affordable process, its efficacy and safety for treatment of hazardous wastes are still open to question.

Biodegradation

Intrinsic remediation has been defined as the combined effect of natural destruction and other nondestructive processes to reduce the mobility and mass, and therefore the associated risks, of a contaminant. The mechanisms that mediate intrinsic remediation of a contaminant include biodegradation by indigenous microbial populations, sorption to, or trapping within matrixes that make the compound non-bioavailable, and loss of mass by dilution or volatilization. Intrinsic remediation of petroleum hydrocarbons is well documented, and it has also been confirmed for chlorinated hydrocarbons, typically under anoxic environmental conditions that favor reductive dehalogenation processes. Before intrinsic processes can be accepted as a cleanup option for a particular site, regulators must agree to accept exposure and risk assessment analyses as the basis for determining cleanup end points. Therefore, in sites undergoing intrinsic remediation, the distributions of contaminants (e.g., the extent of contaminant plume migration in relation to sensitive receptors) must be thoroughly characterized, and it must be possible to confirm that pollutant remediation is

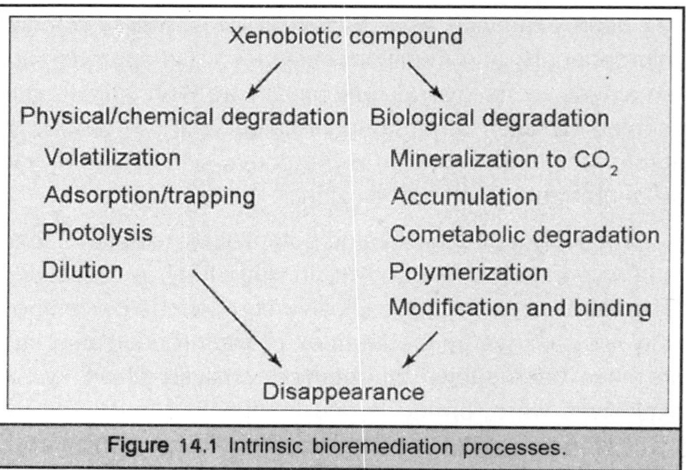

Figure 14.1 Intrinsic bioremediation processes.

proceeding at a rate sufficient to reduce risks to human health and the environment in an acceptable period. To confirm that intrinsic processes are effectively removing contaminants from soil or groundwater, several factors can be periodically measured, including the following:

1. Contaminant mass and its environmental distribution
2. Quantity of contaminant degraders or transformers, as indicated by microbial counts
3. Temperature and pH, to ensure that they are conducive to microbial activity
4. Types of electron donors and acceptors and changes in their concentrations; for example, nitrate, nitrite, metal species (especially iron), methane, ammonia, carbon dioxide, and sulfate
5. Redox potential
6. Other factors determined by site and contaminant

Ideally, the long-term efficacy of intrinsic remediation processes could be predicted at individual sites through modeling. It is obvious that intrinsic remediation requires specific, extensive, and long-term activities at a site and is far from a "do-nothing" option.

Remediate Contamination

Phytoremediation, in which plants are used to remediate contaminated soils or waters, has been studied mainly for the removal of toxic metals but has also been proposed for remediation of some organic contaminants. Phytoremediation can take place through the phytoextraction of contaminants from soil or water, where contaminants are taken up by a plant and concentrated in the plant tissues. The plant materials can then be harvested, removing the sequestered contaminant from the environment. Bacteria associated with the root zones of plants may in some instances degrade organic contaminants, although more study in this emerging area of research is required to confirm which plants promote degradation of which contaminants. Plants may remove contaminants from soil by transpiration of water to the atmosphere, carrying volatile contaminants along. Plants can help in the management and stabilization of polluted sites by reducing soil erosion caused by wind and water and by reducing the infiltration of water into soil, which will in turn slow down or eliminate the leaching of mobile contaminants into groundwater.

An especially promising line of phytoremediation research concerns the use of plants to remove radionuclides from soil or water. After plants or trees have concentrated metals or radioactive compounds, they can be harvested and then burned, which reduces the volume of the collected material by about 95%. Concentration and volume reduction makes disposal of the ash far less expensive than disposal of the original material itself. Alternatively, useful heavy metals might be extracted from the ash as industrial products, especially if the ash contains one or a few valuable metals. It has been suggested that if a plant contained 1% by weight of a metal, then its ash would contain 20% by weight of that metal, a higher metal content than is found in most ores. Any remaining residue after metal recovery would be disposed of as a toxic material.

Phytoremediation does have limitations. It can be slow, so it is not an option for sites that represent an immediate public health or environmental threat. Plant roots also may not penetrate the soil as deeply as necessary to access the targeted contaminants. The roots of most plants generally reach depths of only 30 cm to about 10 m, although a few plants, such as certain varieties of poplar trees, may send their roots somewhat deeper. Thus, deep contamination can be inaccessible to phytoremediation.

Furthermore, as living organisms, plants have specific environmental and nutritional needs. They require a rather narrow range of growth temperatures, grow well only in the presence of favorable soil qualities (texture, water- holding capacity, pH, etc.), must obtain enough water to support growth and metabolism, and usually require the presence of oxygen. Plants may also be inefficient at taking up some pollutants, allowing contaminants to reach groundwater. Some plants proposed for use in phytoremediation include alfalfa (symbiotic with hydrocarbon-degrading bacteria); *Arabidopsis* (genetically engineered to carry a bacterial gene that transforms mercury into a gaseous state); bladder campion (accumulates zinc and copper); *Brassica juncea* (Indian mustard greens; accumulates selenium, sulfur, lead, chromium, cadmium, nickel, zinc, and copper); Buxaceae (boxwood; accumulates nickel); the Compositae family (symbiotic with bacteria; accumulates cesium and strontium); Euphorbiaceae (succulent; accumulates nickel); tomato (accumulates lead, zinc, and copper); poplar (used in the absorption of the pesticide atrazine, and solvents such as carbon tetrachloride); and *Thlaspi caerulescens* (alpine pennycress; accumulates zinc and cadmium).

White-Rot Fungi

White-rot fungi produce enzymes that degrade the plant polymer lignin and oxidize a great variety of chemicals, including many environmental pollutants. Because these fungi degrade such a broad range of anthropogenic compounds, they have been examined extensively as potential bioremediation agents. The peroxidases of the basidiomycete *Phanerochaete chrysosporium* are the most thoroughly investigated of the xenobiotic compound–degrading fungal enzymes. These and other oxidative fungal enzymes such as laccase are thought to be the primary catalysts responsible for initiation of degradation of many xenobiotics. For example, enzyme measurements indicated that both high and relatively stable activities of Mn-dependent peroxidase, Mn-independent peroxidase, lignin peroxidase, and laccase characterized efficient PCB degraders among a selection of white-rot fungi. For application in the field, white-rot fungi grown on a substrate of wood chips or other specially formulated lignocellulosic materials are mixed into a polluted soil. The soil is tilled, watered, and generally managed to encourage the growth and lignin-degrading activity of the introduced organism.

As the fungus degrades the lignocellulosic carrier, it simultaneously degrades the contaminants. This technology is considered experimental, because the fungi so far examined are often inhibited by the contaminant concentrations found in some polluted soils, and even when degradation does take place, it may be incomplete. Clearly, more development work needs to be done in contaminated soil systems. The focus of one recent study was to determine whether *P. chrysosporium* could effectively operate in an actual field sample of oil tar–contaminated soil. The soil was supplemented with [^{14}C]phenanthrene to serve as a representative polynuclear aromatic compound (PAH). This study confirmed earlier work showing that the native soil microflora could acclimate to provide full mineralization of PAHs in some contaminated soils. However, this mineralization could be enhanced by supplementation with *P. chrysosporium*. In another soil study, investigators examined the role of fungal oxidative enzymes in the degradation of soil contaminants by looking at gene expression in the soil matrix. Among the genes implicated in pollutant degradation were two closely related lignin peroxidase genes, whose mRNA transcripts were detected in the soil under study. Thus, the use of white-rot fungi appears quite promising for certain pollutants such as the PAHs and chlorinated phenols. However, more proof of efficacy is required before this becomes a commonly used bioremediation technique. Hundreds of species of white-rot fungi have not yet been examined as bioremediation agents, so the possibility of identifying strains more suitable for fungal bioremediation is good.

CONCLUDING REMARKS

As a biological process, bioremediation has the limitations inherent in any biological system, so it may not always be an appropriate technology for treating a contaminated site. For a site to be treatable by bioremediation, environmental conditions must support biological activity because extremes of pH, temperature, radioactivity, or redox potential may not be tolerated by degradative organisms. Essential nutrients such as nitrogen, phosphorus, sulfur, and trace elements must be present and available. The soil or water to be treated by bioremediation must be free of extremely high concentrations of certain toxic chemicals (e.g., the heavy metals mercury, lead, and zinc) and other antimicrobial substances.

The concentrations of contaminants may be too low to provide sufficient energy for microbial growth, and cometabolism may not be an option. Contaminants may not be bioavailable, or accessible to microorganisms, but they may still be subject to remediation standards required by regulatory agencies. In complex mixtures of contaminants, the feasible bioremediation processes may be unable to remove all the components. The site to be treated, a vast, deep aquifer, for example, may not be accessible to engineers. Finally, some xenobiotic compounds simply are not biodegradable. Thus, bioremediation must be viewed as only a single tool in the environmental restoration toolbox, one that will be used frequently but not for all sites or situations.

15

NEW ANTIBIOTICS

Since the discovery of penicillins in 1928 and the discovery of cephalosporin C in 1945, β-lactams have become the most important class of antibiotics because of their low toxicity and their effectiveness against bacterial infection. In cephalosporins, the four-membered β-lactam ring is fused to a six-membered dihydrothiazine ring rather than to the five-membered thiazolidine ring found in penicillins. The natural product, cephalosporin C, has low activity against gram-negative bacteria, but it is resistant to penicillinase hydrolysis. Many chemical derivatives of cephalosporin C have a higher and broader spectrum of antibacterial activity than cephalosporin C. After 40 years of development, cephalosporins (i.e., chemical derivatives of cephalosporin C or semisynthetic cephalosporins) have become one of the most widely used antibiotics in the world; in 1995, the world market for cephalosporins was valued at $9.3 billion at the dosage form level. Of this total, $6.5 billion represented products made from cephalosporin C, with the remainder derived from penicillins. In this article, we discuss the development and production of cephalosporins and their chemical derivatives.

WORK DONE

Penicillin G (benzylpenicillin) was considered a miracle drug, when discovered in the 1940s, for its effective treatment of the systemic infections caused by gram-positive bacteria such as *Staphylococcus aureus*. Unfortunately, penicillin G was generally ineffective against gram-negative bacteria, and within a few years of its introduction many common gram-positive bacteria had acquired resistance to penicillin G as a result of the expression of bacterial penicillinases. At about this time, a new antibiotic, cephalosporin C, was discovered from a species of mold, *Cephalosporium acremonium*. The *C. acremonium* strain was originally isolated from seawater near a sewage outflow in Cagliari, Sardinia, by Brotzu in 1945. The culture filtrates of this fungus exhibited a broad spectrum of antibacterial activity both in vitro and in preliminary clinical studies. In 1948, Brotzu published his findings, which failed to attract attention. At the British Medical Research Council's suggestion,

Brotzu sent his fungal culture, *C. acremonium,* which was later reclassified as *Acremonium chrysogenum* in 1971 by Gams, to Florey at Oxford.

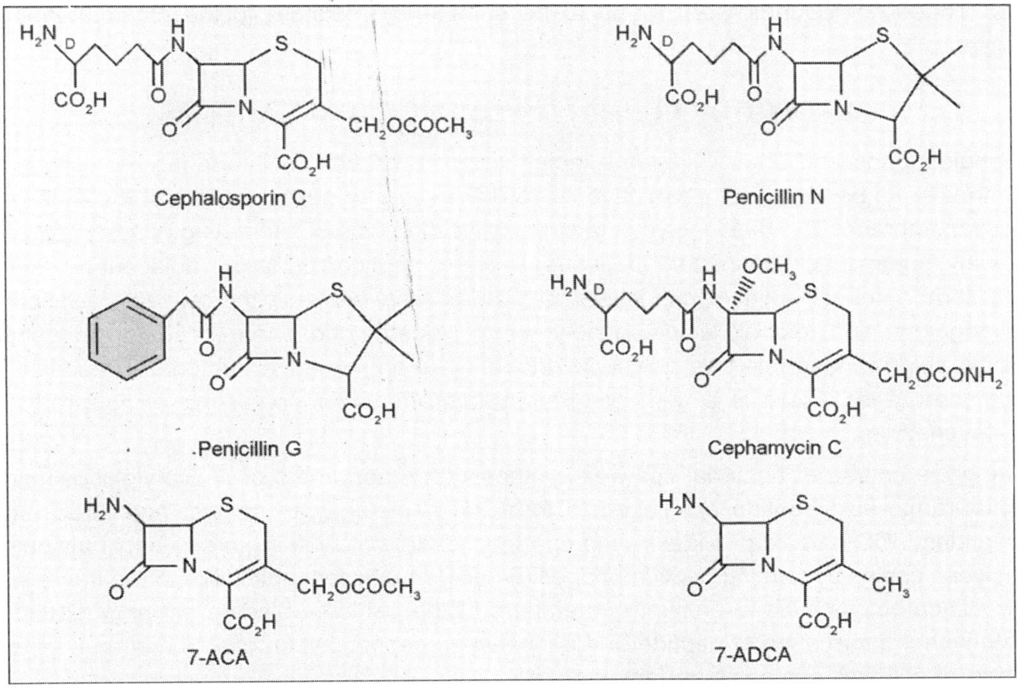

Cephalosporin C

Penicillin N

Penicillin G

Cephamycin C

7-ACA

7-ADCA

Figure 15.1 Chemical structure of β-lactams.

The work of Newton and Abraham at Oxford led to the purification of penicillin N and cephalosporin C in 1955. Both compounds had activity against gram-negative bacteria. Penicillin N was as susceptible to staphylococcal penicillinase hydrolysis as was penicillin G. Cephalosporin C, however, had a high degree of resistance to hydrolysis by staphylococcal penicillinase. Further study also indicated that cephalosporin C had low toxicity in mice, but unfortunately, cephalosporin C had relatively low in vivo antibacterial activity. In 1961, the structure of cephalosporin C was proposed by Abraham and Newton and confirmed by an X-ray crystallographic analysis by Hodgkin and Maslen. It was found that a small quantity of the nucleus of cephalosporin C, 7-aminocephalosporanic acid (7-ACA), could be produced by gentle acid hydrolysis of cephalosporin C. Acylation of 7-ACA yielded a cephalosporin with much higher antibacterial activity than cephalosporin C against gram-positive bacteria.

In 1962, Morin et al. in the Lilly Research Laboratories developed a chemical method for the production of 7-ACA in kilogram quantities, and this opened the way for the extensive exploration of the synthesis of semisynthetic cephalosporins. The extensive use of semisynthetic cephalosporins resulted in the selection in nature of gram-negative bacteria containing R-plasmids that carried a gene encoding β-lactamase. This β-lactamase can hydrolyze most penicillins and all the semisynthetic cephalosporins that had been introduced in the mid 1960s. While searching for new

β-lactam antibiotics that are stable in the presence of β-lactamases, culture filtrates of *Streptomyces* species yielded cephamycins containing a methoxy group at the C-7α position of the cephalosporin nucleus. These compounds were found to be completely resistant to the plasmid-encoded β-lactamase.

EFFECT OF CEPHALOSPORINS AND β-LACTAMS

Cephalosporins and other β-lactams inhibit bacterial growth by interfering with bacterial cell wall synthesis. All bacterial cell walls have a protective layer of peptidoglycan at the outer surface of the cell membrane. The peptidoglycan layer consists of chains of alternating *N*-acetylglucosamine and *N*-acetylmuramic acid connected by peptides. These peptides cross-link adjacent peptidoglycan chains to form a net of peptidoglycan layers. In gram-positive bacteria, the cell walls are composed of peptidoglycan and teichoic acids (a polymer of glycerol phosphate or ribitol phosphate) or teichuronic acids (phosphate-free polysaccharides). In gram-negative bacteria, in addition to the peptidoglycan layer, there is a distinct outer membrane layer consisting of lipoproteins and lipopolysaccharides.

During the course of bacterial cell wall synthesis, the monomers of *N*-acetylglucosamine and *N*-acetylmuramic acid peptide are linked to form the peptidoglycan chains, and three enzymes (transpeptidase, DD-carboxypeptidase, and endopeptidase) catalyze the cross-links between these peptidoglycan chains by forming peptide bridges. Binding of β-lactam antibiotics to the transpeptidase and the DD-carboxypeptidase enzymes, members of the penicillin-binding proteins (PBP) group, interferes with the formation of peptide bridges between peptidoglycan chains and thus inhibits the final stage of cell wall synthesis in bacteria.

MOLECULAR BIOSYNTHESIS

The biosynthetic pathways of cephalosporins have been well characterized. The nomenclature of genes encoding the enzymes involved in each step of the biosynthesis of cephalosporins follows the recommendation by Ingolia and Queener.

Microbial Biosynthesis

The biosynthesis of cephamycin C by *Streptomyces clavuligerus* and *Nocardia lactamdurans* and the biosynthesis of cephalosporin C by *A. chrysogenum* all begin with the condensation of three amino acids, L-α-aminoadipic acid, L-cysteine, and L-valine, to form the tripeptide δ-(L-α-aminoadipyl)-L-cysteinyl-D-valine (LLD-ACV) by the ACV synthetase enzyme. The racemization of valine from the L- to the D-configuration appears to take place during the synthesis of the tripeptide. The LLD-ACV tripeptide is cyclized to form the penam nucleus, isopenicillin N, by isopenicillin N synthetase (or cyclase). Both the LLD-ACV tripeptide and isopenicillin N also serve as intermediates in the biosynthesis of penicillins.

The genes coding for enzymes involved in the biosynthesis of the common intermediates LLD-ACV and isopenicillin N are designated *pcb* (penicillin/cephalosporin biosynthesis). The genes coding for the remaining enzymes involved in the biosynthesis of cephalosporins are designated *cef*. The isopenicillin N epimerase (IPNE) catalyzes the conversion of the L-α-aminoadipyl side chain of isopenicillin N into the D-α-aminoadipyl side chain of penicillin N. The five-membered thiazolidine ring of penicillin N is then expanded to give the six-membered dihydrothiazine ring of deacetoxycephalosporin C by deacetoxycephalosporin C synthetase (expandase). The next enzyme in the pathway, deacetylcephalosporin synthetase (hydroxylase), catalyzes hydroxylation of deacetoxycephalosporin C at the C-3 methyl group to yield deacetylcephalosporin C. In *A. chrysogenum*, both the ring expansion and hydroxylation activities reside on the same protein, which is coded by one gene.

Unlike fungi, the expandase and the hydroxylase are two distinct enzymes in *S. clavuligerus* and *N. lactamdurans* and encoded by two separate genes. In *A. chrysogenum*, the last step in cephalosporin C biosynthesis, catalyzed by cephalosporin C synthetase (acetyltransferase), involves the transfer of an acetyl group from acetyl coenzyme A to the C-3 hydroxymethyl moiety of deacetylcephalosporin C. In the cephamycin C producers *S. clavuligerus* and *N. lactamdurans*, the *O*-carbamoyltransferase catalyzes the transfer of a carbamoyl group from carbamoylphosphate to the C-3 hydroxymethyl moiety of deacetylcephalosporin C. The *O*-carbamoyldeacetyl cephalosporin C is then hydroxylated at the C-7α position by cephamycin hydroxylase. The final biosynthetic step to cephamycin C (7α-methoxy-*O*-carbamoyldeacetylcephalosporin C) is the transfer of the methyl group from *S*-adenosylmethionine (SAM) to the C-7α hydroxyl moiety of 7α-hydroxyl-*O*-carbamoyldeacetylcephalosporin C by cephamycin methyltransferase. The biosynthetic pathways of other cephamycins, such as cephamycin A and cephamycin B, have not been fully investigated.

Genetic Engineering of Cephalosporin

Because the DNA transformation system developed for *A. chrysogenum* is much less efficient than the bacterial systems, cloning of the cephalosporin C biosynthetic genes by complementation of blocked mutants is not feasible. However, the reverse genetics approach has been used successfully in the cloning of a number of cephalosporin C biosynthetic genes. This approach begins with the purification of the desired enzyme, determination of a partial amino acid sequence, and synthesis of oligonucleotide probes based on the amino acid sequence. A genomic DNA library of *A. chrysogenum* is then constructed in *Escherichia coli* and screened for clones that hybridize to the DNA probes. Sequencing of the cloned DNA will confirm that the DNA sequence matches the known amino acid sequence. The reverse genetics approach has also been used in the cloning of the cephamycin C biosynthetic genes despite the existence of highly efficient transformation systems developed for *S. clavuligerus* and *N. lactamdurans*. In 1985, the first β-lactam biosynthetic gene, *pcbC* (encoding cyclase), was cloned from *A. chrysogenum*.

The *A. chrysogenum pcbC* gene has no intron and encodes a polypeptide of 338 amino acids. When the open reading frame of the *A. chrysogenum pcbC* gene was expressed in an *E. coli* expression vector, the purified recombinant cyclase protein was able to convert LLD-ACV to isopencillin N with identical kinetics to the authentic protein, and it also converted ACV tripeptide analog substrates to novel β-lactam antibiotics. Recombinant cyclase is readily produced in the recombinant *E. coli* and can be used for in vitro synthesis of novel β-lactam antibiotics. Other *pcbC* genes have been cloned by reverse genetics or heterologous hybridization from *Aspergillus nidulans, Penicillium chrysogenum, S. clavuligerus, S. griseus, S. jumonjinensis, S. lipmanii, N. lactamdurans,*

Flavobacterium sp., and *Lysobacter lactamgenus*. When these genes were compared at the DNA sequence and the predicted amino acid sequence levels, a high degree of similarity was found to exist among the *pcbC* genes from different sources. The reverse genetics approach was also used to clone the *pcbAB* (encoding ACV synthetase), *cefD* (isopenicillin N epimerase), *cefE* (expandase), *cefF* (hydroxylase), *cefEF* (*A. chrysogenum* expandase–hydroxylase bifunctional enzyme), *cefG* (acetyltransferase), *cefH* (also known as *cmcH*, *O*-carbamoyl-transferase), *cefI* (*cmcI*, cephamycin hydroxylase), and *cefJ* (*cmcJ*, cephamycin methyltransferase) genes. The only gene of the cephalosporin biosynthetic genes that has not been cloned is the *cefD* gene from *A. chrysogenum*.

Most of the cephalosporin C biosynthetic genes cloned from *A. chrysogenum* do not have introns and have a high degree of similarity with their bacterial counterparts, suggesting a bacterial origin for these genes. The *cefG* gene is the exception with two introns. All the fungal genes and some of the bacterial genes encoding for the enzymes involved in cephalosporin biosynthesis have their own promoter for transcriptional control. However, the *Streptomyces cefD* and *cefE* genes are transcribed from the same promoter, with the *cefD* gene located upstream of the *cefE* gene and a 130 base pair (bp) intergenic region separating them. In *N. lactamdurans*, a 7-bp separation between the *cefI* and *cefJ* genes and the presence of a putative ribosome binding site upstream of the *cefI* gene suggests that the two genes are cotranscribed from the same promoter and also cotranslated.

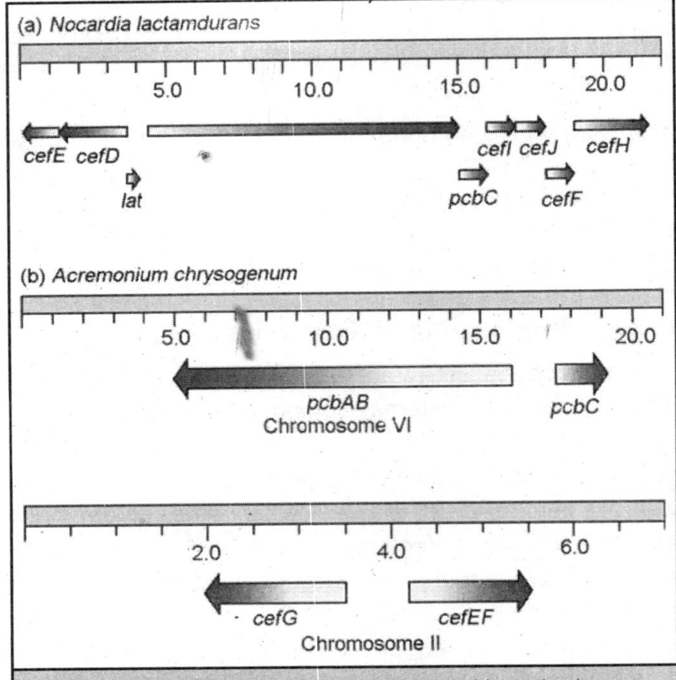

Figure 15.2 Organization of the cephalosporin biosynthesis genes. (a) Physical map of a 25-kb region of *N. lactamurans* DNA carrying all genes related to cephamycin C biosynthesis. (b) Physical map of the clusters of the cephalosporin C biosynthesis genes of *A. chrysogenum*.

Biosynthetic Genes

In 1988, Chen et al. demonstrated that a cloned 29.3-kb DNA fragment from *S. cattleya* carried the entire set of genes encoding enzymes involved in cephamycin C biosynthesis. No further characterization of this cloned DNA fragment was reported. Subsequent work on other bacterial cephamycin C producers also indicated that the cephamycin C biosynthetic genes are clustered in one region of the bacterial chromosome. However, in *A. chrysogenum*, the two early cephalosporin C biosynthetic genes, *pcbAB* and *pcbC*, reside on a different chromosome from the two late biosynthetic genes, *cefEF* and *cefG*. Eight chromosomes have been detected by pulsed- field electrophoresis, an electrophoretic technique that allows separation of chromosomes in an agrose gel system. The *pcbAB* and *pcbC* genes are located on chromosome VI and transcribed in opposite directions. The *cefEF* and *cefG* genes are located on chromosome II and are also transcribed in opposite directions. Because the *cefD* gene of *A. chrysogenum* has not been cloned, it is not known whether it is linked to either of the two clusters.

FURTHER MODIFICATION

Mutant Resistance

Most of the literature describing strain improvement reports the use of the cephalosporin C producer *A. chrysogenum*. However, similar strategies could be applied to the cephamycin C producers (i.e., *Streptomyces* spp.). Improvement in the yield of cephalosporin C has been achieved primarily by increasing the productivity of the *A. chrysogenum* culture by traditional strain mutation and selection techniques. The first significant titer improvement was achieved when in 1959 mutant 8650 was isolated. This strain enabled 100 g of cephalosporin C to be purified for structure determination and is the parent of all the industrial production strains. Mutagenesis, followed by high-throughput screening of the culture, has been the traditional means of strain improvement. Conidia are prepared from the mycelial culture to increase the probability of isolating mutants derived from a single nucleus. Chemical mutagens, such as *N*-methyl-*N*-nitro-*N*-nitrosoguanidine and ethyl-methane-sulfonate, or *UV* light are used to induce chromosomal mutations. Growth of the mutants on selective agents can be used to increase the probability of isolating improved clones. Toxic analogs of cysteine, valine, aminoadipate, lysine, and methionine are used to select mutants that overproduce these amino acids, providing a larger precursor pool for β-lactam biosynthesis. The toxicity to *A. chrysogenum* of heavy metallic ions, such as Cu^{2+}, Mn^{2+}, and Hg^{2+}, has also been exploited for selection purposes. Because the β-lactam antibiotics complex with these ions, one can theorize that mutants resistant to inhibitory concentrations of Cu^{2+}, Mn^{2+}, or Hg^{2+} can overproduce the β-lactam nucleus to detoxify these metallic ions. Elander et al. demonstrated the isolation of superior cephalosporin-C-producing strains from mutants resistant to mercuric chloride.

Technology for Protoplast Fusion

Protoplast fusion has been explored as a means of combining the advantageous characters of different culture lineages (e.g., high titer and fast growth). Mycelia are treated with dithiothreitol (DTT), followed by Novozym 234 (a multienzyme preparation from the fungus *Trichoderma harzianum* available from Novo Biolabs) in the presence of an osmotic stabilizer, usually 0.7 M NaCl. Polyethylene glycol (PEG) is then used to induce membrane fusions between protoplasts; the fusants are identified by regeneration on selective media. Auxotrophic markers have been used to select for fusants, but auxotrophic mutants often have reduced titers and are therefore not suitable for industrial strain improvement. Strains that have been transformed with positive selective markers such as phleomycin and hygromycin resistance are better candidates for protoplast fusion.

Gene Manipulation

In the past 10 years, the cloning of many of the genes involved in the biosynthetic pathway of cephalosporins has led to attempts to improve strains through genetic engineering. Enzymes that have been identified as rate-limiting in the biosynthesis of cephalosporins or their precursors are promising targets for strain improvement through genetic engineering. Transformation of industrial strains with additional copies of the genes coding for these rate-limiting enzymes can result in alleviation of a metabolic bottleneck and increase antibiotic production. Transformation methods for *A. chrysogenum* have been developed using protoplasts and PEG treatment. Based on the kinetic information available for the enzymes in the cephalosporin C biosynthetic pathway in *A. chrysogenum* and the cephamycin C pathway in *S. clavuligerus,* models have been developed to predict the flux of intermediates in these pathways.

The model developed for *S. clavuligerus* predicted that the amino acid precursors, in particular α-aminoadipate, were limiting. Introduction of additional copies of the lysine ε-aminotransferase gene *(lat),* required for conversion of lysine to α-aminoadipate in *Streptomyces,* resulted in a fivefold increase in cephamycin C production. Skatrud et al. have transformed industrial strains of *A. chrysogenum* with a recombinant plasmid containing the hygromycin resistance gene and the *cefEF* gene encoding the expandase–hydroxylase bifunctional enzyme. At the shake flask scale, a 39% increase in cephalosporin C titer was observed, with a significant reduction in the levels of penicillin N and deacetoxycephalosporin C produced. The improvement in cephalosporin C titer at the 150-L pilot plant scale was 15%. Enzyme assays indicated that the expandase–hydroxylase activity in the transformant was twofold higher than in the host strain, and Southern blot analysis revealed a single chromosomal integration at a heterologous position. Interestingly, when the *cefG* gene (encoding acetyltransferase) was identified, it was found to be adjacent to the *cefEF* gene and had incidentally been included on the genomic fragment that Skatrud's group had used for transformation. Chiang and Basch also reported that recombinant *A. chrysogenum* strains transformed with a plasmid containing the *cefEF* gene had a 40% reduction of deacetoxycephalosporin C present in the cephalosporin C fermentation broth as compared to the parental strain. This process was scaled successfully to production tanks.

PROCESS OF FERMENTATION

Biochemistry of Cephalosporin C Fermentation

Cephalosporin C is produced by *A. chrysogenum* as a secondary metabolite, with production of the antibiotic associated with a distinct morphological change. When no nutrient stress is placed on the culture, the organism grows as branched mycelia by extension of the hyphal tips, and little or no cephalosporin C is produced. Under glucose-limiting conditions and in the presence of DL-methionine, the mycelial tips swell and form arthrospores. The mycelial cell walls weaken and fragment, releasing the arthrospores into the media.

Throughout the formation of the arthrospores, cephalosporin C is produced, with the swollen hyphae being the most productive morphology. DL-Methionine addition significantly increases cephalosporin C production. This effect is independent of the methionine role as the sulfur donor for the formation of cysteine. The presence of DL-methionine induces small increases in the enzyme activities of expandase and cyclase and a large increase in the ACV synthetase activity. In the presence of high concentrations of glucose or glycerol in fermentation media, the cephalosporin C productivity decreases dramatically. The high concentration of glucose represses the production of expandase, but not of cyclase or ACV synthetase.

Fermentation at Commerical Level

Cephalosporin C is produced industrially by fed-batch fermentation of *A. chrysogenum.* The pH is maintained between 6.0 and 7.0 in a temperature range of 24 to 28°C. The media must supply carbon and nitrogen for growth, but must also induce the differentiation of the culture required for antibiotic production. The carbon source maybe supplied as separate simple and complex carbohydrates feeds, enabling better control of growth and glucose levels. Monosaccharides, particularly glucose, promote rapid growth of the culture but reduce antibiotic titer. Simple sugars maybe batched into the media or fed at higher rates earlier in the fermentation. Galactose and sucrose support slower growth than glucose but a higher rate of specific productivity. As the fermentation progresses, this sugar feed is reduced and a larger proportion of carbon is supplied as soybean or peanut oil. This limits the availability of glucose and allows the formation of arthrospores and cephalosporin C.

The oil may also serve as a surfactant to reduce potential foaming of the culture. The organic nitrogen is usually supplied by some combination of agricultural by-products, such as soybean and cottonseed meals, and supplemented with an inorganic nitrogen source such as ammonium sulfate. Ammonia gas and ammonium hydroxide can also be used to supply inorganic nitrogen and to provide pH control. Corn steep liquor is commonly used in the medium as an inexpensive source of amino acids, trace elements, and vitamins.

DL-Methionine induces the formation of arthrospores in *A. chrysogenum,* which as mentioned earlier is correlated with cephalosporin C production. Swartz et al. examined the growth and productivity of fed-batch cultures of *A. chrysogenum.* Batch addition of methionine at 3 g/L interfered with sucrose utilization of these cultures but did not affect glucose utilization. Methionine was found to be a competitive inhibitor of the invertase enzyme required for the metabolism of sucrose. Optimal productivity was achieved when sucrose and DL-methionine were fed in an exponentially ramped feed. Surprisingly, feeding of DL-methionine increased the productivity of strains that were considered methionine independent, meaning batched methionine had no effect on titer.

As with all fungal fermentations, dissolved oxygen is an important physical parameter. It is particularly important in the fermentation of cephalosporin C, because the cyclase and particularly the expandase enzymes in the biosynthetic pathway require oxygen for effective catalysis. If the dissolved oxygen level drops below 20%, penicillin N will accumulate and the output of cephalosporin C will be reduced.

Chemical breakdown of cephalosporin C to 2-(D-4-amino-4-carboxybutyl)-thiazole-4-carboxylic acid (compound X) results in the loss of ~40% of the cephalosporin C produced, a significant loss of the biosynthetic potential of the organism. Deacetylcephalosporin C, however, is far more stable in the fermentation broth in comparison to cephalosporin C. Addition of an acetyl esterase produced by *Rhodosporidium toruloides* to the fermentation results in the accumulation of deacetylcephalosporin C and an increase in antibiotic yield of ~40%.

ISOLATION OF CEPHALOSPORINS

After the fermentation is complete, the mycelia and insoluble media components are usually removed by filtration or centrifugation. In the fermentation broth, besides cephalosporin C, there is also present small quantities of penicillin N, deacetoxycephalosporin C, and deacetylcephalosporin C. Cephalosporin C must be extracted from the fermentation broth. Under neutral and slightly acidic conditions, a temperature-dependent chemical conversion of cephalosporin C to compound X occurs in the fermentation broth. At a pH less than 2, cephalosporin C lactone is formed. To minimize this degradation, it is important that the cephalosporin C broth be processed rapidly, avoiding extreme pH conditions and keeping the temperature low. Two main strategies are used for the purification of cephalosporin C.

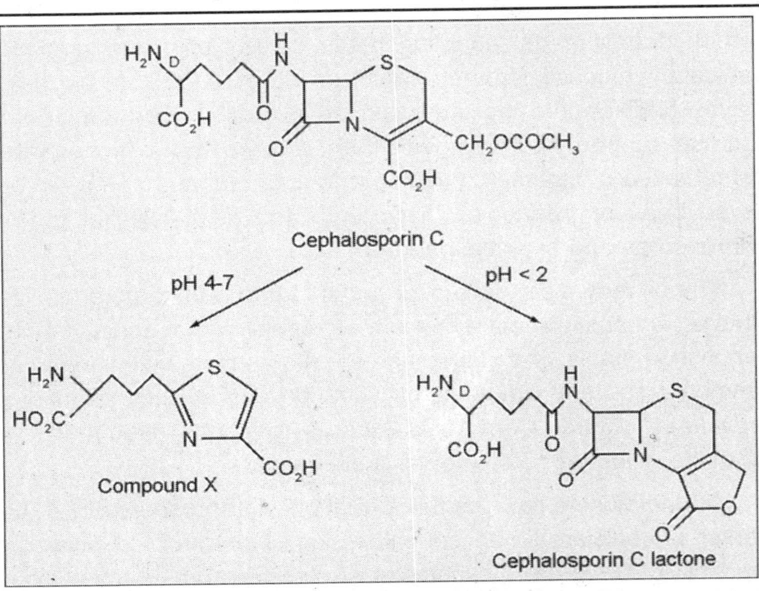

Figure 15.3 Chemical degradation of cephalosporin C.

One strategy uses activated carbon column or nonionic resin column to adsorb cephalosporin C from fermentation broth. Because of the high selectivity of activated carbon and nonionic resin, cephalosporin C is preferentially adsorbed over penicillin N, deacetyl-cephalosporin C, and deacetoxy-cephalosporin C. A subsequent anion-exchange column gives high purity cephalosporin C product. Alternatively, a combination of anion- and cation-exchange columns arranged in series can be used to purify cephalosporin C directly from the fermentation broth.

Because cephalosporin C will be further derivatized to semi-synthetic end products, modifications to the cephalosporin C molecule that do not affect the cephalosporin nucleus can be used to facilitate its purification. Therefore, an alternative strategy for the purification of cephalosporin C is substitution of the amine moiety on the α-aminoadipyl side chain of cephalosporin C. Some of the N-substituted derivatives, such as *N*-2,4-dichlorobenzoyl cephalosporin C and tetrabromocarboxybenzoyl cephalosporin C, can be crystallized in acidic aqueous solutions. Alternatively, salts can be formed between the N-substituted derivatives and organic base, such as dicyclohexylamine and dimethylbenzylamine, and they are solvent extractable.

Several extraction steps are required to achieve an acceptable purity of the end product, although small amounts of penicillin N, deacetylcephalosporin C, and deacetoxycephalosporin C derivatives are still present. N-Substituted cephalosporin C derivatives can be converted to 7-ACA as effectively as cephalosporin C can. The clarified cephamycin C fermentation broth is usually adjusted to pH 2.5 to inactivate most of the penicillin N present. Then, cephamycin C can be isolated with a combination of activated carbon adsorption and anion- exchange resin adsorption. A small amount of deacetoxycephalosporin C and the remaining penicillin N copurifies with cephamycin C. These can be removed by a subsequent silica gel column.

ANTIBACTERIAL ACTIVITIES

Cephalosporin C itself has a relatively low level of antibacterial activity. The semisynthetic cephalosporins with higher antibacterial activity, β-lactamase resistance, and desirable pharmacological properties are derived by modifications of 7-ACA or 7-aminodeacetoxycephalosporanic acid (7-ADCA). Almost one-third of commercial cephalosporins are derived from 7-ADCA. Because of the lower cost of fermenting penicillins, 7-ADCA is most economically produced from penicillin G by ring expansion of a penicillin sulfoxide ester to yield a cephalosporin ester. The ester group is removed, followed by removal of the phenylacetyl side chain to give 7-ADCA. Recently, Merck and Panlabs patented a process for producing 7-ADCA. In this process, *P. chrysogenum* was transformed with the *S. clavuligerus cefE* (expandase) gene. When fed with disodium adipate, this strain produced adipoyl 6-APA, which was then converted by the recombinant expandase to adipoyl 7-ADCA in vivo.

The adipoyl side chain was subsequently cleaved in vitro by a *Pseudomonas* adipoyl acylase to produce 7- ADCA. Two-thirds of the commercial cephalosporins are derived from 7-ACA. 7-ACA is produced from cephalosporin C by chemical or enzymatic deacylation. In the chemical process, after protection of the amino and carboxyl groups of cephalosporin C, reaction with PCl5 in the presence of base yields an iminochloride derivative. Addition of alcohol is used to form the imino ether, which can then be hydrolyzed to form 7-ACA. Alternatively, conversion of cephalosporin C to 7-ACA can be performed enzymatically in two steps. The first step is reaction of the α-aminoadipyl group with D-amino acid oxidase to produce glutaryl-7-ACA. This reaction proceeds through a keto-7-ACA intermediate that undergoes an oxidative decarboxylation in the presence of H_2O_2. Next, a glutaryl acylase is used to remove the glutaryl side chain to produce 7-ACA. An 85% yield of this reaction can be achieved using an immobilized enzyme system. Acai et al. coimmobilized the D-amino acid oxidase and the glutaryl acylase enzymes to perform a single-step conversion of cephalosporin C to 7-ACA.

HYDROPHOBICITY

The outer membrane of gram-negative bacteria is impermeable to penicillins because of their hydrophobicity. Cephalosporins, however, are less hydrophobic, and therefore can pass through the outer membrane to reach their targets. The natural product, cephalosporin C, has low antibacterial activity. Chemical modification at the C-7 position of cephalosporin C yields compounds with a much improved antibacterial activity. Since the development of an efficient method for preparation of 7-ACA, a large number of semisynthetic cephalosporins have been made in the past 30 years. The development of these compounds can be divided into four stages. The first-generation semisynthetic cephalosporins introduced in the mid-1960s include cephalothin, cephaloridine, and cefazoline. These cephalosporins are modified at the C-3 and C-7 positions.

The antibacterial activity of these compounds is still limited because indole positive *Proteus* species and opportunistic pathogens, such as *Pseudomonas aeruginosa*, are resistant because of the presence of β-lactamases.

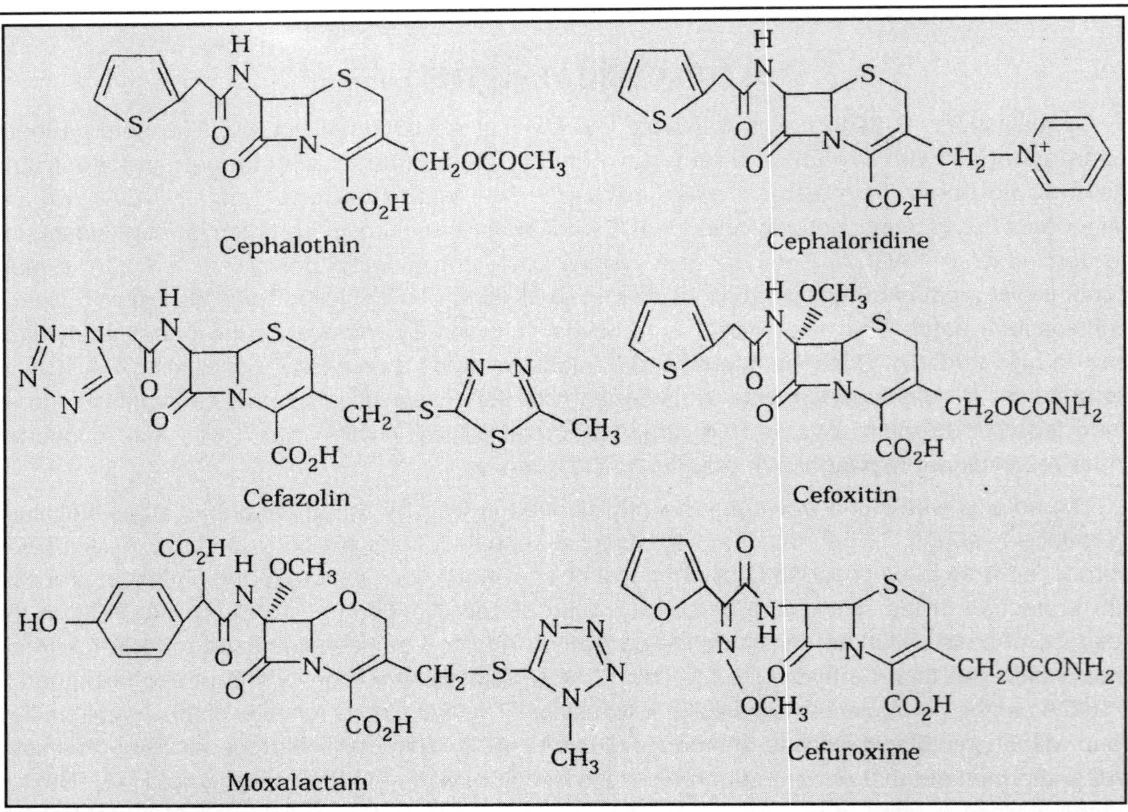

Figure 15.4 Chemical structure of first- and second generation cephalosporins.

Since the late 1960s, an enormous effort has been devoted to the development of cephalosporins resistant to β-lactamases. Out of this research, the second-generation cephalosporins were generated. Among this group, two types of compounds, cephamycins and 7-oximinoacyl cephalosporins, were developed. In 1971, cephamycins were discovered from actinomycetes. Cephamycin C has a 3-carbamoyl and a 7α-methoxy group, which confers a remarkable resistance to β-lactamases. Chemical derivation of cephamycin C led to the development of cefoxitin and moxalactam. The compound developed from 7-oximinoacyl cephalosporin is cefuroxime. Cefuroxime carries the same 3-carbamoyl-oxymethyl group as cephamycin C does. In the search for better cephalosporins with broader antibacterial activity and resistance to different β-lactamases, the addition of a hydrophilic aminothiazole group to oximinocephalosporins produced the third-generation cephalosporins, such as cefotaxime, ceftizoxime, and ceftazidime.

Again, the third-generation cephalosporins have been compromised by the rapid emergence of resistant bacterial strains with plasmid-encoded β-lactamases and by their ineffectiveness against

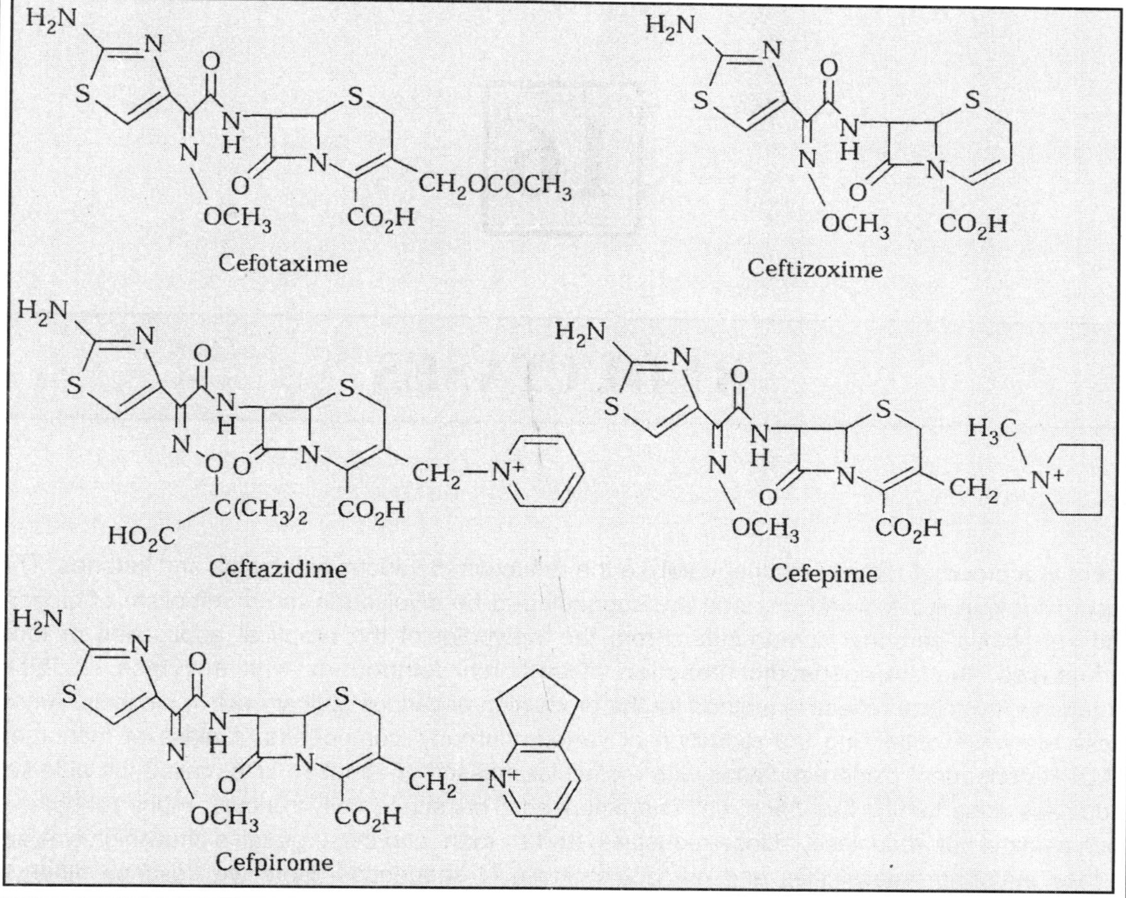

Cefotaxime

Ceftizoxime

Ceftazidime

Cefepime

Cefpirome

Figure 15.5 Chemical structure of third- and fourth-generation cephalosporins.

strains of *Enterobacter cloacae* and *P. aeruginosa* with constitutively expressed chromosomal β-lactamase. By the late 1980s, the fourth-generation cephalosporins, cefepime and cefpirome, had been developed. The new compounds retain the aminothiazole moiety at C-7 with an oxyimino substitution and contain a positively charged quaternary nitrogen at the C-3 position. Since the discovery of cephalosporin C 50 years ago, rapid progress has been made in the strain improvement and fermentation process development for the production of cephalosporin C. Genetic manipulation of *A. chrysogenum* and *S. clavuligerus* allows the potential for engineering of cephalosporin biosynthetic enzymes with broader substrate specificity for in vivo or in vitro production of new β-lactams. Because of the continuous selection of resistant bacteria carrying newly formed β-lactamases, the development of new semisynthetic cephalosporins remains a constant effort to find new compounds effective against these pathogens.

16

REDUCTASES

There is a group of reductases that catalyze the reduction of various aldehydes and ketones. They occur widely in living organisms and are suggested to be involved in the metabolism of biogenic and xenobiotic carbonyl compounds. From the viewpoint of the practical application of these reductases, the stereospecific reduction of carbonyl compounds with enzymes exhibiting stereospecificity is an effective method for the production of various optically active alcohols. Among these enzymes catalyzing the reduction of various carbonyl compounds, a class of monomeric NADPH-dependent oxidoreductases with molecular masses of about 35 kDa, called the aldo-keto reductase superfamily, has been well characterized. This superfamily contains many reductases, such as aldehyde reductase, aldose reductase, and so forth, and these proteins show high similarity in their substrate specificities and primary amino acid sequences. Here we describe aldehyde reductase and the application of microbial aldehyde reductase to the production of chiral alcohols.

FAMILY OF REDUCTASES ENZYMES

The aldo-keto reductase superfamily is a class of monomeric NADPH-dependent oxidoreductases with molecular masses of about 35 kDa. This group includes aldehyde and aldose reductases, prostaglandin F synthase, 2,5-diketogluconate reductase, morphine dehydrogenase, chlordecone reductase, a number of hydroxysteroid dehydrogenases, plant chalcone reductases, and so on. From the strong similarity in their amino acid sequences, ρ-crystallin and a yeast protein encoded by the *GCY* gene, which do not exhibit oxidoreductase activity, are also included in this superfamily. Among them, aldehyde reductase and aldose reductase catalyze the NADPH-dependent reduction of a wide range of aromatic and aliphatic aldehydes, such as *p*-nitrobenzaldehyde, pyridine-3-aldehyde, and glyceraldehyde, to the corresponding alcohols.

These two enzymes are widely distributed in various kinds of mammals including humans, cows, pigs, rats, and mice. Because aldose and aldehyde reductases cause complications in diabetes, the crystal structures of human and pig aldose reductases and aldehyde reductases have been determined by X-ray diffraction methods. As oxidoreductases, the enzymes are unique in that they have a β/α–triosephosphate isomerase (TIM) barrel structure, and are the first oxidoreductases known to possess such a structure instead of a dinucleotide, or Rossmann, binding fold. Elucidation of the tertiary structures and reaction mechanisms of these enzymes will facilitate the design of specific inhibitors that may be used as drugs for the treatment of diabetic complications.

ALDEHYDE REDUCTASE IN RED YEAST

Yamada et al. found a novel aldehyde reductase in a red yeast, *Sporobolomyces salmonicolor*, and succeeded in cloning the aldehyde reductase gene from this strain. Aldehyde reductase was isolated in a crystalline form from cells of *S. salmonicolor* and was characterized in some detail. It should be noted that the cellular content of the enzyme comprised more than 4.5% of the total extractable protein. The enzyme has a monomeric form of about 35 kDa. The enzyme absolutely requires NADPH as a cofactor. The enzyme did not catalyze the reduction of any keto acids or monoketones tested. Several aldehydes such as *p*-nitrobenzaldehyde, *p*-chlorobenzaldehyde, pyridine-3-aldehyde, or D-glyceraldehyde were readily reduced. These aldehydes are typical substrates for aldo-keto reductase family members, as already described. Quinones, such as *p*-benzoquinone, α-naphthoquinone, β-naphthoquinone, and menadione, which are good substrates for carbonyl reductases, were not reduced. Although no detectable oxidation of NADPH was observed with D-glucuronate, D-glucose, D-xylose, or D-galactose as the substrate at a low concentration (0.2 mM), the enzyme effectively catalyzed the reduction of these compounds when they were present at a high concentration (100 mM). The K_m values for these substrates are quite high in comparison with those for other substrates, which suggests that the affinity of the enzyme for these substrates is very low.

Furthermore, the enzyme effectively catalyzed the reduction of 4-haloacetoacetate esters to the corresponding (*R*)-4-halo-3-hydroxybutanoate esters (100% ee). Because the reduction products derived from 4-haloacetoacetate esters are optically active alcohols, the enzyme can recognize the stereo-positions of these substrates. 2-Chloroacetoacetate esters were also effective substrates. Initial velocity and inhibition studies indicated that the enzyme reaction sequence is ordered, with NADPH binding to the free enzyme, and $NADP^+$ being the last product to be released. Quercetin, dicoumarol, and some SH-reagents inhibited the enzyme activity. These results of enzymological studies suggested that the enzyme of *S. salmonicolor* belongs to the aldo-keto reductase superfamily. cDNA coding the aldehyde reductase was cloned from cells of *S. salmonicolor* and sequenced. The aldehyde reductase gene comprises 969 bp and encodes a polypeptide of 35,232 Da. The deduced amino acid sequence shows a high degree of similarity to that of other members of the aldo-keto reductase superfamily.

The identities between *S. salmonicolor* aldehyde reductase and the bovine, rat, human, rabbit, and barley aldose reductases, human aldehyde reductase, and xylose reductase from *Pichia stipitis* are 45.1, 45.0, 43.7, 42.9, 39.5, 43.9, and 37.7%, respectively. In general, aldo-keto reductases exhibit greater similarity in their N-terminal regions than in their C-terminal regions. Analysis of the genomic DNA sequence indicated that the aldehyde reductase gene was interrupted by six introns (two in the 5′ non-coding region and four in the coding region).

PRODUCTION OF OPTICALLY ACTIVE COMPOUNDS

The asymmetric reduction of carbonyl compounds with microorganisms or enzymes is a useful method for the synthesis of optically active compounds. The aldehyde reductase produced by *S. salmonicolor* effectively catalyzes the asymmetric reduction of 4-haloacetoacetate esters to the corresponding (*R*)-isomers, which are promising chiral building blocks for the chemical synthesis of L-carnitine and so forth.

Enzymatic Reduction of Ethyl 4-Chloroacetoacetate

Ethyl (*R*)-4-chloro-3-hydroxybutanoate formed through the reduction of ethyl 4-chloroacetoacetate with cells of *S. salmonicolor* showed low optical purity (8.5% ee) because the cells contained a significant amount of an enzyme(s) catalyzing the reduction of ethyl 4-chloroacetoacetate to ethyl (*S*)-4-chloro-3-hydroxybutanoate in addition to the aldehyde reductase forming ethyl (*R*)-4-chloro-3-hydroxybutanoate. Because heat treatment and acetone fractionation of a cell-free extract of *S. salmonicolor* remove the (*S*)-isomer-forming enzyme(s) from the cell extract, the cell extract after these treatments was used as the source of aldehyde reductase.

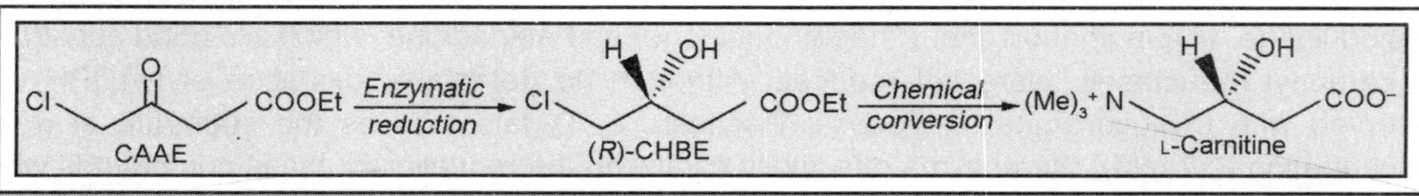

Figure 16.1 Chemicoenzymatic route for the synthesis of L-carnitine, CAAE, ethyl 4-chloroacetoacetate; CHBE, ethyl 4-chloro-3-hydroxybutanoate.

Commercially available glucose dehydrogenase, NADP$^-$, and glucose were added as a cofactor regenerator. A practical preparation can be carried out in an organic solvent–water two-phase system, because the substrate is unstable in an aqueous solution, and both the substrate and product strongly inhibit the enzyme reaction. In this system, most of the substrate is present in the organic phase; consequently, the ethyl 4-chloroaceto-acetate is quite stable and cannot inhibit the enzyme reaction occurring in the aqueous phase. *n*-Butyl acetate is regarded as the most suitable organic solvent for such a two-phase system because it shows high partition efficiencies with regard to both the substrate and product, and both enzymes are stable in the presence of this organic solvent. In a bench-scale

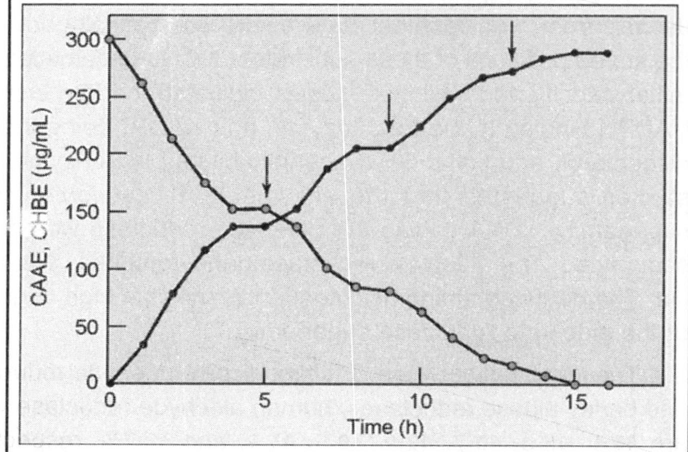

Figure 16.2 Conversion of ethyl 4-chloroacetoacetate to ethyl (R)-4-chloro-3-hydroxybutanoate in a two-phase reaction system by *E. coli* JM109 [pKAR, PKKGDH] cells.

two-phase reaction, 83.8 g/L of ethyl (*R*)-4-chloro-3-hydroxybutanoate (86% ee) was produced from ethyl 4-chloroacetoacetate, with a molar yield of 95.4%.

Enzymatic Production of Ethyl (*R*)-4-Chloro-3-Hydroxybutanoate

The two-phase reaction was improved by using an *Escherichia coli* transformant. The aldehyde reductase gene from *S. salmonicolor* and the glucose dehydrogenase gene from *Bacillus megaterium* were transformed into the same *E. coli* strain; consequently, a transformant over-producing both aldehyde reductase and glucose dehydrogenase, *E. coli* JM109 [pKAR, pKKGDH], was obtained. The reduction reaction was carried out in an organic solvent-water two-phase system containing ethyl 4- chloroacetoacetate, glucose, and $NADP^+$, and *E. coli* JM109 [pKAR, pKKGDH] cells as the catalyst. When the *E. coli* cells were incubated in the *n*-butyl acetate–water two-phase system, 300 mg/mL of ethyl 4-chloroacetoacetate was almost stoichiometrically converted to ethyl (*R*)-4-chloro-3-hydroxybutanoate (92% ee) in 16 h. Because the use of *E. coli* transformant cells as the catalyst is simple and does not require isolation of the aldehyde reductase, which is necessary for *S. salmonicolor* cells, it is highly advantageous for the practical synthesis of ethyl (*R*)-4-chloro-3-hydroxybutanoate.

17

AEROBIC TREATMENT OF WASTEWATER

The most common type of bioprocess, at least in the United States, is the aerobic treatment of wastewater, in which contaminants in the aqueous phase are biodegraded with the help of oxygen transferred from the gas phase (air) to generate a metabolic product (carbon dioxide) that, being volatile, is then stripped out of the aqueous phase into the gas. The familiarity of this arrangement should not be allowed to limit our imaginations about the possibilities of bioprocessing. It is perfectly possible to construct a process in which a gas-phase contaminant (hydrogen sulfide) is biodegraded with the help of a nonvolatile oxidant (nitrate), generating a product (sulfuric acid) that remains dissolved in the aqueous phase.

The bioreactor for such a process must share many of the features found in aerobic wastewater treatment, such as efficient gas–liquid contacting, cell recycling or immobilization to create a high concentration of biomass, control of temperature and pH, and addition of nutrients in order to create a comfortable environment for microbial metabolism. However, the different objective, the treatment of a gas rather than a liquid, must dictate differences in the details. This point can be illustrated by thinking of a conventional trickling filter not as a device for treating sewage, but as one for removing oxygen from the air that flows through it. It would work, but terribly inefficiently, with a fractional removal on the order of 1%.

This chapter presents a three-step study of the use of bioprocesses for the removal of undesirable components from gas streams. The existing and potential applications are described first, followed by the types of bioreactors that may be employed. A simple process analysis is then presented to illustrate how the choice and design of the bioreactor is dictated by the particular application, the type of microbial metabolism involved, the concentration and solubility of the contaminant, and so on. It should be noted that there are also a number of potential bioprocesses in which the objective is not the removal of a contaminant from a gas stream, but the bioconversion of gaseous substrates into useful products. They include the partial oxidation of methane to methanol, and the production of ethanol and acetic acid from mixtures of carbon monoxide and hydrogen. Work on these processes is not described in detail, except when doing so illustrates the common problems of bioreactors with gaseous substrates.

ROLE OF MICROORGANISMS

Due to inherent difficulties in achieving the sterile conditions necessary for prolonged maintenance of pure cultures in field-scale bioreactors, most gas treatment bioreactors contain a mixed culture or consortium of microorganisms. These consortia may be derived from a number of different inocula, from common sources such as sewage-treatment facilities, biofilms from established bioreactors used for similar applications, or soils and waters from areas contaminated with the substrate of interest. In some instances, it is possible to enrich for the desired microbial consortium directly from the bioreactor bed medium, particularly when media components such as soil, compost, peat, or bark chips are used, because these materials naturally contain a mixture of microbes with wide-ranging physiological capacities. Regardless of the inoculum source, the bioreactor must be operated under conditions that select for and maintain, over the life of the bioreactor, microorganisms with the physiological capabilities necessary to catalyze the desired bioconversion.

Typical selective factors include the electron donor (i.e., the metabolizable energy source), available terminal electron acceptors, supplemental nutrients, pH, Eh, and temperature. Regardless of the selection strategy used to achieve a microbial consortium with the desired physiological properties, the use of a consortium versus the attempted use of a pure culture of microbes has a number of practical advantages. As mentioned above, it is difficult to achieve the sterility standards necessary to maintain a pure culture of microbes in a gas treatment bioreactor under field conditions. In addition, a pure culture of microbes is often metabolically incapable of fully degrading a contaminant, so hazardous intermediates may be created and build up within the bioreactor or in effluent streams from the bioreactor. Because mixed consortia inherently tend to be composed of a number of different microbial species with a gamut of physiological capabilities, undesired intermediates that may be created by pure cultures are often fully degraded by consortia. Similarly, in many applications, the concentrations and chemical compositions of gas streams to be treated may be transient. Consortia generally have a higher probability than pure cultures of adapting to this transience. The contaminant gases and vapors that can be successfully biodegraded can be classified by their role in microbial metabolism.

Metabolic Energy

Microorganisms require a constant supply of metabolic energy, which is normally derived from the oxidation of either organic or inorganic compounds, with these compounds serving as electron donors to the microorganisms. Many environmentally or industrially significant gases and vapors are metabolically oxidizable by microorganisms and can serve as electron donors. Thus, many different types of gas and vapor streams can be and have been, treated microbiologically. Microbially oxidizable gases and vapors that have been treated in bioreactors include hydrocarbons, ketones, ammonia, xylene, alcohol, terpene, and carbon disulfide vapors, as well as gases such as methane and hydrogen sulfide. In the case of carbon-containing substrates (hydrocarbons, carbon monoxide, etc.), the metabolism of these compounds provides heterotrophic microbes not only with an electron source, but also with a source of cellular carbon. Non-carbon-containing substrates such as ammonium, while providing a metabolic electron source, must be metabolized by autotrophic microorganisms capable of obtaining their cellular carbon via the fixation of atmospheric carbon dioxide.

Cometabolic Activities

Some substrates cannot serve as sole energy sources for microorganisms, but are nevertheless potentially bioconvertable. Their biodegradation is achieved in the presence of another compound that serves as the microorganism metabolic electron donor, a process that is termed cometabolism. Trichloroethylene (TCE) provides a well-documented example of a substrate that is oxidized cometabolically by microorganisms. Typical of cometabolic TCE degraders are the methanotrophic bacteria, aerobes that oxidize methane as their sole carbon and energy source. This oxidation takes place in a sequential manner, with methane first being converted to methanol, which is in turn oxidized to formaldehyde, formate, and finally, carbon dioxide. Along this oxidative pathway, the methanotrophs generate cellular energy as well as fixing methane-derived carbon into additional biomass. The first step, in which methanol is formed from methane, is catalyzed by the enzyme methane monooxygenase (MMO).

Although MMO is selective for its natural substrate, it can, under certain conditions, catalyze the oxidation of a variety of other compounds, including partially chlorinated aliphatic solvents such as TCE. Two distinct forms of MMO have been identified, membrane MMO (mMMO) and soluble (sMMO). sMMO, expressed by methanotrophs growing under copper-limited conditions, supports much higher rates of TCE oxidation than does mMMO, which is expressed under conditions of copper sufficiency. Air streams contaminated with TCE and similar chlorinated solvents are generated by industrial operations and the remediation of contaminated ground and water by air stripping. They can be treated in a bioreactor that is started up with methane as the sole carbon and energy source, while supplying the necessary oxygen and supplemental nutrients under copper-limiting conditions.

After the methanotrophic consortium is established, the methane load to the reactor can be decreased or fed on an intermittent basis concurrently with TCE vapors. The methanotrophic microbes then will be maintained by the lessened methane feed, while catalyzing the oxidation of the TCE via their sMMO. The reduced or intermittent methane feed not only reduces the amount of methane that must be supplied, but is essential in achieving optimal TCE removal rates since methane itself is an effective competitive inhibitor of TCE oxidation. An alternate, but more expensive, strategy used to avoid this type of competitive inhibition is to supply a catabolic intermediate such as formate in order to provide energy to the methanotrophs, because formate does not compete with TCE for access to the sMMO.

It should be noted that a number of other microbial oxygenases have been shown to be capable of catalyzing cometabolic TCE oxidation. These include toluene oxygenases from a number of *Pseudomonas* spp., propane monooxygenases from propanotrophs, and ammonia monooxygenase from *Nitrosomonas europea*.

Role of Oxygen as Electron Acceptors

Most gas treatment bioreactors remove contaminants from air streams and oxidize them to innocuous end products (CO_2, Cl^-, etc.), using the oxygen in the air as the electron acceptor. However, there are also gas streams containing little or no oxygen where the objective is to remove a compound that can act as an alternative physiological terminal electron acceptor. This approach has been shown to work for compounds like sulfur oxides (SO_x) and nitrogen oxides (NO_x). NO_x biofilters have been developed using a consortia of denitrifying microbes that have the physiological capacity, as a part of their normal metabolism, to use a variety of NO_x compounds as terminal electron acceptors, reducing them to innocuous nitrogen gas (N_2). This process generally requires low-oxygen conditions because denitrifiers preferentially use oxygen, if present, as their terminal electron acceptor. As a result, selective conditions for denitrifying microbes include a suitable carbon and energy source with a NO_x compound as a terminal electron acceptor, a lack of oxygen, and the presence of supplemental nutrients necessary for microbial growth.

The potential application to stack gases raises the question of the maximum temperature at which the microbes will function. Research has shown that naturally occurring thermophilic microbes selected for from compost are capable of reducing nitric oxide (NO) to nitrogen gas (N_2) at temperatures up to 60°C. Other processes are far more sensitive than denitrification to trace amounts of oxygen. The sulfite-reducing bacteria responsible for the removal of SO_2 are obligate anaerobes, although they may be active in anoxic micro- niches, for example, deep in a biofilm. Chlorinated aliphatic solvents will also accept electrons under anaerobic conditions, a degradative mechanism that works best on the completely chlorinated compounds that are not touched by the oxidative enzymes already discussed. Carbon tetrachloride, for example, will be progressively reduced to chloroform, dichloromethane, and methyl chloride. This mechanism is often categorized as cometabolism, implying that the microbes derive no metabolic benefit from it, but this remains uncertain.

TYPES OF BIOREACTORS

Gas treatment bioreactors are sometimes described as *gas-phase bioreactors*, but this term is inexact in the sense that the actual reactions happen inside a microbial cell, which is necessarily an aqueous phase. Some reactions carried out in immobilized enzyme reactors do seem to involve direct interaction between the gas and the enzyme, although even here the humidity remains a critical variable and the enzyme may be covered with a layer of water a few molecules thick. For our purposes, the bioreactor is divided into a gas phase, a solid phase consisting of any solid packing or biofilm support particles, and a biophase, which contains the water and the microorganisms.

The fraction of the bioreactor volume occupied by each phase is called its holdup, and the three values, e_g, e_s, and e_b, must add up to 1. For convenience, we think in terms of a bioreactor of unit volume and express the gas flow rate as the volume per volume per minute (VVM) flow rate, g, at the effluent gas conditions, taken as temperature T and atmospheric pressure. Others choose to work in terms of the molar outlet gas flow rate (g/RT, where R is the gas constant) or a gas residence time, either the empty bed residence time, $1/g$, or the average time an element of gas actually spends in the reactor, e_g/g. The flow rate of the influent gas is not the same as g because its temperature may not be T, it is at a higher pressure, (1 + h atmospheres, where h is the pressure drop through the reactor), and because various components (contaminant, oxygen, metabolic products, water vapor) are added to, or removed from, the gas stream as it flows through the reactor. When these effects are significant, it is essential to specify exactly what is meant by "gas flow rate." The objective of some processes is to remove a gaseous contaminant by converting it into a soluble, nonvolatile metabolic product.

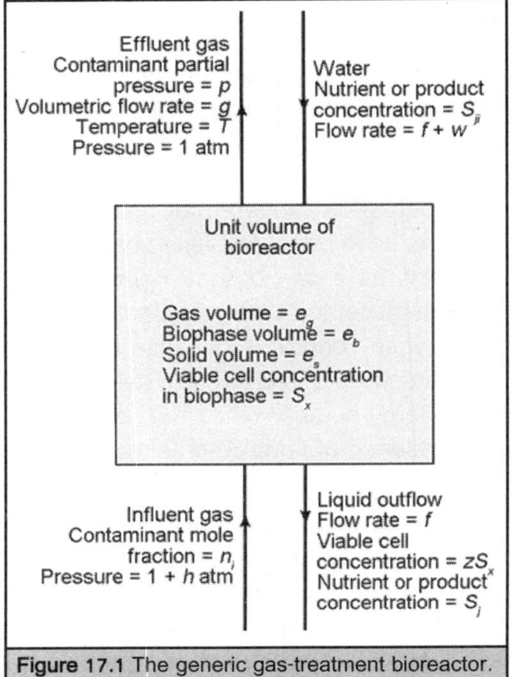

Figure 17.1 The generic gas-treatment bioreactor.

Some such products, H^+ for example, can be neutralized in the bioreactor, but others, such as the Cl^- from the degradation of chlorinated organics, must be removed in the liquid if they are not to accumulate to concentrations that inhibit the microorganisms. There may therefore be a liquid outflow, f per unit bioreactor volume, and an inflow that must be greater than f to allow for the evaporation rate, w. This outflow tends to wash microorganisms out of the bioreactor, and it is usually beneficial to retain them either by cell recycle or immobilization, making the parameter z less than 1. The question is what type of bioreactor should be put in the "black box". Almost every possible configuration by which a gas can be contacted with a biofilm or a suspension of microorganisms has been employed in some application, and these can be roughly divided into four categories.

Process of Biofilteration

The earliest gas-treatment bioreactors were used to control odors in air coming from slaughterhouses and rendering plants and consisted of porous pipes buried in the ground, the noxious compounds being removed by a combination of adsorption and biodegradation as the air flowed up through the soil. Applications of the idea have since expanded to include the removal of hazardous, volatile compounds from chemical plants, paint shops, foundries, and a variety of agricultural and food-processing operations. Although often effective, the operation of these simple soil beds depends critically on the local soil and weather conditions. Unless they are covered and/ or periodically sprayed with water, they may stop working, either due to becoming waterlogged or to drying out. They have largely been replaced by custom-made biofilters, which are essentially boxes made of plastic, wood, sheet metal, concrete, or even stainless steel, containing a bed of media and a gas distribution system to ensure uniform flow of the gas through the bed. In terms of the variables defined, biofilters are the small e_b, small f, and small z approach. The biophase consists of individual microorganisms attached to the media, perhaps enough to constitute a thin biofilm. The amount of moisture in this layer is critical, and is controlled by humidification of the inlet gas and/or by spraying water directly over the bed. Little or no water drips out of the bed, making it difficult to provide soluble nutrients and pH control chemicals, or to wash out nonvolatile metabolic products.

Biofilters are thus best suited to applications that generate no such products and involve no pH swings, the treatment of air streams containing low levels of volatile organics being the obvious example. The great advantage of biofilters over the early soil beds is that they allow the choice of the nature and particle size of the bed media. This media must support a high density of attached microorganisms suited to the particular application, perhaps even providing some of the nutrients needed for their growth. The ability to adsorb the contaminant is an advantage because adsorption can provide some contaminant removal during the start-up period, before a large population of well-acclimated microorganisms has developed (sometimes called "the bed-ripening period"), and during any sudden slugs of concentrated contaminant in the influent gas. Also, although the interactions between such adsorption and biodegradation are complex and poorly understood, they are generally favorable to biofilter performance.

Finally, the media must be mechanically strong and resistant to disintegration, compacting, and the resulting channeling of gas flow. Soil, compost, peat, and wood-chip mixtures are all inexpensive media that have been used successfully. Materials such as activated carbon and limestone, although more expensive, may be mixed in to provide extra adsorption and pH buffering. The particle size of the media is critical. Small particles give a huge gas–biophase interfacial area per unit bed volume (a quantity subsequently called "*a*"), thus eliminating concerns about mass-transfer limitations between the two phases. But beds of small particles have a higher resistance to gas flow, are more prone to waterlogging and plugging by excess biofilm, and if too light, may get blown out of the reactor altogether.

A general guideline is to choose the smallest available particles that avoid the latter problems, and then ensure that the shape of the bed is optimized to give the desired contaminant removal with a reasonable pressure drop. If the life of the bed is found to be finite due to compaction, loss of pH buffering capacity, accumulation of salts from the evaporation of added water and so on, one solution is to employ two beds in series. Bed 1 can provide most of the contaminant removal, while bed 2 polishes the effluent, mainly by adsorption, while developing its own population of well-acclimated microorganisms. When bed 1 is exhausted it can be replaced by bed 2, which is itself replaced by a bed of fresh media. The arrangement of the media on a series of trays in the bioreactor facilitates this mode of operation.

Bubble-aerated Fermenters

Conventional bubble-aerated fermenters make poor gas treatment bioreactors because they are designed for a high rate of transfer of gaseous nutrients and products (O_2 in and CO_2 out), rather than a high fractional removal of a component from the air stream. This latter objective requires taller bioreactors with longer gas residence times, but tall reactors mean higher pressure drops due to the hydrostatic head. Nevertheless, tall, mechanically agitated bioreactors with multiple impellers have been proposed for removing H_2S from industrial gas streams, and bubble columns with no mechanical agitation have been demonstrated for the removal of TCE from air using a toluene-oxidizing organism. These bioreactors are in many ways at the opposite extreme from biofilters. The biophase occupies most of the reactor ($e_b \rightarrow 1$), and it is dilute and well-mixed rather than dense and immobilized.

The large volume of water makes it easy to add soluble nutrients, control the pH, and remove water containing nonvolatile products. However, care must be taken to not wash out the biomass, because $z = 1$ unless there are extra surfaces in the bioreactor for the attachment of a biofilm, or if a cell separation and recycle loop is added. Another major difference is that the gas–biophase interfacial area, *a*, is not only much smaller than in a biofilter, but is not even a constant, varying with the gas flow rate. Bubble gasified bioreactors are therefore best suited to contaminants like NO_x that are very soluble in water (thus reducing concerns about the gas–liquid mass-transfer rate), whose dissolution in water cause pH swings, and whose metabolism requires large amounts of dissolved nutrients.

Secondary Treatment

Trickling filters for gas treatment are similar to, but usually taller than, those in use for decades for the secondary treatment of municipal and industrial wastewaters. Like biofilters, they consist of a bed of media through which the gas flows either upward or downward, but unlike biofilters, the microbial culture is recirculated continuously over the bed from a reservoir beneath it. Stone aggregate or wood chips are the traditional media, but even seashells have been used. Most modern filters employ ceramic or plastic media, Pall rings, Raschig rings, and so on specifically designed to make a bed with a high surface area to maximize gas transfer, and a high porosity to minimize the pressure drop and the chance of clogging or flooding. These bioreactors can be thought of as intermediate between biofilters and bubble-gasified systems and are best suited to applications between those already described. If the characteristics of the media and the microorganisms are such that most of the biomass stays attached, then a trickling filter is essentially a biofilter with a higher liquid flow and a smaller gas–biophase interfacial area, *a*, due to the larger media. If on the other hand, most of the biomass stays suspended and the recycle ratio (RR = flow of recycled liquid/f) is high enough to keep the biophase well mixed, then it is functionally similar to a bubble-gasified bioreactor. The main differences are that the pressure drop, *h*, is usually much smaller, and the interfacial area, *a*, is a constant that is independent of the gas flow rate.

Shifting from Gas to Water

The systems just described essentially combine two functions, stripping the contaminant from the gas into the aqueous biophase and then subjecting it to biodegradation. But what if the physicochemical conditions in the biophase resulting from the stripping are not suitable for the microbial metabolism? The removal of sulfur dioxide from stack gas is a good example because stack gases contain a small percentage of oxygen that although relatively insoluble, inhibits the bacteria responsible for the reduction of the sulfite formed by the dissolution of SO_2. One solution is the *bioscrubber*, which consists of an aqueous stripper in which the contaminant is transferred from the gas to water, and a separate bioreactor, in which the liquid effluent from the stripper is essentially treated as wastewater. The bioreactor should be a completely mixed, immobilized-cell type so that a large population can grow up in a well-controlled environment without continuously being recirculated through the potentially toxic environment in the stripper. Several configurations are possible, the simplest being that in which the liquid reservoir at the bottom of a packed-bed stripper is used as the bioreactor. Bioscrubbers are not considered further here because building a separate bioreactor and stripper is necessarily more expensive than combining them in a single unit, and when it is necessary, the two unit operations can be designed by conventional methods.

OPERATIONAL DESIGN

A process variable is any quantity that is under the direct control of the designer or operator of the process. For gas treatment bioreactors, the variables to be fixed by the designer include the bioreactor type, shape, and size, whereas those fixed by the operator include the addition of water, nutrients, and chemicals for pH control. The objective in fixing these variables is to satisfy a set of process requirements that specify the flow rate and composition of the gas stream, the nature and concentration of the contaminant, and the fractional removal required. The difficulty is that many designs, that is, many combinations of the process variables, will meet the process requirements for a particular case. The proper goal of engineering design is not simply to pick one of these designs at random, but to find the single design that meets the requirements at the lowest possible cost. The design of gas treatment bioreactors, like that of any other process, is an economic optimization problem, but one that is so complex that the exact solution cannot be found within the constraints of time and money normally imposed on the design process.

The best solution for a particular case will certainly depend on the scale of the problem, the solubility and possible inhibitory effects of the contaminant, the type of microbial metabolism (aerobic or anaerobic, growth associated or cometabolic, etc.) needed to degrade it, and the temperature and pressure of the gas stream. Even coming close to the best solution requires a careful scale-up program in which the insights available from education and experience are used to integrate experiments at different scales with mathematical models that can extrapolate the results from one scale to the design of the equipment at the next scale. The bioreactor analysis presented in this section is intended as a guide to this program, suggesting which bioreactor types be chosen for experimental evaluation in each case, specifying the operating conditions around which the experiments should be designed, and showing how performance may be expected to vary with scale. Because the design goals are essentially economic, the relationship of the variables shown in Figure 1 to the process costs must be kept in mind.

The volume of the commercial-scale bioreactor equals the actual gas flow to be treated (a process requirement) divided by g; a large value for g translates directly into a smaller and, all other things being equal, cheaper bioreactor. The liquid flow, f should be kept as small as possible because providing clean water and disposing of any process wastewater both cost money, as do any nutrients and chemicals that must be added. The pressure drop, h, through the bioreactor may be significant, particularly on a large scale, and it determines the capital cost of the gas compressors and the energy costs for running them. Compressing the feed gas is not only expensive in itself, it heats the feed such that heat exchangers may be needed to cool it to the desired temperature.

Bioprocess Models

It will be assumed that the biophase is completely mixed and that the gas approximates plug flow through the bioreactor. This provides a realistic description of some bioreactors (bubble columns, some biotrickling filters), and a reasonable starting point for quantitative thinking about the others. More detailed mathematical models are available in the literature for specific types of bioreactors and metabolism. The equilibrium solubility of the contaminant in the biophase, S^*, will be described by Henry's law, $S^* = p/H$, where p = partial pressure of the contaminant in the gas and H = Henry's law constant. Although this is adequate for relatively insoluble and dilute contaminants, it is a considerable oversimplification for others. For example, the dissolution of oxides of nitrogen (NO_x: a mixture of NO and NO_2) from stack gases into water is an extremely complex process that involves the gas-phase chemical oxidation of NO to NO_2 and a dissociation reaction that generates H^+, NO_3^- and NO_2^- when the gas dissolves.

In such cases, Henry's law can only be interpreted in the purely qualitative sense that low H means a soluble gas. Any complete bioprocess model combines the mass conservation equations that describe the bioreactor with equations that try to approximate the complexities of microbial metabolism. Because the objective here is to compare different bioreactor configurations, it is essential to have consistent descriptions of the metabolism. In most gas treatment applications the contaminant is a major microbial nutrient, either the electron donor or the electron acceptor. Its consumption provides metabolic energy for growth and cell maintenance, so its specific consumption rate, q is given by the yield equation

$$q = \frac{\mu}{Y} + K \qquad \ldots(1)$$

where μ is the specific growth rate of cells, Y is the cell-yield coefficient, and k is the cell-maintenance coefficient.

In a well-operated bioreactor, the pH and temperature are carefully controlled and all required nutrients are provided in adequate amounts, making the specific consumption rate a function only of the concentrations of contaminant, S, and any inhibitory product, S_m, dissolved in the biophase. The exact form of this function, written $q(S, S_m)$, will not be specified because it varies among cases, but it must describe the effects on the metabolic rate of substrate limitation, product inhibition, and possibly (when n/H is large) substrate inhibition. It is more common to describe the specific cell growth rate, μ, by a Monod-type function of the concentration S, but this is clearly incorrect at the low concentrations found in gas treatment bioreactors, because it predicts that growth stops and substrate uptake continues when no substrate is available ($\mu = 0$ and $q = k$ when $S = 0$). In fact, when a major nutrient is exhausted, its consumption must stop, and the microorganisms go into an endogenous state in which the viable biomass declines, a phenomenon correctly approximated by requiring that $q(S, S_m) = 0$, and thus $\mu = kY$ (equation 1) when $S = 0$.

Cell growth now stops at a nonzero "stationary phase" concentration, S_s, corresponding to the point where substrate uptake is just sufficient to satisfy the maintenance requirement, and defined by $q(S_s, 0) = k$. For cometabolic processes, the contaminant is neither the main electron donor nor the acceptor; two rate equations are needed, and equation 1 contains an extra parameter whose value may be found from knowledge of the degradative pathway. The design of cometabolic bioreactors follows many of the principles described here but is too complex and case specific to be analyzed in detail. At steady state the conservation of mass requires that the rate of growth of microorganisms in the bioreactor, $\mu e_b S_x$, equals the rate at which they flow out, fzS_x. The loss of microorganisms in mist suspended in the effluent gas may be significant in some bioreactors without demisting devices but is not included explicitly in the analysis.

The specific growth rate, μ, a measure of the physiological state of the microorganisms, thus equals zf/e_b, the reciprocal of the mean cell-residence time (the number of cells in the bioreactor divided by the cell outflow rate), which is a quantity under our direct control. This remarkable result, although well known for chemostats, has not been fully appreciated for gas treatment. It follows from equation 1 that the concentrations in the biophase, S and S_m, are constrained by

$$q(S, S_m) = k(1 + D) \quad \text{...(2)}$$

where $D = zf/Yke_b$ is the dimensionless dilution rate or reciprocal mean cell-residence time.

The steady-state conservation of mass for the contaminant requires that whatever is removed from the gas phase be transferred to the biophase, where it is either consumed by the microorganisms or flows out in the liquid. For a bioreactor, as opposed to a gas stripper, the latter term should be negligible, but the flows of dissolved nutrients and nonvolatile metabolic products must be included

$$\frac{g}{RT}(n_i - p) = \frac{k_1 a(n_i - p)}{H \ln \dfrac{(n_i - HS)}{(p - HS)}} = e_b q S_x = \pm \frac{(fS_j - (f + w)S_{ji})}{y_j} \quad \text{...(3)}$$

where y_j is the amount of component j produced (+ sign) or consumed (– sign) per unit of contaminant consumed.

The description of the driving force for mass transfer in terms of the log–mean concentration difference in the second term of this equation follows from the assumption of plug flow of the gas. The expression is not strictly correct when the pressure drop through the bioreactor is large, as in a tall bubble column, because the contaminant partial pressure in the influent is then $(1 + h)n_i$; but this complication is rarely included in practice. The essential points here are that the mass-transfer factor, k_1a, is an empirically determined parameter, that it can be based on different forms of the driving force, and that the best choice depends on the reactor type and must be clearly understood by everyone involved in a project. Equation 3 can be rearranged to give the fractional removal of the contaminant from the gas, only the product (subscript m) being considered in the last term:

$$X = 1 - \frac{p}{n_i} = (1 - e^{-G})\left(1 - \frac{HS}{n_i}\right) = \frac{S_x}{X}(1 + D) = FD\frac{S_m}{S'_m} \quad \text{...(4)}$$

where $G = k_1aRT/gH$ is the dimensionless mass transfer coefficient; $F = YS'_m/zXy_m$ is the dimensionless product inhibition factor; $X = gn_i/ke_bRT$ is the maximum possible viable cell concentration in the biophase; and S'_m is the product concentration that completely inhibits cell growth.

Rather than solving equations 2 and 4 exactly, which would require the form of the inherent kinetic function $q(S, S_m)$, consider some limiting cases in order of decreasing liquid outflow rate.

These cases are used to summarize the bioreactor performance, a graph that can be universal, at least within the limitations of the assumptions, because the parameters describing the contaminant (H, n_i), the reactor (f, k_1a), and the metabolism (k, S'_m) in individual cases have been incorporated into the dimensionless numbers x, D, F, and G.

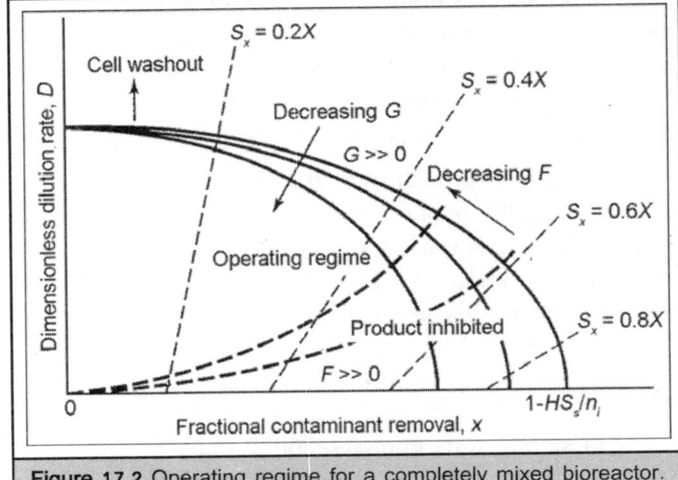

Figure 17.2 Operating regime for a completely mixed bioreactor.

1. *Washout.* $D > (q(n_i/H,0)/k) - 1$: Because the dissolved concentration, S, must obviously be less than that in equilibrium with the feed gas, n_i/H, equation 2 now has no solution, meaning that all of the biomass is washed out of the reactor and there can be no contaminant removal.
2. *Substrate limitation.* $D > (1 - e^{-G})/F$: Equation 4 shows that under this condition, $S \to 0$, while $S_m < S'_m$, meaning that microbial kinetics are limited by the availability of substrate rather than product inhibition and the exact value of the inhibition number, F, is irrelevant. If gas-biophase mass transfer is very rapid, the dissolved contaminant is in equilibrium with the outlet gas, giving the best possible bioreactor performance. The shape of this $G = \infty$ line is simply a reflection of the relationship between $\mu(= zf/e_b)$ and $S(= p/H)$ from the inherent microbial kinetics. Mass-transfer limitation ($G < \infty$) leads to a progressive reduction in the contaminant removal, x.
3. *Product inhibition.* $D < (1 - e^{G})/F$: Now the concentration of metabolic products, S_m reaches inhibitory levels, S'_m, while dissolved contaminant is still available, meaning that metabolism is limited by product inhibition rather than substrate limitation, and the exact value of G is irrelevant. Increasingly stronger inhibition (smaller F) produces a series of curves whose exact shape depends on the details of the inherent inhibition kinetics. All of these curves pass through the origin because, with no liquid outflow ($D = 0$), any soluble, nonvolatile metabolic products must eventually accumulate to inhibitory levels.

Problematic Details

Any pair of values of F and G define a regime within which the bioreactor can operate. The most important result is that for any such pair of values there exists a value of D that maximizes the contaminant removal, x. If the form of the rate equation $q(S, S_m)$ were known, the pointed maximum would be replaced by a smooth curve that defined this optimum design more accurately. In the absence of such detailed modeling the optimum must be found experimentally, the analysis providing a starting point and overall guidance. Because the gas flow rate, g, appears in the denominator of both F and G, the first step is to establish the maximum g for which the desired contaminant removal, x, falls safely within the operating regime. It is clear from equation 4 that a high removal rate (e.g., $x = 0.95$) with reasonable values of $S(\gg Hn_i)$ and $S_m(<0.5\ S'_m)$ requires both $G \sim 3$ and $FD \sim 2$, and such values are reasonable, if approximate, general design goals. Much larger values represent inefficient design because they imply excessive consumption of energy and water. For processes that produce large amounts of very inhibitory, nonvolatile, soluble metabolic products, the problem is to wash the products out of the bioreactor without washing out the biomass. Mathematically, $y_m/S'_m \rightarrow \infty$, making it difficult to keep $FD > 2$ without D exceeding the washout criterion. These are the processes that require cell retention ($z \rightarrow 0$), either by cell recycle or by immobilization as a biofilm.

For processes with more moderate values of y_m/S'_m, the analysis shows that there exists an optimum set of values of the process variables g, f, and z. Some processes, the aerobic mineralization of hydrocarbons for example, generate only innocuous (H_2O) or gaseous (CO_2) products, so $y_m/S'_m \rightarrow 0$, and the analysis suggests operating with no liquid outflow. This not only gives the best effluent-gas quality ($F = \infty$) but, because the biomass is in a maintenance condition ($\mu = 0$ when $D = 0$), it also avoids the cost of adding nutrients, such as phosphates, that are needed for microbial growth. In practice there are reasons to avoid this condition. Even if y_m/S'_m is too small to be observed in short-term laboratory experiments, any nonvolatile product must eventually accumulate to inhibitory levels in a continuous bioreactor with no liquid outflow. Salts in the water added to compensate for evaporation or for pH control will also accumulate in the biophase. Also, it must not be forgotten that complete mixing of the biophase is an abstraction never achieved in practice and sometimes, as in the biofilter, not even attempted. The discussion prior to equation 2 shows that any bioreactor with $D = 0$ must have a cell growth rate $\mu = 0$ as an average, but it will contain microniches where cells are dying because conditions are worse than average, and others where conditions are slightly better and cells are growing.

Although the growing cells may assimilate some of the material released by the lysis of the dying cells, there must be some long-term accumulation of dead biomass, with consequent reduction in bioreactor performance. Keeping D on the of order of 1, a very low flow rate in most cases can wash out the dead biomass and ensure stable operation for the cost of a few growth nutrients. The analysis assumed the usual case in which the cell concentration in the bioreactor is kept as high as possible, with sufficient amounts of other nutrients provided to ensure that the contaminant is the limiting nutrient. The quantity X, defined under equation 4, is then the maximum possible viable-cell concentration in the biophase, corresponding to the situation where all of the contaminant flowing into the bioreactor, gn_i/RT, is being used to satisfy the cell maintenance requirement, ke_bX. The actual concentration at any given operating condition, S_x, can be calculated from the fourth term of the equation and as a function of D and x.

There is, however, no guarantee that this concentration corresponds to a dilute, easily mixed cell suspension. If the maintenance requirement, k, is small, and the contaminant loading gn_i is large, it may be a paste of cells too dense to be, for example, pumped around a biotrickling filter. Although it is possible to reduce S_x by restricting the amounts of growth nutrients provided, such artificial reductions in the microbial "catalyst" concentration are generally poor bioprocess design. The alternative is to choose the right type of bioreactor, recognizing that, although the mass of microorganisms in the bioreactor is fixed by the process requirements and the microbial metabolism, the amount of water is a process variable that can be controlled. The concentration, X, is an order of magnitude higher in a trickle-bed type bioreactor where $e_b \approx 0.1$, than in a bubble column where $e_b \approx 1$.

Aerobic Mineralization

In some processes, the aerobic mineralization of organics in beds of compost for example, the air provides the electron acceptor and the media provides what little growth nutrients are needed. In other processes these substances, and sometimes the electron donor, must be provided in the liquid inflow. The amounts can be estimated from the "stoichiometry" of metabolism, a procedure best illustrated by example. Ongcharit et al. have shown that hydrogen sulfide can be removed from industrial gas streams by the acidophilic, autotrophic, sulfur-oxidizing, denitrifying bacterium *Thiobacillus denitrificans*. Even such esoteric metabolism can be approximated by a pseudo-chemical reaction in which all of the main nutrients and products are written in the ionic form in which they enter or leave the reactor. After performing all of the element and charge balances the result is

$$H_2S + (1.6 - 0.2y\gamma\gamma)NO_3^- + yCO_2 + ypH_2PO_4^- \rightarrow yCH_hN_oS_sP_p + (0.8 - 0.1y(\gamma + 5n))N_2$$
$$+(1 - ys)SO_4^{2-} + (0.4 + y(0.2\gamma - p - 2s))H^+wH_2O \quad \ldots(5)$$

The first term on the right-hand side represents the dry weight of the biomass produced. $\gamma = 4 + h - 2o + 6s + 5p$ is a measure of its oxidation/reduction state and is remarkably consistent between species (note that the base oxidation state for nitrogen here is N_2). The actual cell yield is y, the ratio of cells produced to H_2S consumed. For a completely mixed biophase this equals μ/q or, from equations 1 and 2,

$$y = \frac{DY}{(1+D)} \quad \ldots(6)$$

The "stoichiometric coefficients" on the other components in equation 5 then provide the y_j values needed to calculate the concentrations of nutrients and products from the last term in equation 3. These simple calculations can provide a surprising amount of approximate information about process design and development. A feature of the example in equation 5 is the large amount of acid (H^+) produced by the metabolism. Even with the high liquid flow rate allowed by cell immobilization or recycle its neutralization requires a high pH in the liquid inflow, so the biophase must be very well mixed if metabolism is not to be inhibited by pH gradients, suggesting some type of bubble-gasified bioreactor. If all of the NO_3^- and OH^- are provided by their sodium salts, then the salinity in the biophase would be high, suggesting a need for halophilic strains of bacteria. A reasonable alternative would be to explore the use of calcium salts, which would precipitate some of the product sulfate as gypsum, reducing inhibitory effects. Like all the previous analyses, equation 6 applies only to steady-state operation.

Gas treatment bioreactors, like any other bioprocess, are generally started with a relatively small inoculum of acclimated microorganisms, the objective of the start-up period being to increase the population to the steady-state value, S_x. This requires additional growth nutrients, and the initial feed concentration, S_{ji}, can be calculated as above, but with setting the observed cell yield, y, equal to the growth yield coefficient, Y. These concentrations can then be decreased slowly as steady state is approached. Start-up of biofilters is sometimes called ripening of the bed, and it is not always necessary to add growth nutrients, because the microorganisms capable of growth on the contaminant may obtain nutrients from the support material (compost, soil, etc.), or may grow at the expense of cells that die and lyse under conditions in the biofilter.

If these sources are not sufficient to produce the biofilm thickness needed for maximum contaminant removal, some nutrient addition after steady- state contaminant removal has been reached will further improve performance. However, excessive nutrient addition must be avoided because modeling suggests that, under contaminant limited conditions, a biofilm will grow much thicker than is needed. The inlet end of the biofilter may plug with useless biofilm. Just because a steady state exists in theory, it does not follow that it can be reached by any start-up procedure. Of particular concern are those contaminants that are so concentrated (high n_i) and soluble (low H) that they will cause substrate inhibition of the microbial metabolism if their dissolved concentration in the biophase ever reaches equilibrium with the feed (n_i/H). The steady state analyzed above still exists, but $S \ll n_i/H$ only as a consequence of the metabolism of the large numbers of microorganisms present. Starting with a small inoculum, the dissolved concentration in the biophase will be much higher, metabolism will be inhibited, and the biomass may wash out. For similar reasons, batch experiments may produce the erroneous conclusion that the gas cannot be treated biologically. A proper start-up procedure may involve dilution of the feed gas and careful control of the dilution rate and nutrient addition, until a dense, well-acclimated microbial culture has developed.

Mass Transfer

The data shows that high levels of contaminant removal require high values of the dimensionless mass-transfer factor, G, which is actually the product of two dimensionless numbers: H/RT, an inverse measure of the contaminant solubility, and k_1a/g, a measure of the efficiency with which the gas flow generates mass transfer. This is not usually a difficulty for biofilters with their thin biophase and very large a but how can it be achieved for other bioreactor types? Measurements of gas–liquid mass transfer factors are often given as correlations of the form

$$k_1a = AU_g^c \qquad \ldots(7)$$

The superficial gas velocity U_g is related to the flow rate g by $U_g = gL$, where L is the bioreactor height. Consider first an idealized bubble column, a hypothetical device in which all the bubbles are the same size, and doubling U_g simply doubles their number. The exponent c must then be 1, making $k_1a/g = AL$, a constant dependent upon scale but independent of gas-flow rate. In real bubble columns, with their imperfect bubble dispersion, c is closer to 0.7, but this still implies that G varies only slightly with g. The copious data available for oxygen transfer from air shows that, at normal gas flows (g on the order of 1 VVM), k_1a/g is on the order of 1 at the laboratory scale, rising toward 10 in commercial-scale equipment.

Assuming for the purposes of illustration that k_1 for a gas is proportional to its diffusivity in water raised to the 2/3 power (strictly valid only for relatively insoluble gases, for which most of the mass-transfer resistance is on the liquid side of the gas–biophase interface), this can be used to estimate G and the maximum possible fractional removal, $1 - e^{-G}$, (equation 4 with $S \rightarrow 0$) for contaminant gases of low, medium, and high solubility. The results shown in Table 1 suggest that bubble columns would be useless for insoluble contaminants like carbon monoxide, but worth consideration for moderately soluble contaminants like hydrogen sulfide. Most hydrocarbon vapors fall into the former category, but many alcohols and chlorinated aliphatic solvents are in the latter. For very soluble contaminants, such as sulfur dioxide, G tends to be unnecessarily large in bubble columns, and while acceptable removal with a moderate pressure drop could be achieved in very short, wide columns, this configuration has not been explored in practice. The addition of a mechanical agitator can increase k_1a by roughly one order of magnitude, mainly by breaking up the gas flow into many small bubbles with a large interfacial area. Equation 7 still applies, but the A parameter becomes a function of the mechanical power input per unit reactor volume.

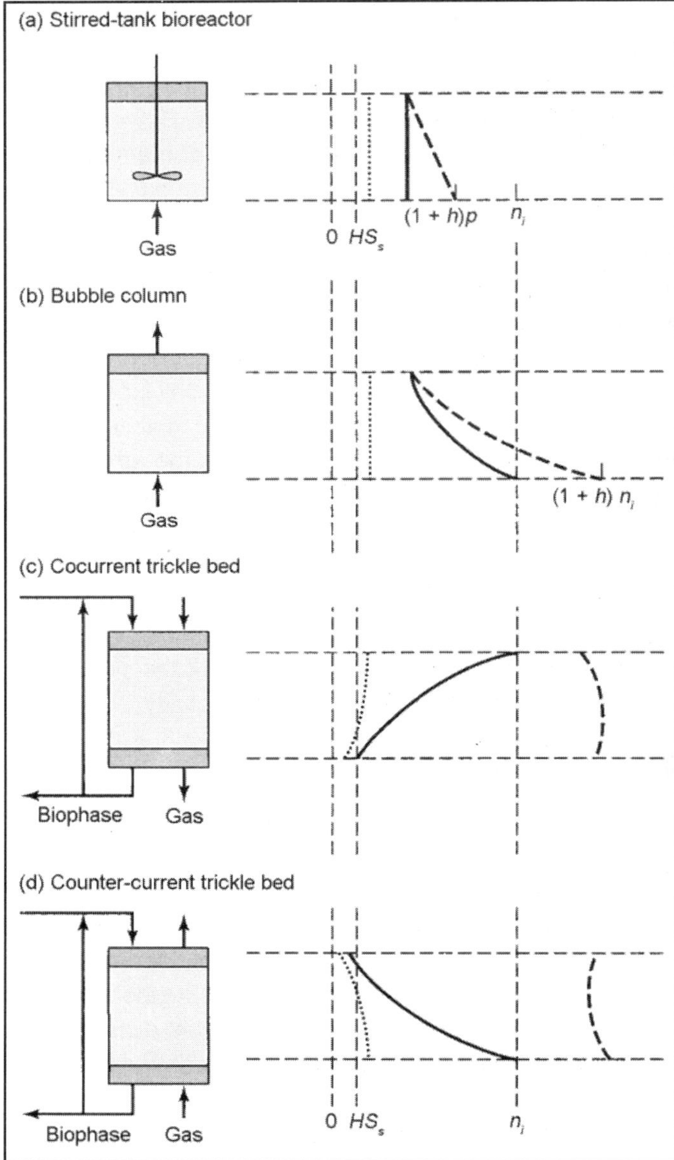

Figure 17.3 Concentration profiles in different bioreactors. (a), (b) Partial pressure in gas phase, dot-dash line (c), (d) Viable biomass concentration, dot-dash line, Dotted line, partial pressure in biophase = HS; solid line, mole fraction in gas phase.

Unfortunately, agitation also undermines the assumption that the gas moves though the reactor in plug flow. In a mechanically agitated tank, the gas bubbles are assumed to recirculate several times so that the gas phase is completely mixed; a sample of gas containing many bubbles taken from anywhere in the tank would have the same composition as the outlet gas. The effect on the driving force for mass transfer can be seen by comparing the concentration profiles. Mathematically, the long-mean concentration difference used in equation 3 must be replaced by the more familiar effluent partial pressure driving force $(p/H - S)$, which makes the maximum possible removal not $(1 - e^{-G})$, but $G/(1 + G)$. These two functions are virtually identical at small G, which explains why the exact definition of the driving force is not an issue for aeration (oxygen is an insoluble gas, $H/RT = 30$ at 20°C, and its small fractional removal from the air stream in conventional fermenters is unimportant). However, a goal of 95% contaminant removal now requires not $G > 3$, but $G > 19$, which gives some idea of the lower effectiveness of a completely mixed gas phase. The numbers in Table 17.1 show that this lower effectiveness cancels out the increased k_1a leading to conclusions similar to those for bubble columns. The extra expense of mechanical agitation is rarely justified in gas treatment bioreactors, although reactors designed for a high k_1a/g with large aspect (height: diameter) ratios and multiple impellers to prevent bubble coalescence have been proposed for removing hydrogen sulfide. Stirred tanks have been used successfully for the microbial production of ethanol and acetic acid from carbon monoxide and hydrogen (another insoluble gas; $H/RT = 51$ at 20°C), but the objective there was high volumetric productivity, rather than complete gas consumption. Several tanks were used in series, and they were pressurized to several atmospheres to improve gas transfer.

Table 17.1 Performance of bioreactors for which $k_1a/g = 1$–10 for O_2

			Maximum possible removal	
			Mixed gas phase	*Plug flow gas*
Contaminant gas	*H/RT at 20°C*	*C*	*G/(1 + G)*	*1 – exp(–G)*
CO	37	0.023-0.23	0.022-0.19	0.023-0.20
H_2S	0.41	1.8-18	0.65-0.95	0.84-1
SO_2	0.0088	88-880	0.99-0.999	1

In contrast to bubble-gasified reactors, the gas–biophase interfacial area in biotrickling filters is almost constant, fixed by the choice of packing media and only slightly influenced by U_g and the superficial velocity at which the liquid is pumped over the bed U_1. This makes it possible to achieve the necessary value of G for contaminants of almost any solubility by appropriate choice of the packing media and the gas flow per unit volume, g. Many types of packing are available and the best choice in a particular situation is a complex trade-off between the following considerations:

1. Less soluble contaminants will need the large a provided by smaller media and the small g implied by the lower permeability of such media to gas flow. For these contaminants, most of the gas–biophase mass- transfer resistance is on the biophase side of the interface, so in correlations of the form of equation 7, $c \to 0$, but A is a fairly strong function of U_1.
2. More soluble contaminants can achieve the same G with the smaller a and larger g given by larger media. In this case most of the mass-transfer resistance is on the gas side of the interface, so $c > 0$ and A is less dependent on U_1.
3. The conservation of mass shows that the change in concentration of a nonvolatile metabolic product in the biophase as it moves down the bed is $U_g x n_i y_m / U_1 RT$. If this number is small then the biophase approximates complete mixing. If it is large, then the concentration gradients, including that of pH, may be large enough to inhibit metabolism.
4. Although a high liquid flow improves mixing and mass transfer, there is a definite upper limit of U_1 at which the bed floods. This limit is predictable and is much smaller for small media.

5. It is difficult to generalize about the attachment of the microorganisms to the media because this depends on both the nature of the media and the type of biomass involved. However, it is clear that a bed of small media with attached biomass may quickly become plugged if the contaminant is sufficiently concentrated and the biofilm is allowed to grow. It is reasonable to suppose that the larger the media and the higher the liquid velocity, the more cells will be suspended in the liquid rather than being attached.

6. A biophase containing suspended and attached biomass may, if too thick or too dense, have its own mass-transfer resistance. While such resistance is usually to be avoided, it may be beneficial. For example, in trying to separate one electron acceptor, NO_x, from stack gases that contain another, O_2, which is preferred by the microorganisms, it should be possible to take advantage of the much higher solubility of the NO_x to create locally anaerobic environments in the biophase.

Somewhere among all of these considerations is a best choice of media for any situation. For dilute, less-soluble contaminants whose metabolism does not produce any nonvolatile metabolic products, the biotrickling filter will look more like a biofilter, with small media, little liquid flow, and attached biomass. For more soluble contaminants which do generate such products, it will tend toward the other extreme of high liquid flows and a completely mixed biophase. If there seems to be no sensible solution, the answer may be to adopt a bioscrubber. Although more expensive, it greatly simplifies the design problem by separating gas–liquid mass transfer from the microbiological considerations. It is certainly preferable to think in terms of this continuum of possibilities and to choose the one best suited to a particular application, rather than considering the different types of gas treatment bioreactors as completely separate technologies, each of which can be forced to fit any job.

Mixed Biophase Analysis

The single bioreactors with a completely mixed biophase analyzed here have an obvious drawback: all of the microorganisms are exposed to the same physicochemical environment, it is the worst possible environment for metabolism, with the lowest nutrient concentrations and the highest concentrations of metabolic products. Even with no mass transfer resistance ($G = \infty$), no product inhibition ($F = \infty$), and no wastage of biomass ($D = 0$), the analysis suggests that complete contaminant removal is still impossible. With no biomass outflow there is no net cell growth, and all of the contaminant is used for maintenance metabolism (equation 2 with $\mu = D = 0$ gives $q = k$). Under these conditions, the dissolved contaminant concentration is not zero, but the stationary phase concentration, S_s, so the partial pressure in the effluent gas cannot be less than HS_s, and the fractional removal, x, can never exceed $(1 - HS_s/n_i)$, even with perfect mass transfer (equation 4 with $G = \infty$).This maximum removal also exists for completely mixed biological wastewater treatment systems, but in that case it does not matter because S_s is much less than common effluent standards.

It may matter for some insoluble gas-phase contaminants because the corresponding minimum-achievable partial pressure, HS_s, may be significant, even when S_s is small. Bioreactors with a completely mixed biophase are a poor choice for such contaminants. The simplest improvement would be two bioreactors through which the gas flows in series and between which the microbial culture is continuously recirculated. Reactor 1 would have a relatively high effluent partial pressure and dissolved contaminant concentration, providing a good environment for metabolism and cell growth. If the process variables are fixed correctly (a complete analysis is beyond the scope of this chapter) reactor 2 could run with a dissolved concentration $S < S_s$. The biomass in reactor 2 would then be in an endogenous state, but this would no longer matter because fresh, viable cells would be continuously supplied from reactor 1, and the dying cells would either flow back there to recover or be lost in the liquid effluent.

The liquid inflow would go to reactor 1, where the growth nutrients are needed, and where it would dilute the concentrations of any metabolic products. The more bioreactors in series, the better the average physicochemical environment for metabolism. Columns containing many sieve trays or bubble-cap trays have long been used for gas–liquid contacting in the chemical process industry and should be considered for the biological treatment of gases in cases where the biomass grows best in suspension and the contaminant is relatively dilute, insoluble, and a poor microbial substrate (i.e., HS_s/n_i is large). These columns are inherently countercurrent contacting devices with the gas flowing upward while the microbial culture falls down from tray to tray. An individual microorganism experiences a series of environments with ever- increasing dissolved contaminant concentration, S, until it is suddenly pumped back to the top tray where conditions are endogenous ($S < S_s$), but where it can continue metabolizing contaminant and survive long enough to start through the cycle again. Nutrient addition and pH control can be done on each tray as needed, and the liquid effluent can be taken from near the top of the column where cell viability is lowest. A similar effect, with the gas closer to the theoretical optimum of plug flow, can be achieved in an up-flow biotrickling filter with a small liquid recycle ratio.

The only drawbacks are that pH control and nutrient addition can now only be done at the top of the bed, and the liquid can only be removed from the bottom, which is not necessarily the best arrangement. In real trickle beds some of the biomass remains attached, which is a great advantage over tray columns when a nonvolatile metabolic product is generated, because it allows a large liquid outflow to remove the product without the danger of washing out the biomass. (Mathematically, D is small because z is small, even though f is large.) Biotrickling filters can also be operated in the gas-down-flow, cocurrent mode, which gives concentration profiles (again assuming suspended biomass). Now, the microorganisms moving down the bed experience an ever-worsening environment for metabolism, with lower S and higher product concentration S_m. They leave the bottom of the bed in an endogenous, low-viability state, which is very suitable for the liquid outflow stream, but which may cause a significant lag phase when cells are recycled to the better environment at the top of the bed. The question of whether the cocurrent mode works better can only be answered definitively by experiments on a particular gas stream.

In biofilters the biophase is far from being completely mixed. It consists of an immobilized mass of up to xXe_b viable cells per unit bioreactor volume (less if cell growth has been limited by some nutrient other than the contaminant), and one of the goals of media selection is to provide sufficient surface area so that these cells can be distributed as a biofilm thin enough to avoid mass transfer limitation due to diffusion through the film. However, simple models suggest that biofilters should share the minimum effluent partial pressure, HS_s, predicted earlier for bioreactors with a completely mixed biophase, because biofilms exposed to any lower partial pressures near the biofilter effluent would be in an endogenous state and eventually die. Experience suggests otherwise, probably because the adsorption of the contaminants on the media increases their effective concentration in the microorganisms' immediate environment. It follows that highly adsorptive support particles, such as compost or activated carbon, will work best, and that biofilters can treat very insoluble, hydrophobic contaminants if they adsorb well.

Further Development

First-stage scale-up experiments can be done in a reactor of any size that is convenient or available, as long as it is large enough so that the flows of gas and liquid are not significantly altered by entrance effects, channeling, or wall effects. Ideally, several different bioreactor types or packing materials are studied, but this is not always possible. The objective is to vary the gas flow, g, the liquid flow, f, nutrient addition, Sji, and so on, to determine how the specified contaminant removal can be achieved most economically. The scale-up problem is then to extrapolate the results in order to calculate the necessary volume (volume = gas flow to be treated/g) and shape (specifically the height, L) of the commercial-scale bioreactor, which must satisfy the process requirements without further experimentation immediately after start up. The best combination of process variables found in the experiments must be close to the optimum identified by the point at the right end of the operating region.

To preserve this solution at the commercial scale, the dimensionless numbers D, F, and G that define this point must be kept the same between scales. Scaling up a gas treatment bioreactor is thus an example of Reynold's similarity principle, states that two physical situations are similar only if the all of the relevant dimensionless numbers are identical. Consider first the case where there is no inhibitory metabolic product and the process needs little or no liquid outflow. $F(\rightarrow\infty)$ and $D(\rightarrow 0)$ both drop from the list of relevant dimensionless numbers, and the scale-up problem reduces to finding combinations of gas flow, g, and bioreactor height, L, that keep G constant between scales. Given a mass-transfer correlation of the form of equation 7, G can be written $ARTL^c/Hg^{(1-c)}$. The problem is illustrated graphically in Figure 17.4, where lines of constant G are drawn on a log–log plot of g versus L. If the small-scale experiments have found different combinations of g and L that produce the specified contaminant removal, x, then this data should fall on such a line, labeled *empirical* in Figure 17.4, because the theory suggests that "constant x" implies "constant

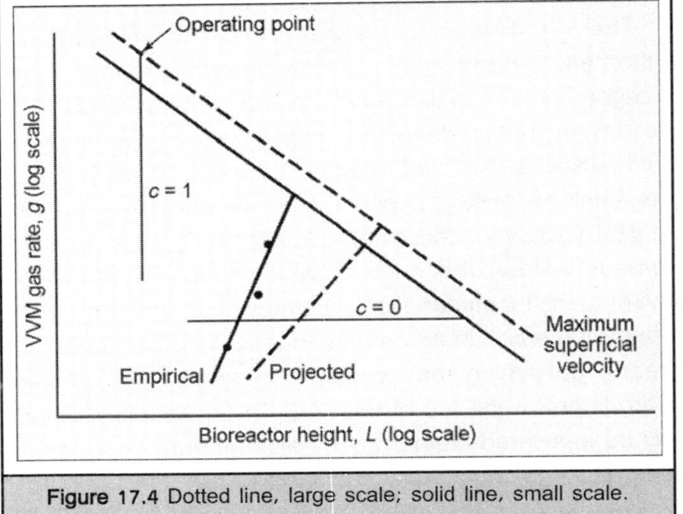

Figure 17.4 Dotted line, large scale; solid line, small scale.

G." Also shown in Figure 17.4 is an upper limit on the gas superficial velocity, $U_g = gL$, caused by physical phenomena such as slugging of bubble columns, fluidization of the media in upflow biofilters, or flooding of the impeller in stirred tanks. This limit may be slightly different at different scales. Two extreme cases of the scale-up problem can be identified. For a biofilter or a biotrickling filter treating sparingly soluble contaminants, the mass-transfer factor is little affected by the gas velocity, implying that $c = 0$ and making the empirical line horizontal.

Several important phenomena are lumped into the parameter A, but there is no reason to expect them to vary with scale because microorganisms cannot "know" if they are in a small or large bioreactor. Consequently, performance on the large scale is defined by the same horizontal line, meaning that scale-up is done at constant g, or equivalently at constant gas-residence time (=1/g). Because $g = U_g/L$, constant g can be achieved either with a deep bed and a high gas velocity, the right end of the $c = 0$ line in Figure 17.4, or with a shallow bed and a low velocity, the left end of the line. Although the former are easier and cheaper to build, they have a much higher pressure drop though the bed, meaning higher costs for gas compression and possibly subsequent cooling to return the gas to a temperature tolerable to the microorganisms. The latter are much cheaper to operate and work well as long as the beds are deep enough (at least several hundred times the particle size) to avoid problems of channeling and short-circuiting of the gas flow. Many biofilters consist of shallow beds of media on trays arranged in various configurations in a reactor designed to ensure an even flow of gas into the beds. At the other extreme is the ideal bubble column described in the section "Gas Biophase Mass Transfer" for which $c = 1$.

The value of G, and thus the contaminant removal, is now independent of gas flow but proportional to bioreactor height. Scale-up must therefore be done by keeping L constant but increasing the reactor width, the large-scale bioreactor operating at maximum g in order to minimize its volume (the *operating point*). Although this ideal bubble column is an abstraction, it illustrates a paradox that can arise in scaling up practical bubble columns (for which $c \cong 0.7$); if the desired removal can be achieved in a small-scale column, then a bubble column is probably not the best solution at the commercial scale. For very soluble contaminants, the L needed to obtain the desired fractional removal may be in the range of laboratory reactors, around 1 m. Scale-up would produce a very unusually shaped large-scale bioreactor, a similar height but several meters wide. There would be no point in making it any deeper, except to provide a safety factor in the design, and it could not be made any narrower and still meet the process requirements. For insoluble contaminants the necessary L may be several meters.

The conclusion from laboratory experiments would then be that a bubble column is not a feasible bioreactor, even though the desired contaminant removal may be achievable (albeit with a high pressure drop) with the reactor heights used at the commercial scale. All practical situations fall between these two extremes, so neither simple scale-up rule, constant g or constant L, is necessarily appropriate. The empirical line will have a positive slope, and a similar line for the same G at the larger scale can be projected from it if any variations of the parameters A and c with scale can be estimated from the published literature, experience, consultations with equipment manufacturers, and so on. The design of the large-scale bioreactor can be anywhere along this projected line in Figure 17.4, the exact design point again being constrained by the maximum U_g line, but essentially an economic trade-off between the large bioreactors demanded by a small *g* and the large pressure drops associated with a large *L*. The essential point is that the shape, and the size of the large-scale bioreactor are fixed by this procedure and should not be chosen based simply on guesswork or convenience. Some bioreactor types allow design flexibility because they have variables that can be adjusted when drawing the projected line. Both *A* and *c* tend to decrease with scale in any bubble-gasified reactor due to poorer gas dispersion, but in mechanically agitated tanks, *A* is a function of the power input per unit volume, which can be varied to some extent.

For trickle beds and biofilters, both *A* and the bed permeability are predictably related to packing size, allowing the packing to be chosen to give the large-scale reactor both a convenient shape and a reasonable pressure drop. Note that, whether or not such changes are made between scales, the estimates on which the projected line in Figure 17.4 is based should always be conservative in order to provide a factor of safety in the large-scale design. When product inhibition is significant, the liquid outflow, *f*, becomes a critical variable, and all three dimensionless numbers *D, F,* and *G* must be kept constant if the physicochemical environment of the biomass (S, S_m) and the optimum bioreactor performance established in the small-scale experiments are to be extrapolated to the large scale. An exact solution is possible only when the retention of the biomass by cell immobilization or recycle (the *z* parameter) can be controlled, although when retention is almost complete ($z \rightarrow 0$) only the product *DF* (from which *z* cancels) is important, implying that the liquid flow, *f*, must be kept proportional to the gas flow, *g*. Biotrickling filters have an extra dimensionless number, the liquid recycle ratio, *RR*, which introduces the extra difficulty that increasing the bed height, while keeping *f* and *RR* constant increases the liquid superficial velocity [$U_l = Lf(1 + RR)$], thus changing the mass-transfer situation and possibly flooding the bed.

The procedure just described will not work in these more complex situations (constant *x* no longer implies constant *G*), but the general considerations still apply, including the physical limits on the gas superficial velocity, the desirability of tall small-scale bioreactors to reduce uncertainties, the need for compromise between capital and energy costs, and the possibility of predicting a very inconvenient aspect ratio for the large-scale bioreactor. Judgment and calculation are needed in each case because the trivial solution to the scale-up problem— "build a bigger bioreactor with the same shape, *f,* and *g*as the small one"—is unlikely to be the best. Gas treatment bioreactors are one of the few devices whose performance tends to improve with scale, but taking advantage of this tendency in practice is not straightforward.

18

BIOACCUMULATION OF METALS

It has been known for decades that some types of microbial biomass tend to accumulate metals rather effectively. This is the case even when the biomass is dead, that is, metabolically inactive, since it is the chemical composition that is responsible for sequestering metal ions and/or metal complexes from solutions. This fact obviates the necessity to maintain special growth-supporting conditions to make the biomass sequester metals. This metal-sorption phenomenon can find enormous applications wherever metals would need to be removed from solution and concentrated.

Rather than searching among thousands of microbial species for biomass with this particular metal-sequestering feature, feasibility dictates us to look first at biomass types that are readily available and in quantities to support the potential demand. Copious quantities of sea- weeds, which could easily be collected, grow in the oceans and, indeed, some of them have been identified for their especially outstanding metal-sorbing properties. Other quantities of biomass are generated as by products of large-scale industrial fermentation processes. Often posing disposal problems, some of these readily available industrial-biomass types have been found to be excellent metal sorbers. A whole new family of metal (bio)sorbing materials appears to be on the horizon, serving as a basis for the new technology of biosorption aimed specifically at metal removal and recovery from metal-bearing industrial solutions and/or effluents.

METAL REMOVAL TECHNOLOGY

By far the largest demand for particularly cost-effective metal-removal technologies from dilute solutions is in ecoapplications. Metal ions in the environment are biomagnified in the food chain and accumulated in tissues. Cd, for example, has a half-life of 10–30 years in the human body. Therefore, their toxic effects are especially pronounced in animals of higher trophic levels, particularly humans. From the ecotoxicological point of view, the most dangerous metals are the so-called big three: Hg, Pb, and Cd. Following well-documented and publicized disasters and problems with Hg and Pb, the large-scale use of these metals by various industries has been either eliminated or very curtailed. Technological processes making use of these metals have recently been phased out, particularly in developed countries, and replaced by environmentally safer technologies.

A major step in limiting the amount of lead discharge into the environment, for instance, has been made by the elimination of leaded gasoline. On the other hand, the use of Cd has been steadily increasing in growing applications such as in electroplating, pigments, and batteries. Cd toxicity causes kidney damage, bone disease, and cancer. Other common heavy metals, such as Cr, Cu, Zn, Ni, and V are less toxic, but are among the most widely used metals (after Fe and Al). Cu is mainly employed in the electrical industry (e.g., wires) and brass production. The major applications of Zn are galvanization and production of brass and other alloys. Mine tailings and effluents from nonferrous metal industry are the major discharge sources of different combinations of dangerous heavy metals.

Therefore, it would be beneficial to devise separate pretreatment facilities for the elimination of heavy metals from wastewater, ideally at the site of those plants that discharge significant amounts of heavy metal ions with their wastewater. Such on-site treatment of waste streams contributing a load of heavy metals would be more efficient than treating large volumes of mixed wastewater in a general sewage plant.

Additionally, from a policy viewpoint it seems reasonable that those who cause metal pollution should be responsible for its elimination. While industrial enterprises are presently obliged to meet certain discharge standards, drinking-water quality cannot be considered since such treatment would be prohibitively costly. Natural-metal recycling in the environment makes metal recovery the only effective way of preventing heavy metals from poisoning the environment. Yet if a cost-effective way of completely eliminating heavy metals could be found, this goal could be attainable. Current research activity in the field attempts to evaluate whether biosorption may eventually provide such an effective and economical treatment-process alternative.

Table 18.1 Drinking water standards and discharge limits of heavy metals

Metal	*Drinking water standards WHO (mg/L)*	*Discharge limits Canada (mg/L)*	*United States (mg/L)*
Cd	0.003	1.5	0.1
Cu	2	1	3.3
Cr	0.05	1	2.8
Ni	0.02	2	4
Pb	0.01	1.5	0.6
Zn	3	2	2.6

While biological treatment is reasonably effective in removing organic pollutants, heavy metals tend to accumulate in biological sludges, which are considered unsafe as fertilizer and require incineration. The accumulation of metals by the microbial biomass is a result and typical example of combined metabolism-dependent active bioaccumulation and passive biosorption of metals.

Utility of Metal Biosorption

The term *biosorption* commonly refers to the passive binding of metal ions by biomass, which may even be dead. It must be distinguished from bioaccumulation, which is usually understood to be an active, metabolically mediated process occurring in living organisms. It has been known for decades that different types of microbial biomass bind trace metal ions, achieving very high concentration factors. Black and Mitchell reported a 1,000-fold increase of metal concentration in biomass of some brown marine algae as compared to the seawater. The biomass-metal-uptake phenomenon has been made use of in, for example, electron microscopy, to visualize cellular components. The focus in early studies of microbial-metal uptake has been almost exclusively on the nutritional and toxicological aspects of metal presence.

Recently, however, efforts have been made to develop the process of metal accumulation by biomass into a technique for the treatment of toxic metal-bearing industrial effluents, or for the recovery of precious elements from processing solutions, or even seawater. The process of gold sorption by *Sargassum* has been patented, and the immobilized biosorbents AlgaSORB and AMTBIOCLAIM are being commercially developed. Biosorption removal of toxic heavy metals is especially suited as a "polishing" water-treatment step because it is possible to reach drinking water quality of the treated water (initial metal concentrations, e.g., 1 to 100 mg/L, final concentration <0.01 to 0.1 mg/L), especially in packed-bed flow-through sorption column applications. To prevent unnecessarily rapid exhaustion of the sorption capacity when the metal concentrations in wastewater to be treated are high (>100 mg/L), it may be desirable to use a different pretreatment technique, such as precipitation or electrorecovery, for removal of the bulk of the metal content. However, generation of toxic sludges during pretreatment must be considered, since these represent another type of hazard and the eventual recovery of metals from them may not be feasible. The metal-laden biosorbent can be regenerated, incinerated, or stored in landfills. The biosorption process basically serves to reduce the waste volume.

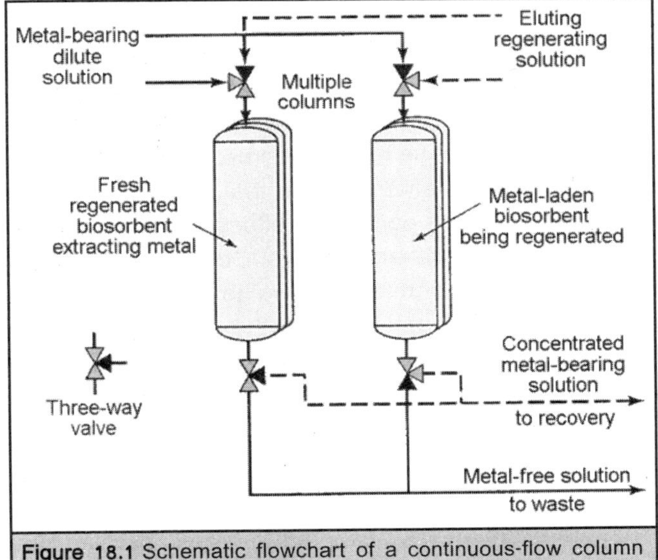

Figure 18.1 Schematic flowchart of a continuous-flow column biosorption process.

Alternatively and preferably, regeneration of the biosorbent material or multiple (re)use is desirable to increase the process economy. It can be accomplished by metal desorption with, for example, acids or salt solutions. The resulting highly concentrated metal solution can be processed by other techniques such as precipitation or electro winning to remove or concentrate the metal, which could be the recovered and resold. The latter process, in particular, is aimed at recuperation of the metal. The overall achievement of the biosorption (complete adsorption + desorption cycle) process is to concentrate the metal solution, possibly by at least a factor of 100 or more. The flowchart of the biosorption process with in situ regeneration of the biosorbent is quite similar to the use of ion-exchange resins.

Compared to conventional wastewater "polishing" techniques such as ion-exchange, activated carbon treatment, or membrane technologies (electrodialysis, reverse osmosis), the advantage of biosorption is not only that it can be operated under a broad range of conditions (pH, temperature), but especially that it may be economically attractive due to the cheap raw materials that can be used as biosorbents. These include waste products from other industries (e.g., fermentation by-products) or naturally abundant biomass (e.g., ubiquitous marine algae). Recent studies show that for certain types of seaweed biomass, for example, virtually no pretreatment may be necessary: Untreated *Sargassum* biomass was successfully applied in packed-bed column biosorption of Cu. This means that the sole costs of the seaweed-based biosorbent may be only those of collection and transport. To evaluate the expected performance of biosorption for treating typical waste waters, it is desirable to reliably predict how much metal could be bound under different operating-process conditions. Computerized process simulations could effectively and economically provide this relevant information, which is essential for the engineering process design and scale-up.

Application of Metal Biosorption

Ecological approaches

A multitude of biomass types, comprising fungal biomass, bacterial biomass, algae, peat, and so on, have been studied for their biosorption of metals. It was observed that not only the species but also growth conditions such as culture medium, as well as physiological state and age of the organism, influence the biomass-sorption performance. In many cases, the sorption by dead material proved to be more effective than that by living organisms. Biomass of some marine brown algae has been found to be particularly effective in metal-ion binding, showing similar or better performance than industrial ion-exchange resins. Gold accumulation by the brown alga *Sargassum* can constitute up to 40% of the biomass dry weight. The particular amount of metal bound in the biosorbent depends, however, not only on the chosen biosorbent but also on the type of metal ion, its concentration, and other physicochemical parameters of the solution (e.g., pH, ionic strength).

Thermal effect

Simple physical sorption processes are generally exothermic (i.e., the equilibrium constant decreases with increasing temperature). Biosorption ion exchange among alkaline earth metals in binding to alginate was observed to be exothermic. However, when Cu was the replacement, the reaction exhibited a positive enthalpy change (i.e., it was endothermic—the equilibrium constant rose with temperature). The driving force in this case was attributed to a large positive entropy change, possibly caused by a larger ordering effect of Cu (than Ca) on the water molecules in the hydration sphere. This is illustrated by the following equation:

$$K = \exp\left(\frac{-\Delta G^0}{RT}\right) = \exp\left(\frac{-\Delta H^0 + T\,\Delta S^0}{RT}\right) = \exp\left(\frac{-\Delta H^0}{RT}\right)\exp\left(\frac{\Delta S^0}{R}\right) \quad ...(1)$$

A rise in K with increasing T means that $\Delta H^0 > 0$, that is, that the reaction is endothermic. But since a reaction can only proceed when $\Delta G^0 = \Delta H^0 - T\Delta S^0 < 0$, it follows that DS^0 must have been >0, that is, the system's entropy must have increased.

Similar results were obtained by Weppen and Hornburg in calorimetric studies of the binding of divalent metal ions by potassium-saturated microbial biomass. The reactions were slightly endothermic and therefore driven by an increase in the entropy of the system. For most metals, the heat of reaction was constant, independent of the degree of site occupation. For Cu, however, the heat of reaction decreased with increasing degree of site occupation from 27 kJ/mol to 14 kJ/mol, which may indicate the involvement of different binding sites or the formation of different types of Cu complexes with the biomass.

For other heavy metals, the heat of reaction was between about 7 and 11 kJ/mol, for light metals between about 2.5 and 6 kJ/mol. Overall, however, the effect of temperature in biosorption is relatively small and perhaps only slightly more pronounced in the lower temperature range: the binding of Co by the brown alga *Ascophyllum nodosum* increased by 50–70% when the temperature was raised from 4°C to 23°C. With further temperature increase to 40°C the binding increased only slightly, whereas temperatures of 60°C or more caused a change in the texture of the sorbent and a loss in the sorption capacity. The distribution ratio (metal bound/metal in solution) for biosorption of Cd, Zn, Pb, Ni, and Cu to *Spirulina* algae increased by only about 20% when the temperature was raised from 4°C to 55°C. It is reasonable to expect that possible application of biosorption will be in a rather narrow temperature range (~5 to 50°C) to avoid decomposition of biosorbent materials.

Anions in bioprocess

Theoretically, the presence of ligands (at levels that do not cause precipitation) can lead to the following:

1. Formation of complexes that have a higher affinity to the sorbent than the free metal ions (i.e., an enhancement of sorption)
2. Formation of complexes that have a lower affinity to sorbent than the free metal ions (i.e., a reduction of sorption)
3. Interaction of anions with the biomass, changing the state of the active sites such that the binding is either enhanced or reduced

There are no indications of the third kind of interaction reported in the biosorption literature. Additionally, these are unlikely to occur in brown algal biomass since the probable binding sites (negatively charged carboxyl and sulfate groups) are not expected to interact with anions such as sulfate, nitrate, and so on. It has been reported for sorption of Cu^{2+} and Ag^{+} to inorganic oxide surfaces that metal binding at low pH was enhanced in the presence of certain ligands. Since the ligand itself was able to bind to the oxide, and since a 1:1 stoichiometry of ligand and bound metal was noted, it was concluded that some metal complexes may bind more strongly than the respective free ions. Negatively charged metal–EDTA complexes were bound to oxide surfaces (probably releasing OH^{-}), with sorption decreasing with increasing pH. Chen et al. report that up to 50% of

Cu binding by peat occurred as complexation of copper nitrate [$Cu(NO_3)_2$] with the binding sites. The remaining metal binding was due to ion exchange such that the overall charge neutrality was conserved. The data allow, however, another interpretation as well: Instead of up to half of the Cu being bound as $Cu(NO_3)_2$ and the other half as Cu^{2+}, up to 100% could have been bound as $Cu(NO3)^+$.

Glaus et al. found that the binding of Co and uranyl (UO_2^{2+}) to humic substances could be modeled better when binding of metal–ligand (oxalate) complexes was included in the model. In most cases of biosorption, however, metal binding tends to be reduced in the presence of ligands (exception: OH^-). This means that the biomass apparently has less affinity for many metal–ligand complexes than for free hydrated metal ions. Therefore, the influence of ligands in solution can be understood as a competition with the biomass for binding of the metal ions. This effect is, however, not very pronounced unless these anions show a strong complexing power for the metal ions of interest (e.g., EDTA), which at only 1.5 times excess caused 100% inhibition of metal binding. The addition of 0.1% EDTA resulted in >90% reduction of binding for most cases of Cd, Cu, Zn, and Pb biosorption by seaweeds (also *Sargassum*). A large excess (1,000 times) of the ligands glutamate, sulfate, phosphate, carbonate, or chloride resulted in the maximum reduction of, for example, Cd binding to *Rhizopus arrhizus* biomass of only 21%. Since significant concentrations of strong complexing agents are rare in typical metal-bearing industrial effluents, the effect of the presence of ligands is of secondary importance.

Effect of pH

Of great importance in sorption is the pH value of the solution. It is commonly agreed that the sorption of metal cations (e.g., Cd, Cu, Zn, Pb, Ni, Mn, Al, Co) increases with increasing pH as the metal ionic species become less stable in the solution. Only those metal ions that can occur as negatively charged complexes or that have a strong "b" character (i.e., tendency to form strong covalent bonds), such as Ag, Hg, or Au (e.g., as tetrachloroaureate), may show either a decrease in binding with increasing pH or may have no significant pH effect at all. There are three ways that the solution pH can influence metal biosorption, as described in the following.

First, the state of the active sites maybe changed. When the binding groups are acidic, the availability of free sites depends on pH: At lower pH the active sites are protonated, therefore, competition between protons and metal ions for the sorption site occurs. At a low enough pH, virtually all sites become protonated and complete desorption of the bound metal ions is possible, which is why acid treatment is a method for metal elution and regeneration of the sorbent material. Decreasing the pH value by 2 units can in some cases result in a ~90% reduction of metal binding.

Second, extreme pH values, as they are employed in the regeneration (desorption) of the sorbent, may damage the structure of the biosorbent material. Microscopic observations have shown distorted cells; significant weight loss and decrease in the sorption capacity have been observed.

Third, the speciation of the metal in solution is pH dependent. Whereas metals in aqueous solutions occur as hydrated cations in solvation shells when the pH is low, hydroxides may form at a higher pH, especially for cations of high charge and small size. The formation of metal oxide and hydroxide complexes and precipitates is often called hydrolysis (i.e., decomposition or conversion by water). Adsorption depends not only on the attraction of the sorbate to the solid surface but also on its lyophobic behavior. This means that sorption increases with decreasing solubility.

Since the solubility of many metal complexes in solution decreases with increasing pH, this provides an additional possible explanation as to why sorption increases with increasing pH. In the narrow pH range where the metal ions are hydrolyzed, sorption is especially enhanced, so that even a charge reversal of the surface to positive values can occur. Further possible explanations of increasing sorption with increasing pH are that hydrolyzed species have a lesser degree of hydration, that is, less energy is necessary for removal or reorientation of the hydrated water molecules upon binding.

Furthermore, OH shows affinity for some sorbents. With further increase of pH, the solubility of metal complexes decreases enough for precipitation to occur. Although precipitation may contribute to the overall removal of metals from solution (and therefore be desirable for metal-removal applications), it renders the study of the biosorptive binding more difficult. For scientific purposes it is therefore recommended to study biosorption at pH values where precipitation does not occur.

Cations in bioprocess

Other sorbable ions in the solution may compete with the metal ion of interest for sorption sites. The binding of this metal ion is then decreased. The amount of inhibition depends on the binding strength of the respective ions to the biomass. For example, Mg and Na are effective in competing with Zn for binding sites of green algae [90% reduction of Zn binding at 200 mM NaCl or $Mg(NO_3)_2$], but they interfere less strongly with the binding of Pb or Cu [<50% reduction of Pb or Cu binding at 200 mM NaCl or $Mg(NO_3)_2$]. The inhibition by alkali metals (K, Na) of heavy-metal- ion binding to microbial biomass was much less pronounced than the inhibition by heavy metals (Zn, Cu, Fe) of uranium or radium binding.

For *R. arrhizus* fungal biomass, the presence of uranium reduced the binding of Cd, Zn, and Ag more strongly than vice versa; Ag in particular was a weak competitor, but nevertheless its binding was unaffected by the presence of Na. In the sorption of Cd, Cu, and Zn on formaldehyde cross- linked biomass of the brown alga *A. nodosum,* the presence of each ion reduced the binding of each other ion, with Zn being the weakest binding, and Cu the most strongest binding, of the three cations.

From the different literature sources it can be generally concluded that the light metals (alkaline and alkaline earth metals) bind less strongly than the heavy-metal ions or radioactive elements. Therefore, the former do not strongly interfere with the binding of the latter. There are indications that among the heavy metals, Zn binds rather weakly and its binding is therefore more strongly affected by other ions.

In order to understand why some ions bind more strongly than others, it is useful to look at the properties of different metals, to be discussed later. Such knowledge can also serve to estimate how strongly an ion may bind that has not yet been investigated. This knowledge serves as a basis to better understand the metal-binding mechanism.

Biological Accumulation of Metals

The effect of any of the previously mentioned influencing factors can be best quantitatively estimated when the mechanism of biosorption for different ions is known and when a mechanism-oriented mathematical model is used for predicting the effect of these factors.

Sequestering mechanisms

The metal binding in biosorption has been attributed to a number of different sequestering mechanisms, such as complexation, coordination, chelation, adsorption, ion exchange, or microprecipitation (as metal or metal complex). It is a recognized fact that a combination of several mechanisms, each functioning independently, can contribute to the overall metal uptake. In the studies of biosorption conducted so far, very little attention has been paid to examination of a well-defined metal uptake (by a specific mechanism) as opposed to the overall uptake where several types of sequestration may be taking place simultaneously. Systematic understanding of the metal-uptake mechanisms and their relationships may greatly clarify the otherwise confusing broad definition of biosorption mechanisms in the literature. For example, some mechanisms can be submechanisms of other overall mechanisms. The claim of an ion-exchange metal-binding mechanism for a specific system may not necessarily be contradictory to reports of chelation being the relevant mechanism.

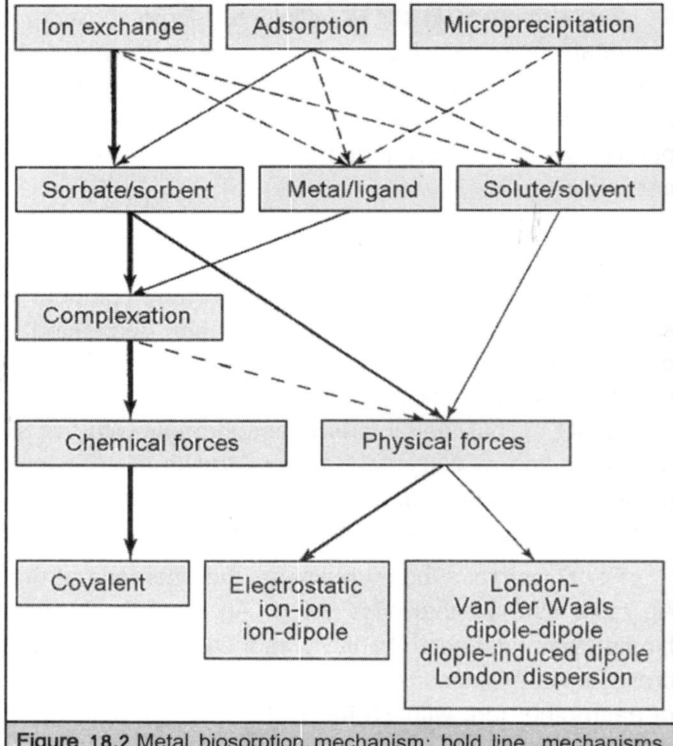

Figure 18.2 Metal biosorption mechanism: bold line, mechanisms probably important in biosorption by Sargassum biomass; dashed lines, biosorption binding relations of secondary importance.

The overall mechanism of ion exchange can be based on the submechanism of complexation. This means that one complexed ion exchanges against another one. The classification of bond types as described by Myers was used, as well as information on sorption processes from Westall and Pagenkopf. In the context of this work, the term *sorption* refers to binding of a solute (usually a metal cation) to free sites that previously were not occupied by another cation. If the sites are initially occupied by another cation (regardless of the submechanism, i.e., whether covalent or electrostatic binding is involved), and if this second ion is released on the binding of the first ion, then the term *ion exchange* is used. Microprecipitation is the deposition of electrically neutral material (metal or metal salt) at the surface of the biomass, and does not necessarily involve a bond between the biomass and the deposited layer. Microprecipitation may, however, be facilitated by initial binding of metal ions to reactive sites of the biomass, which serve as nucleation sites for further precipitation. This process is not limited to a monolayer (or saturation of sites): Cells can accumulate several times their dry weight in metal.

Microprecipitation is based on interactions between the solute (dissolved solid) and the solvent, and occurs when the local solubility is exceeded. Ion exchange and adsorption can be the result of three different interactions (which can act in combination): The main contribution for free metal ions (which are highly soluble in water) is usually attraction of the sorbate (metal ion) to the sorbent (biomass). In the case of binding of metal–ligand complexes, or in the case of ligand exchange (where the metal ion in solution is complexed with a ligand that has to be released when the metal ion is bound to the biosorbent), interactions between ligands and metal cations must be taken into account. Additionally, hydrophobic (or, generally, lyophobic) expulsion (which is an interaction between sorbate and solution) may also play a role, especially in the adsorption of organic dipoles and large organic ions because of their low affinity to the aqueous phase.

Among inorganic ions, the less hydrated ones are more readily accumulated at the interface. The hydration effect is further discussed later. Complexation plays an important role in both metal–ligand and sorbate–sorbent interactions. A complex (also referred to as a coordination compound) is a polyatomic molecule that consists of one or more (then called polynuclear complex) central atoms (usually metal cations) surrounded by ligands (other atoms or groups, usually of negative or neutral charge) that are attached to it. Complexes can be neutral, positively charged, or negatively charged. The number of coordinating atoms in the ligands that are directly attached to the central atom is called the coordination number, and can be larger than the valence of the central atom. (The most common coordination numbers are 4 and 6, but 2 and 8 also occur frequently.) If one ligand is attached to the central atom through two or more coordinating atoms, then the complex is called a chelate. The bonds between the central atom and the coordinated groups can be arising from principal or auxiliary valence forces.

Physical and chemical forces

In general, the forces between atoms or molecules can be classified into chemical and physical ones. Chemical forces extend over very short distances (0.1–0.2 nm). This type of bond is rather strong: Stumm and Morgan report an energy of $\gg$40 kJ/mol, Myers 150–900 kJ/mol, Smith for chemisorption 20–400 kJ/mol and Pagenkopf 40–400 kJ/mol. Covalent bonds are formed by merging electron clouds such that a nonionic molecule is formed. These bonds are directional (characteristic bond angles and lengths) and localized. Physical forces can be subdivided into electrostatic and London–van der Waals forces. The energy of physisorption is reported as 2–20 kJ/mol and 20–40 kJ/mol by Smith and Pagenkopf, respectively. In the resulting bonds, the electrons stay in their original systems.

Electrostatic (or coulombic) forces between ions or between ions and dipoles extend over a long range and are the strongest among the physical bonds, with energies $\gg$40 kJ/mol. The interaction is repulsive for ion charges of the same sign and attractive for unlike charges. The magnitude of the force is proportional to the charge of each ion and inversely proportional to the square of the distance between the ions. London–van der Waals forces can be divided into three categories: dipole–dipole interactions (creating orientational energy), dipole-induced dipole interactions, and the London dispersion force. The first two are closely related to coulombic forces, while the last is of quantum– mechanical nature and acts over a long range, of up to about 10 nm. The energy of the dispersion force (8–40 kJ/mol) is larger than the one of orientational or induced dipole (or polarization) energy (<8 kJ/mol).

An example of untypical strong (almost ionic) dipole interactions is hydrogen bonding (which may even display some covalent bond character). It occurs between molecules in which H is bound to a very electronegative atom such as O (e.g., in H_2O). The mechanisms and forces that are probably the most important ones in the case of metal binding by, for example, brown algal biomass, that is, complexation by chemical covalent bonds based on ion exchange (for which the driving force is the affinity of the sorbate [metal ion] to the sorbent [biomass]) supported by physical electrostatic interactions between positively charged metal ions and negatively charged biomass.

Role of ion exchange

There have been some indications that ion exchange plays an important role in metal-ion sorption by algal biomass. Kuyucak and Volesky noted that the amount of ions (K, Na, Ca, Mg) released from the marine brown alga *A. nodosum* was much more pronounced in metal (Co) bearing than in metal-free solutions. A linear correlation between Ca release and Co uptake (2:3) was found when the biomass had been previously washed with CaCl and HCl. It was concluded that ion exchange was responsible for the sorption. If ion exchange was responsible, an exchange ratio of 1:1 should, however, be expected for two divalent ions such as Ca and Co. Since it can be expected that some sites would have been protonated due to the pretreatment with a CaCl and HCl mixture, and since some proton release was noticed, it would have been more appropriate to relate the Co binding to the sum of Ca and H released (in milliequivalent not mole). This would have led to an exchange ratio much closer to 1:1. Crist et al. included protons in the quantitative study of ions (Ca, Mg, H) released during metal (Cu) uptake by the freshwater alga *Vaucheria* and found that the amount of divalent metal ions taken up equaled the amount of other divalent ions plus protons released (in milliequivalent terms).

Treen et al. reported that in column application of *Rhizopus* fungal biosorbent 2 mol of protons were released for 1 mol of uranyl sequestered. According to Chen et al., the Cu binding to peat at low metal concentrations was predominantly ion exchange (with H, Ca, Mg), while at higher concentrations, binding of copper–nitrate complexes occurred. The latter mechanism was dominant above 6 mM of Cu. It appears reasonable that in many cases ion exchange rather than sorption to free sites is the relevant overall mechanism for the binding of metal ions in biosorption. Since the overall charge of the biomass particle has to be neutral, any binding of one cation must be accompanied by either a stoichiometric release of other cations or by the binding of anions. Since the charge of algal biomass is negative, it appears unlikely that free anions will be bound by the biomass. If complexes in solution occur, double exchange of cations and anions or binding of the complex may take place. Generally, sorption of neutral (metal cation + anion) complexes would be possible. However, it is unlikely to be the case at the moderate pH and metal, as well as the ligand concentrations, relevant to the biosorption process.

Desorption of the metal

Due to the weak nature of metal-binding forces, it is possible to wash the bound metal from the sorbent in a small volume of solution offering another cation (e.g., H^+, Ca^{2+}) in higher concentrations. From the process point of view, this enables the reuse of the biosorbent in multiple uptake/desorption cycles allowing for a greater process economy. Desorption can be achieved by acids (e.g., HCl, H_2SO_4) and/or salt solutions (e.g., $CaCl_2$). The effectiveness of desorption depends in this case on the binding strength of the added cation to the biosorbent. A different approach to desorption is the addition of a complexing agent (e.g., EDTA, KSCN) that binds free-metal ions and thereby lowers their concentration in solution to the extent that they are released from their binding sites on the solid. In this case the driving force is the complexation of the metal ion. The efficiency of the desorption depends on the binding strength of the desorbed ion to the complexing ligand. Nevertheless, a charge balance of the sorbent particle has to be maintained, that is, ion exchange still occurs. That means some cation (e.g., the one that was added along with the complexing ligand, often Na) has to replace the de- sorbed ion at the binding sites, even if its affinity toward the binding sites is lower.

Physiochemical binding

The proportion of protons among the released species varied with the metal taken up and corresponded to the binding strength: More protons were released by the stronger binding ions. It could be concluded that in those cases where no protons were involved, as for most alkaline and alkaline earth metals, electrostatic attraction was the binding mechanism, whereas covalent binding occurred when protons were released. It appears plausible that the number of protons released increases with the binding strength of the metal ion because it can more effectively compete with protons. Also, as a general rule, the force of electrostatic binding alone is weaker than binding with covalent contribution (especially since alkaline earth ions have similar radii to those of many divalent heavy metal ions, so that large differences in the binding strength are due to different amounts of covalent binding). However, it does not necessarily follow that proton release indicates covalent binding; part of the protons (depending on the relationship between pH and $p^{BH}K$) may also be bound by electrostatic attraction and therefore be easily replaceable by electrostatically bound ions.

Specific sites of binding

Biosorption has been attributed to different types of groups such as carboxylate, carbonyl, hydroxyl, amine, amide, imidazole, phosphate, thio, and thioether groups. Which binding mechanism applies in any specific case depends, of course, largely on the type of biosorbent used and on the types of binding sites it contains. Crist et al. proposed that, for example, carboxylate, sulfate, and amino groups may be responsible for metal binding by freshwater algae. Indications for carboxyl- and sulfate-group participation were found by titration of marine algal biomass. The chelating ability of polysaccharides from green freshwater alga *Chlorella* was connected to their content of uronic acids. Their carboxyl groups would be negatively charged and could bind metal ions. Tobin et al. concluded that in *R. arrhizus* fungal biomass at pH 4, carboxylate and phosphate groups offered the primary metal-binding sites, supported by hydroxyl groups. Greene et al. noticed a complete loss of available sulfhydryl groups (determined by polarimetrictitration) after the binding of Au(III) to the freshwater alga *Chlorella,* thus giving evidence for the involvement of this group, but it could only account for a small amount of total gold binding. Cell modifications with succinic anhydride indicated the relevance of amine groups, since the sorption of tetrachloroaureate decreased by 50% after their modification. The modification of carboxyl groups by esterification with acidic methanol resulted in a significant decrease in the binding of Cu and Al, but a slight increase in Au binding by five different algae.

The decrease in sorption was more pronounced for Al, which as a hard metal is expected to bind preferably to hard sites, such as carboxyl groups, whereas Cu as a softer metal may bind to nitrogen- or sulfur-containing group instead. The binding of the negatively charged Au complex increases, probably due to a decrease in the negative surface charge. For brown marine algae it is assumed that alginates play a key role in metal-ion binding. Fourest and Volesky investigated the binding of H, Cd, and Pb by alginic acid and *Sargassum fluitans* biomass before and after modification of the carboxyl groups using acidic methanol or propylene oxide. A linear correlation between the binding capacity for Cd and the content of weak acidic groups (probably carboxylate) resulting from different degrees of active-site blocking was noted. The maximum metal-binding capacity of different algae was linearly correlated with their titer of weak acidic groups, with a stoichiometric relation between metal bound and acidic titer of 0.36 to 0.5 mol/mol. This means that about two acidic sites are necessary per bound metal ion.

Electrostatic attraction

The relevance of electrostatic attraction in biosorption depends on the types and amounts of sites present in the biomass and on whether they are ionized or occupied by protons or other ions. That, in turn, depends on the pH and on the pK_a of the respective group. Usually, a good agreement between the pK_a values of inorganic acids and their corresponding acidic groups in biomolecules exists. Amine groups are positively charged in their protonated form and neutral when deprotonated. Carboxyl, sulfate, and phosphate groups are neutral when protonated and negatively charged when deprotonated.

Table 18.2 K_a values of inorganic acids and acidic groups in biomolecules

Inorganic acid	*pK_a*	*Acidic group in biomolecules*	*pK_a*
Sulfuric acid	1.9	Sulfate	~1.5
Phosphoric acid	2.1, 7.2	Phosphate	2
Benzoic acid	4.2	Carboxyl	3–5
Acetic acid	4.8	Carboxyl	3–5
Ammonium	9.2	Amine	8–10
Phenol	10.0	Phenolic OH	10

If the freshwater alga *Chlorella* could be taken as an example, its isolelectric point (zero overall charge) is at pH 3. At lower pH, the positive charge (possibly of amine groups) leads to an attraction of anions; at higher pH, the negative charge (probably of carboxyl groups) attracts cations and repels anions. Biomass of the marine brown alga *Ascophyllum nodosum* bears a negative charge whose magnitude sharply increases from pH 3 upward. It is probable that this surface charge originated from deprotonated carboxyl groups of the alginate in the cell wall.

Location of uptake metal

Knowledge pertaining to the location of metal uptake in the cell can provide some indication about the relevance of binding groups if it is correlated with information about which biomolecules occur in the respective parts of the cell. Electron micrographs of bacterial, fungal, and algal metal-bearing cells in the literature indicate that the cell wall plays an important role in metal-ion biosorption. While the basic biomass metal-sequestration process is usually very fast, some time-lapse electron micrography indicates slower metal penetration into cell interiors and/or building up of external metal deposits. For a better understanding of the relevant mechanisms in metal binding by microbial biomass, more detailed and accurate knowledge about the biomolecules and binding sites offered by different types of cells is necessary.

APPLICATION

Studying metal-ion binding to alginic acid from *Laminaria digitata* Haug reported that the amount of protons released into solution decreased in the order Pb > Cu > Cd > Ba > Sr > Ca > Co > Ni > Zn > Mn > Mg. This was explained by a better ability of strongly binding ions to compete with protons for organic binding sites. The affinity sequence for metal-ion binding to alginate from *L. digitata* followed a similar trend Cu > Ba > Ca > Co, and for binding of alkaline earth metals to polymannurate and polygulurate the binding strength decreased in the order Ba > Sr > Ca > Mg. According to Haug and Smidsrod, the preferential binding of the heavier ions could have been due to stereochemical effects, as larger ions might better fit a binding site with two distant active groups. The amount of proton displacement by metal ions decreased in the order Pb > Cu > Ba > Sr > Ca > Cd > Zn > Ni > Mg > Co, Mn for polyguluronic acid, and in the order Pb > Cu > Cd > Ba > Ni > Sr > Ca > Mg > Zn, Co, Mn for mannuronic acid. Again, the trend was concurrent with the one in binding strength. Schweiger found a different order of complex stability for binding to alginate as being Ba > Cd > Cu > Sr > Ni > Ca > Zn > Co > Mn > Mg, but still the binding of alkaline earth metals increased with their weight. For the elements Cu, Ni, Co, and Mn of the third period, an increase of binding strength with increasing atomic number was noticed.

In binding to fucoidan, the affinity sequence Pb > Ba > Cd > Sr > Cu > Fe > Co > Zn > Mg > Cr > Ni > Hg > Ca was reported. For *R. arrhizus* biomass, the maximum metal-binding capacity decreased in the order UO^{2+} > Cr^{3+} > Pb ~ Ag^{+} >Ba > La^{3+} > Zn > Hg > Cd > Cu > Mn. The metal-binding capacity increased linearly with the crystal ionic radius, and no correlation to the charge was noticed, except that the binding of alkaline metal ions was negligible. The observations of higher binding (both capacity and binding constant) with increasing ionic radii were interpreted as a consequence of the binding-group conformation: Assuming that binding groups are distant from each other, large ions could not only be more easily dehydrated, but could also bind to several groups simultaneously. This could explain the binding strength; however, it does not explain the higher binding capacity.

To the contrary, if several groups are involved in the binding of one large ion, the binding capacity of small ions, each binding to one site, could be higher. Avery and Tobin found that the amount of proton displacement decreased in the order Cu > Cd > Zn > Mn > Sr. It was assumed by these authors that proton release occurred only when the metal was covalently bound, and that the release of Ca and Mg indicated ionic binding. The ratio of (Mg + Ca) released per H^+ released was used as an index for the relative amount of ionic/covalent binding. It was, however, not considered that, for example, Cu might bind covalently to a site from which Ca, which had been bound electrostatically, was released. This index should therefore not be used in a quantitative way. Evans states that the affinity sequence in alkali- metal binding to soil (Cs > Rb > K > Na > Li) reflects an increase in the hydrated radii of these ions; for divalent ions the binding strength to soil organic matter at pH 5.8 decreased in the order Hg = Pb = Cu > Cd > Zn > Ni > Co > Mn.

For freshwater algae, Crist et al. observed a trend in binding strength Cu > Sr > Zn > Mg > Na being similar to the one for proton replacement Cu > Zn > Mg > Sr > Na, and concluded that a trend from covalent to ionic bonding was responsible. As pointed out earlier, however, electrostatic effects could have contributed to proton binding on the one hand and, on the other hand, covalent binding to sites previously occupied by a light metal could be possible. For binding of light metal ions to freshwater algae, the affinity sequence was Na > K > Li > Cs and Mg > Ca > Ba > Sr. It was concluded that if the affinity between biomass and metals was high, the binding should increase with increasing charge density, that is, with decreasing radius. If binding was weak, it would be controlled by dehydration energy and the larger ions would be preferably bound because they could be dehydrated more easily.

The observed sequence, which does not show a clear preference for large or small ions, was interpreted as resulting from these opposite trends, indicating medium to strong bonds. It is questionable, however, whether the use of crystal ionic radii for characterizing the ionic binding of the hydrated ion is appropriate. Leusch et al. determined the affinity sequence for untreated *Sargassum fluitans* as Pb > Cu > Mi ~ Cd > Zn. Fourest and Volesky observed that the binding capacity for divalent metal binding to *Sargassum* decreased in the order Pb > Cu > Cd > Zn > Ca, and related this to decreasing electronegativity. However, the binding strength, rather than the maximum binding capacity, should be influenced by the electronegativity. Possibly, the maximum binding was not reached under the conditions employed by these authors.

Factors Affecting the Binding Strength

There are three factors that can determine the affinity of a metal ion (because they contribute to the free energy of binding of hydrated metal ions to biosorbent binding groups).

Hydration effects

A change in the orientation of the hydrated water molecules occurs in conjunction with electrostatic binding, but in such "outer-sphere" complexes both partners retain their hydration spheres. Complete or partial dehydration takes place for "inner-sphere" complexes resulting from covalent binding. Both types of changes in the hydration state can occur more easily if the hydrated water molecules are not held strongly by the metal ion.

Williams and Hale showed that a good correlation (Born equation) between the ion hydration energy and the term (z^2/r) exists. For estimating the strength of hydration, the radius of the nonhydrated cation should be used, of course. This is usually taken as the ionic crystal radius. Taking into account the hydration sphere, the effective hydrated radius is larger than the crystal-ion radius. In general, the larger the hydration energy of an ion (i.e., the smaller the crystal radius) the larger the effective hydrated radius.

Consequently, the hydrated radii follow a reverse trend as the ionic crystal radii, both for the groups of alkali and alkaline earth ions. It is difficult to quantify the radius of the hydrated ion, but some values have been compiled. According to Jain, hydration effects can be dominating when the electric field of the sorbent is weak (i.e., weak binding). In this case, larger ions (comparing crystal radii of ions of the same charge) that exert a smaller ordering influence on their hydrated water molecules (i.e., that are less strongly hydrated) are preferably accumulated at the interface.

Electrostatic binding

If the binding groups are negatively charged, they can attract metal cations. The interaction is stronger the higher the charge density of both biosorbent and metal ion. If a strong electric field is present, electrostatic effects may become the dominant factor such that small ions that have a higher charge density are bound more strongly. In the past, ionic-crystal radii have been used in estimation of the ionic binding strength, as well as in aqueous solution. However, since the cation retains its hydrated water molecules, the hydrated cation radius is more characteristic for electrostatic attraction (as opposed to hydration). According to Marcus and Kertes, the selectivity increases with increasing charge and decreasing hydrated radius.

Covalent binding

This type of bond involves sharing of the electrons. Therefore, the more similar the electronegativities of the metal ion and the coordinating atom of the ligand, the higher the covalent character of the bond. The selectivity (or binding strength) increases with increasing polarizability of the ion (i.e., softness).

Overall, the binding strength increases with increasing ionic radius and decreasing charge if binding is weak and largely due to hydration effects, decreasing hydrated radius and increasing charge if binding is intermediately strong and due to electrostatic effects, and decreasing electronegativity difference if binding is strong and covalent.

Since the first two criteria are almost opposed to each other, the affinity sequence can be reversed with changing electric field strength of the sorbent, except that the ionic and hydrated radii can exhibit opposite trends. Therefore, it is possible that for the alkaline and alkaline earth metals, for example, the affinity sequences Cs > Rb > K > Na > Li and Ba > Sr > Ca > Mg > Be observed in industrial ion exchangers are due both to decreasing crystal and increasing ionic radii.

Behavior of Metal Ions

In order to classify the behavior of metal ions, a division into so-called hard and soft ions was introduced by Pearson. Hard ions are characterized by predominantly electrostatic bonds where the entropy gain in the change of the orientation of the hydrated water molecules makes an important contribution to the free energy change ΔG. Hard ions have a weak polarizing power and retain their electrons strongly. Williams used the charge density (z^2/r) (with z being the charge and r the cation radius) as a measure for the strength of ionic binding. The characteristics of hard ions resemble the ones of class "a" cations in the "a"- "b" categorization of metal ions. Type "a" ions form their strongest bonds with the first element of each group, "b" ions with the second or higher. Type "a" ions tend to form strong bonds with highly electronegative donors that are difficult to polarize (i.e., hard). The difference between the electronegativity (x) of the metal ion and that of oxygen has been used as a measure for "a" (i.e., ionic bond) character, which increases with $\Delta x^2 = ({}^{M}x - {}^{O}x)^2$.

Table 18.3 Parameters characterizing binding strength

Metal	z (–)	r_{cryst} (Å)	r_{hyd} (Å)	x (–)	z^2/r_{cryst} (1/Å)	z^2/r_{hyd} (1/Å)	"b" type (Å)	Δx (–)	ionic total (–)	χ (1/Å)
H	1.	—	2.82	2.2	—	0.35	—	1.3	0.34	1.03
Li	1	0.76	3.82	1.0	1.32	0.26	1.61	2.5	0.79	0.33
Na	1	1.02	3.58	0.9	0.98	0.28	1.51	2.6	0.82	0.34
K	1	1.51	3.31	0.8	0.66	0.30	1.51	2.7	0.84	0.36
Rb	1	1.61	3.29	0.8	0.62	0.30	1.57	2.7	0.84	0.36
Cs	1	1.74	3.29	0.7	0.57	0.30	1.27	2.8	0.86	0.35
Mg	2	0.72	4.28	1.2	5.56	0.93	2.26	2.3	0.73	1.27
Ca	2	1.00	4.12	1.0	4.00	0.97	1.85	2.5	0.79	1.23
Sr	2	1.26	4.12	1.0	3.17	0.97	2.11	2.5	0.79	1.23
Ba	2	1.42	4.04	0.9	2.82	0.99	1.84	2.6	0.82	1.21
Mn	2	0.67	4.38	1.5	5.97	0.91	3.42	2.0	0.63	1.44
Co	2	0.65	4.23	1.8	6.15	0.95	4.86	1.7	0.51	1.84
Ni	2	0.69	4.04	1.8	5.80	0.99	4.99	1.7	0.51	1.92
Cu	2	0.73	4.19	2.0	5.48	0.95	6.32	1.5	0.43	2.22
Zn	2	0.74	4.30	1.6	5.41	0.93	4.07	1.9	0.59	1.56
Cd	2	0.95	4.26	1.7	4.21	0.94	5.20	1.8	0.56	1.69
Pb	2	1.19	4.01	1.8	3.36	1.00	6.61	1.7	0.51	1.94
Al	3	0.54	4.75	1.5	16.67	1.89	3.60	2.0	0.63	3.00
Ag	1	1.15	3.14	1.9	0.87	0.29	7.22	1.6	0.47	0.62

According to Pauling, the fraction of ionic bond character is $1 - \exp(-\Delta x^2/2)$, that is, equal contributions of covalent and ionic binding occur when $\Delta x = 1.7$. As seen in Table 18.3, the ionic bond character is larger than 70% for alkaline (earth) metals and about 50% for most heavy metals. Dean describes a similar criterion: The ionic bond character increases with Δx and 100% ionic bonding occurs for $\Delta x > 2$ (but according to Pauling's criterion the ionic-bond character at $\Delta x = 2$ is 63%, not 100%). Table 18.3 shows that the Δx criterion is fulfilled for alkaline and alkaline earth ions. Soft ions participate in more covalent bonding where the free energy is mostly enthalpic. The characteristics of soft ions are opposite those of hard ions, resulting in a reversed affinity sequence. The affinity sequence of soft, or class "b", ions for ligands is I > Br > Cl > F and S > N > O. A combined affinity sequence was described by Buffle: S > I > Br > Cl 39 N > O > F. Therefore, the difference between complexation constants for F and Cl, respectively, has been introduced by Turner et al. to characterize softness, which increases with decreasing $\Delta K = \log({}^{F\,M}K) - \log({}^{Cl\,M}K)$. Type "a" ions are characterized by $\Delta K < 2$ and type "b" ions by $\Delta K > 2$; among the border elements, tendencies for "a" or "b" character are present for $\Delta K <$ or > 0, respectively.

Typical hard, or class "a," ions are the alkaline and alkaline earth metals, transition elements up to class VI, and, for example, Zn. As opposed to the original classification of Pearson, who considered Cd a soft ion and H a typical hard ion, these two are now regarded as borderline ions. Table 18.3 confirm that light metals and Mn can be classified as hard, and that most of the heavy metals may be listed as intermediate, except for Zn, which is just on the threshold to being hard, and Ag, which is soft. For any donor (and especially for soft ones) the binding strength increases with cation softness. Soft cations form more stable complexes with soft donors (S, P, or As), while hard cations prefer hard donors (F or O). N donors have intermediate characteristics. For O or N donor atoms, the Irving–Williams order of complex stability for divalent ions in the first series is Mn <Fe <Co <Ni < Cu > Zn. This can be explained by the electronegativity differences with respect to the donor atoms O or N. Examples of hard ligands are carbonate, phosphate, sulfate, or hydroxide ions as well as sulfate, carboxylate, alcohol, or hydroxyl groups. Soft ligands include sulfhydryl, thioether, and amino groups.

The relevant parameters for different metals are listed in Table 18.3. The hardness parameter (z^2/r_{hyd}) for different metal ions is plotted over the softness parameter $X^2(r_{cryst} + 0.85\ \text{Å})$ in Figure. Since the parameter (z^2/r_{hyd}) expresses the ionic, binding strength, and the parameter $1 - \exp(-\Delta x^2/2)$ is supposed to characterize the relative contribution of ionic bonding to the total binding strength, one can obtain an indicator for the total binding strength by dividing (z^2/r_{hyd}) by $[1 - \exp(-\Delta x^2/2)]$. This parameter is called ζ; it is listed in Table 18.3. The lower part represents the ionic binding strength; the upper part represents the covalent binding strength. The graph should, however, be used only for illustrative purposes, rather than in a quantitative way, because the exact magnitude of the contribution of ionic bonding is disputable.

According to the Δx criterion, for example, there is no covalent bonding for the light metals, while this plot, using Pauling's criterion, shows –20% covalent binding. The binding trends summarized in the section "Need for Heavy-Metal-Removal Processes" can now be explained: The tendency of stronger binding for the heavier elements in both the alkaline and alkaline earth metals can be explained as being due to lower charge density (z^2/r_{cryst}) (i.e., weaker hydration) of the ions themselves and higher charge density (z^2/r_{hyd}) of the hydrated ions (i.e., stronger electrostatic bonds) in the heavier elements of each series. Since these ions are hard, there is little contribution of covalent bonding; therefore, the electronegativity differences between them are irrelevant. The trend among the elements in the second period (Cu > Ni > Co > Mn) and the trend Pb > Cd > Zn are probably mainly due to a decrease of the covalent binding [characterized by $x^2(r_{cryst} + 0.85)$], but a decrease in the electrostatic binding (z^2/r_{hyd}) (except for the relation between Ni and Cu) and increasing hydration strength (z^2/r_{cryst}) (except for Mn) may also contribute. Since the binding trends in biosorption by algal biomass or its constituent molecules indicate that there is significant electrostatic and covalent binding, the latest biosorption models described at the end of the next section ("Modeling") attempt to incorporate these stronger interactions and neglected hydration effects, which are weaker.

BIOSORPTION BINDING MODELS

In order to evaluate the feasibility and effectiveness of biosorption in wastewater treatment, it is essential to make predictions of the sorption performance (e.g., for facilitating process design). Therefore it is necessary to develop appropriate mathematical models of biosorption. Modeling the biosorption-binding equilibrium is a prerequisite for all further work involving batch kinetic studies and, in particular, for the continuous-flow sorption-column applications that represent the most effective configuration of the sorption-based process. The amount of metal M (or in general, sorbate) bound per mass of sorbent is called the binding (or uptake), ^{M}q. The binding is not only dependent on the sorbent material but also on the equilibrium concentration $[M]$ of the sorbate in solution and on other parameters, such as pH and equilibrium concentration of other ions in solution.

The relationship between the equilibrium binding and the concentration of an ion (at constant temperature) is depicted in an isotherm plot of ^{M}q versus [M]. With increasing metal concentration in solution its binding increases from zero to the maximum. It is desirable for a sorbent to possess a high sorption capacity (maximum binding obtained at high equilibrium concentrations) and a high affinity for the sorbate species, which is reflected in a steep slope of the isotherm curve at low equilibrium concentrations.

Langmuir and Freundlich Isotherms

Most mathematical biosorption models used in the literature describe simple Langmuir or Freundlich sorption isotherms, where the metal-ion binding is determined as a function of the equilibrium concentration of that metal ion in the solution, without reference to pH or other ions in the same solution system. The Freundlich isotherm is originally of empirical nature, but has later been interpreted as sorption to sites with an affinity distribution; that is, it is assumed that the stronger binding sites are occupied first and that the binding strength decreases with increasing degree of site occupation. More specifically, the Freundlich isotherm is obtained when a log-normal affinity distribution [i.e., a normal distribution of log (K)] is assumed (57p.319;100). The Freundlich isotherm is defined by the following expression:

$$^{M}q = k[\mathrm{M}]^{1/n} \qquad \ldots(2)$$

where k and n are empirically determined constants, with k being related to the maximum binding capacity, and n related to the affinity or binding strength.

The Langmuir isotherm is based on considering sorption as a chemical phenomenon. It is assumed that the forces exerted by chemically unsaturated surface atoms (total number of binding sites B) do not extend further than the diameter of one sorbed molecule, and that therefore sorption is restricted to a monolayer. In the simplest case the following assumptions are made.

1. All sorption sites are uniform (i.e., constant heat of adsorption).
2. There is only one sorbate.
3. One sorbate molecule reacts with one active site.
4. There is no interaction between sorbed species.

The rate of adsorption is proportional to the rate constant of the forward reaction ${}^{BM}k_f$, the number of free sites ($B = {}^{t}B - {}^{M}q$), and the number of sorbate molecules hitting the surface per unit time (i.e., proportional to their concentration [M]). Under the same conditions, the rate of desorption is proportional to the rate constant of the backward reaction ${}^{BM}k_b$ and the number of occupied sites ${}^{M}q$. The equilibrium is attained when the rate of adsorption equals the rate of desorption:

$$ {}^{BM}k_f B[M] = {}^{BM}k_b {}^{M}q \text{ (mol/s)} \quad \ldots(3)$$

With the equilibrium constant ${}^{BM}K$, indicating the affinity for the sorbate, defined as the ratio of adsorption and desorption rates ${}^{BM}K = {}^{BM}k_f / {}^{BM}k_b$, it follows:

$$ {}^{BM}q = \frac{{}^{t}B \cdot {}^{BM}K\,[M]}{1 + {}^{BM}K\,[M]} \text{(mmol/g)} \quad \ldots(4)$$

Both Langmuir and Freundlich isotherms have been used to describe the equilibrium of metal biosorption. While it is possible to obtain good agreement between predictions of these models and the experimentally determined metal binding, their simple hyperbole-based equations have significant limitations: They cannot describe a more complex sorption behavior often reflected by nonhyperbolic relationships and they cannot predict the effect of pH or of other ions in the solution, among other shortcomings.

Effect of pH

Although the crucial role of protons in biosorption is generally known, it is usually neglected in the mathematical description of the process. The complex nature of biosorbent materials that may contain a multiplicity of binding groups makes the modeling of the pH effect in biosorption difficult. As a consequence, it has been common practice to determine a separate isotherm and a new set of parameters, for each pH value or for each initial biomass saturation state (e.g., protonated or loaded with ions from the seawater). This has been necessary because the most frequently used Langmuir or Freundlich sorption models do not take into account the fact that metal-ion biosorption is largely an ion-exchange phenomenon. They do not allow the prediction of the remaining binding of those ions (e.g., protons or sodium) that were initially loaded onto the biosorbent, and neither do they incorporate the concentration of the exchanged species (e.g., protons) as a parameter, normally not allowing reliable calculated predictions of the biosorbent performance. Some attempt has been made, assuming a simplified linear relationship between the metal uptake and pH difference, to empirically model the biosorptive binding of metal ions by a Freundlich isotherm, which was modified by a factor proportional to the pH difference. This approach could not represent the change of proton binding in the presence of metal ions.

Though protonation reactions for two acidic sites (alga) have been considered, the model simplification by lumping pH and proton- and metal-binding constants into one parameter for each resulted in a standard two-site Langmuir isotherm. A rare example of a Langmuir model including protons and one or two binding sites assumed a 1:1 stoichiometry of metal binding and proton release, although it was recognized that metal binding with a 1:2 stoichiometry probably occurred. A modified Langmuir model for competitive Cu and H binding to two sites on alginate required iterative calculations of each cation uptake as a function of the binding of the other one. Another two-site model based on a 2:1 proton–metal exchange constant allows direct calculation of the metal binding as a function of the concentration ratio $[H]^2/[M]$, but it does not account for free sites and it cannot easily be extended to include more species.

Sorbed ions

Often, the prediction of metal-sorption equilibria is complicated by the presence of several sorbed ions, requiring the use of multicomponent isotherm equations. This is especially relevant when the sorption of one ion influences the binding of another when different ions compete for the same binding site. The use of multicomponent Langmuir isotherm models was demonstrated in describing the equilibrium metal biosorption. However, biosorption was not regarded as an ion-exchange process; the ions with which the biomass had originally been saturated were not considered. Instead of using multicomponent isotherm models, more complex systems involving different cations and ligands can be modeled by incorporating binding constants for metal-ion–biomass complexes into a computer program for the calculation of the chemical equilibrium. Reviews of different computer programs for calculating the equilibrium speciation have been presented. These programs take acid/base reactions, coordination, and dissolution/precipitation into account. Their use is advantageous if complexation in solution is expected or if final concentrations are to be predicted from the initial state.

Ion-Exchange Biosorption Mechanism

There have been indications that ion exchange is an important biosorption mechanism. However, this does not preclude an adequate use of simple sorption isotherms instead of appropriate ion-exchange models with equilibrium constants taking into account the reversibility of the ion-exchange reactions: The ion that is displaced into the solution can compete with the sorbed metal ion for the sites.

The Langmuir equation and the ion-exchange constant for the binding of a metal ion M (for simplicity here a monovalent ion) replacing a proton H on a complexing site B are related as follows:

Langmuir: $B + M \rightleftharpoons BM$

$$^{BM}K^* = \frac{BM}{B[M]}[B]_t = [B] + [BM] \qquad \ldots(5)$$

Ion exchange: $BH + M \rightleftharpoons BM + H$

$$^{BM}K = \frac{BM[H]}{BH[M]} [B]_t = [BH] + [BM] \quad \text{...(6)}$$

therefore:

$$^{BM}K^* = {}^{BM}K/[H] \quad \text{...(7)}$$

The difference between the two approaches is that the first one assumes that all sites are initially free and does not consider any reverse reaction of a displaced ion, in this case a proton, with the site. The second model assumes that all sites to which metal ions are sorbed are initially occupied, that is, the number of free sites stays constant. Crist et al. compared the fit of the Langmuir sorption isotherm model and the one using ion-exchange constants. The differences between the two models were especially pronounced at low metal-ion concentrations because the reverse reaction involving the displaced ion is most noticeable if the concentration of the bound cation is low.

The ion-exchange approach is probably somewhat closer to reality than the simple Langmuir model, but it is not completely satisfying either. Although the constant number of free sites [in other words, the number of sites occupied by the bound ion is equal to the number of sites from which the exchanged ion is released so that $^tB = (B^1M) + (B^2M)$ can be assumed] may be a reasonable working assumption for a constant pH system, it may not hold for systems with changing pH values. According to Stumm and Morgan, the cation-exchange capacity increases with increasing pH above the isoelectric point. Therefore, one cannot simply model the competitive binding of metals and protons by using only a metal–proton-ion-exchange constant. It is necessary to include at least one reaction where a cation reacts with a free site. In the modeling of industrial ion exchangers, protons have been considered. However, these synthetic resins are chemically rather simple compounds that mostly contain only one active group. Biosorbents, on the other hand, can be heterogeneous and contain different acidic groups active in metal ion binding. The $p^{BH}K$ value, the quantity of binding sites, and the metal-ion binding constant must be known for each of these sites.

Ionic Strength Models

Ubiquitous ions such as sodium may interfere with heavy metal biosorption through the effect of ionic strength. In biosorption, no modeling of the effects of ionic strength and electrostatic attraction has been done. Several models of ionic-strength effects, however, have been published concerning metal binding to humic and fulvic acids. The physical characteristics of these molecules, which are degradation products of organic matter, are quite different from microbial biosorbents. Humic and fulvic acids are small water-soluble molecules. However, since these molecules contain carboxyl groups as one of their main functional groups they may in some aspects be comparable in particular to those biosorbent materials based on seaweeds. Buffle systematically described possible models for complexation equilibria, including site protonation, multiple sites, and secondary effects such as the influence of the electric field in polyelectrolytes on the local concentration.

The simplest approach to modeling the influence of Na in the biosorption system consists of the introduction of a binding constant for Na, as done by Westall et al. for the binding of H, Co, and Na to humic acids. This approach enables derivation of a multicomponent sorption isotherm model so that the binding of each species can be directly calculated. This was done for biosorbents only recently in the work of Schiewer and Volesky. Although the Na binding constant can easily be incorporated into computer programs for the calculation of chemical equilibria in aqueous solutions, such as MINEQL+, the disadvantage of this approach can be that the Na binding constant is purely empirical and does not reflect the electrostatic nature of the Na binding. Another possibility is the use of the Donnan model. An application of this approach for different charged polymers has been pioneered by Marinsky and coworkers, who established the use of Donnan models for interpreting protonation equilibria in organic polyelectrolytes including humic and fulvic substances and some metals and alginic acid.

The latter contribution also involves an estimation of the water uptake (swelling) using osmotic coefficients. Unfortunately, this modeling work did not include model predictions of proton or metal binding under different conditions. More Donnan-based model work done with alginate considered competition of Cu and H for the same sites. Guntelberg activity coefficients have been used to account for ionic-strength effects in conjunction with fulvic acid and Cu–alkaline earth metal binding competition. Although this method may be suitable for rather small molecules, it would not be appropriate for polyelectrolytes, such as the alginic acid found in seaweed-based biosorbents. While the isotherm sorption model did not include a term for free sites, it was noticed that increasing ionic strength (Na) or presence of Ca or Mg can reduce Cu binding. A model that is suitable for both oligo- and polyelectrolytes was used only for modeling pH titrations and Cu binding of humic acid molecules. Two of the sites were assumed to be diprotic and able to bind metal ions and one was assumed to be monoprotic and not to bind metals. In this approach, the humic acid molecules were assumed to be penetrable or impenetrable spheres of two size classes (assuming that the total charge is proportional to the volume of the sphere).

The advantage of this approach is its generality concerning the size of the molecule. It is particularly suited for oligoelectrolytes that show a behavior intermediate between simple ions and polyelectrolyte gels. Most biosorbent materials could be considered true polyelectrolytes. However, the semiempirical correlation used in this approach required three fitting parameters for the electrostatic effects for each of the two particle sizes. Moreover, the fitting parameters had to be determined individually for each ionic strength. This means that the effect of ionic strength could not be predicted. A Helmholtz metal-sorption model for the charged interface requires only one fitting parameter, the constant capacitance, for modeling the electrostatic effects, but not the ionic strength effects. Another model considering eight sites of two general types for light-metal binding to humic acids used relatively few fitting parameters for the prediction of a wide range of experimental conditions. Unfortunately, it also relies on an empirical relationship involving the ionic strength. A Poisson–Boltzmann model for proton binding to humic substances incorporates the effect of ionic strength on the concentration factor in a realistic manner and requires only one fitting parameter for electrostatic effects (the particle radius).

The assumption of small impermeable spheres or cylinders, however, does not appear applicable to the large, penetrable particles as in the preferable case for biosorbents. Recently, Schiewer and Volesky presented a series of equilibrium biosorption models gradually incorporating the metal-ion concentration, pH, and ionic strength. The models proposed had only one fitting parameter, instead of up to six as in some of those already mentioned. The formation of $BM_{0.5}$ complexes was assumed, and a combined Donnan–biosorption–isotherm equation was derived that allowed for direct calculation of cation binding without iterations when swelling proportional to the number of free sites was assumed. Using the parameters (number of binding sites, proton binding constant, and specific particle volume) obtained from pH titrations, it was possible to predict the effect of pH, ionic strength, and Ca concentration on Cd binding for a seaweed biosorbent.

MECHANISM OF BIOSORPTION

Because biosorption is just another form of a basic sorption operation, the same principles apply for the design and sizing of the equipment. The approach is always a combination consisting of basic equilibrium sorption tests (batch) and dynamic continuous-flow apparatus studies, which can become very sophisticated if empiricism is to be minimized. Invariably, numerical computer methodologies provide a powerful tool for the design of a continuous-flow (bio)sorption process.

Balancing Activities

When contacting the solid-phase biosorbent with the sorbate species (i.e., metal ions) in solution, a sorption equilibrium will become established after an appropriate length of time. When equilibrium between the species sorbed on the solid and the portion remaining in solution is established, the solids are separated and the supernatant analyzed for the residual, final, equilibrium concentration of the remaining sorbate. The metal uptake by the solid sorbent is calculated (according to the formula) and used for the sorption isotherm plot, which serves to compare the sorption performance of different systems. This is easily done for one-sorbate (one-metal) systems. The situation becomes more complicated when there are more than one sorbed species in multisorbate systems. Of interest is often the case of "interference" assessment when the uptake of one key metal (e.g., toxic Cd) may be affected by the presence of a ubiquitous metal of no major interest (e.g., Fe). This situation can be still depicted by three types of three-dimensional sorption isotherm surfaces, depending on what particular metal uptake appears in the plot: (i) Cd uptake, (ii) Fe uptake, and (iii) combined (Cd and Fe) uptake.

The base two axes are always the equilibrium concentrations of the two metals, respectively. Since the experimental equilibrium (final) concentrations are the product of the experiment, there is no control over their values and they are generated more or less randomly. Consequently, the sorption isotherm surface needs to be fitted to the date points. Finding a suitable mathematical model capable of representing reasonably well the two-metal sorption behavior enables this and more; it can facilitate the summarizing of the effect of one metal on the uptake of the other for the desirable conclusion. For many metal biosorption cases an equilibrium model considering only one type of binding site would be adequate assuming the following reaction:

$$M^{2+} + 2B = 2BM_{0.5} \quad H^{+} + B = BH \qquad \ldots(8)$$

The model provides explicit formulas for metal uptakes as functions of final concentrations of the species present in the system, even for the case of heterovalent exchange, which makes it especially suitable for use in numerical simulations of dynamic sorption systems. The model equation for the equilibrium uptake of a bivalent metal ion M1 present in a binary system with hydrogen protons can be written as follows:

$$q^*_{M1} = \frac{0.5M_{WM1}C_t(K_{M1}C_{fM1})^{1/2}}{1+K_HC_{fH}+(K_{M1}C_{fM1})^{1/2}} \quad ...(9)$$

Similarly, for a binary system containing bivalent species M1 and M2, the expressions for uptakes of each of the species can be written as follows:

$$q^*_{M1} = \frac{0.5M_{WM1}C_t(K_{M1}C_{fM1})^{1/2}}{1+(K_{M1}C_{fM1})^{1/2}+(K_{M2}C_{fM2})^{1/2}} \quad ...(10)$$

$$q^*_{M2} = \frac{0.5M_{WM2}C_t(K_{M2}C_{fM2})^{1/2}}{1+(K_{M1}C_{fM1})^{1/2}+(K_{M2}C_{fM2})^{1/2}} \quad ...(11)$$

The values of the parameters K_H, $K_{M1,}$ and K_{M2} are normally determined by fitting of the model to the experimental data using, for example, the method of weighed least squares. The constant C_t in equations 9-11 may be obtained from pH and/or conductometric titration of protonated biomass with NaOH. Metal biosorption can often be viewed as a simple ion exchange; hence, the classical ion-exchange concept based on exchange-equilibrium constants this case. For the ion-exchange reaction:

$$bA^{+a} + aB^{+b} = bA^{+a} + aB^{+b} \quad ...(12)$$

the equilibrium constant K_{AB} and the separation factor r_{AB} are defined as

$$K_{AB} = \frac{q_A^b c_{Br}^B}{C_{Af}^b q_B^a} = \frac{y_A^b x_B^a}{x_A^b y_B^a} \cdot \frac{C_0^{a-b}}{Q^{a-b}} \quad ...(13a)$$

$$r_{AB} = \frac{y_a x_B}{x_a y_b} \quad ...(13b)$$

for the case of ideal behavior of the exchanging species in both of the phases. As can be seen from equation 13a, the total normality C_0 of the solution affects the equilibrium of a heterovalent exchange (i.e., when $a \neq b$), but it does not affect the equilibrium when the valence of the exchanging species is the same (i.e., $a = b$). Consequently, the ion-exchange isotherms of a species in a binary heterovalent system vary in shape depending on the total normality of the solution.

Corelative approaches

The term *equilibrium sorption isotherm* is commonly used in biosorption to refer to a line connecting equilibrium metal uptakes plotted against the final concentrations of the metal. Typically, biosorption equilibrium data are obtained under the same pH. However, this is not the case in ion exchange where the sorption isotherm represents data points obtained at a constant normality of the solution.

Furthermore, in compliance with the established convention, ion- exchange isotherms for binary systems usually display dimensionless uptakes as functions of dimensionless concentrations for the species of interest:

$$\frac{q_M}{Q} = f\left(\frac{C_{ML}}{C_0}\right) \quad ...(14)$$

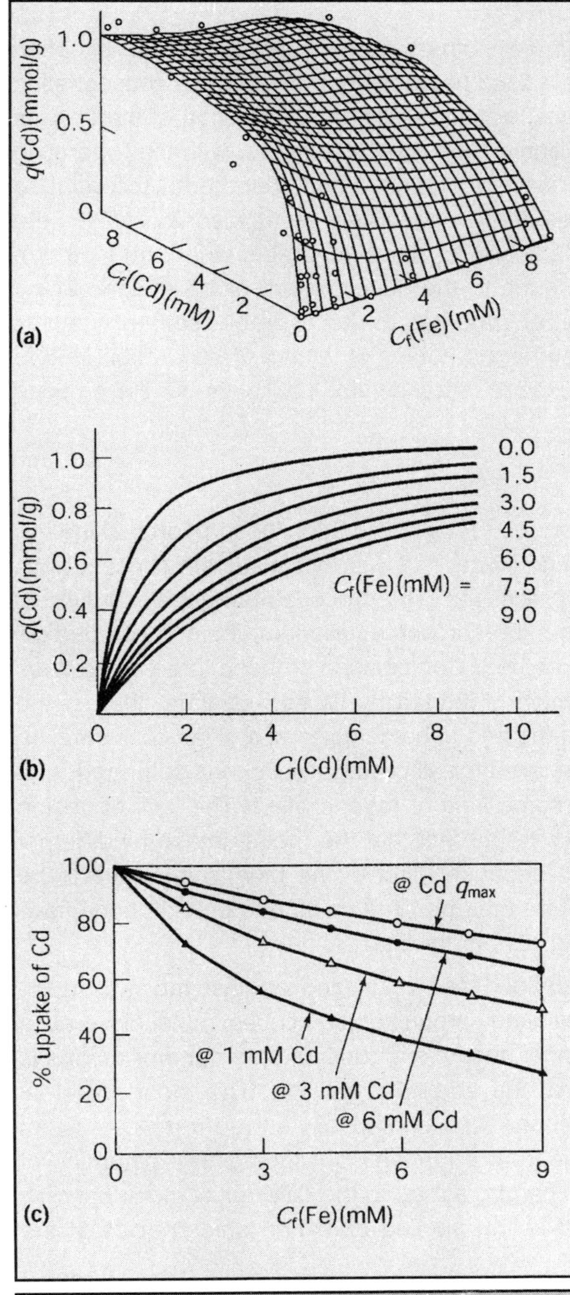

Figure 18.3 (a) A 3-dimensional sorption surface for the Cd-Fe-Sargassum biosorption system, cadmium uptake at pH 4; (b) The effect of Fe on the equilibrium uptake of Cd by Sargassum biomass at pH 4.5; (c) The summary of the effect of Fe presence on the Cd uptake by Sargassum biomass at pH 4.5.

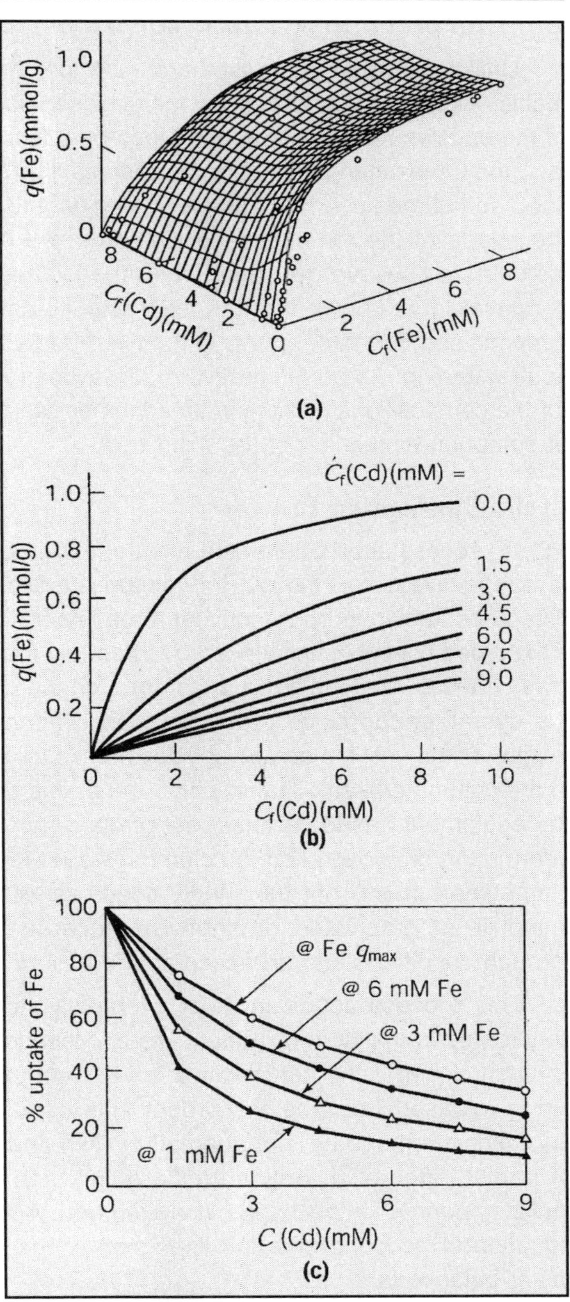

Figure 18.4 (a) A 3-dimensional sorption surface for the Cd-Fe-Sargassum biosorption system, iron uptake at pH 4.5; (b) The effect of Cd on the equilibrium uptake of Fe by Sargassum biomass at pH 4.5; (c) The summary of the effect of Cd presence on the Fe uptake by Sargassum biomass at pH 4.5.

Equilibrium and dynamic sorption

Unlike the biosorption isotherm, the ion- exchange isotherm depicts an equilibrium situation that is relevant to the case of the metal biosorption in a fixed-bed adsorber in which the concentration of the species released from the biosorbent increases monotonically in the direction of the flow. In fact, the ion-exchange isotherm represents the “equilibrium line” that is, together with the “operating line,” sometimes used for assessing design and performance of fixed-bed adsorbers. According to the shape of the ion-exchange isotherm, the equilibrium is classified as (i) favorable when the isotherm is concave, (ii) linear when the isotherm is linear, or (iii) unfavorable when the isotherm is convex. It is known that the favorable isotherm leads to the development of a self-sharpening dynamic sorption zone (front) in the fixed-bed adsorber, whereas the unfavorable isotherm causes its broadening. As an alternative to ion-exchange isotherms, tables of values of separation factors for the compositions existing in an adsorber can be used to represent the equilibrium for the purpose of computer simulation of the fixed bed.

Cyclic Biosorption Process

In process applications, the most effective apparatus for cyclic sorption/desorption making the most effective use of the reactor volume is a fixed-bed column that also optimizes the concentration difference known to be the driving force in sorption processes. The process of metal biosorption in a fixed-bed column is influenced by three key regimes: the sorption equilibrium, the sorption particle mass transfer, and the flow pattern through the packed bed. Combination of these three determines the overall performance of the sorption column, which is judged by its service time, that is, the length of time until the sorbed species breaks through the bed to be detected at a given concentration in the column effluent. At that point, the bed is considered for all practical purposes saturated, and the equipment has to be taken out of operation for some kind of regeneration. The lack of design information on biosorption in columns is due, in part, to the fact that the metal uptake by different biomaterials is still not fully understood, as well as because most of the biomasses have to be immobilized, granulated, or otherwise processed before their use in a column to provide good flow-through conditions and an acceptable pressure drop across the bed.

This process condition must eventually be balanced and optimized against the maximum-sorption-performance requirement. As opposed to activated carbon-sorption cases, metal biosorption is much closer to the heterovalent ion-exchange model because a considerable number of ions is being released from the biosorbent in exchange for the sorbed metal ion. The most universal approach seems to be that adopted by Tan and Spinner, who numerically solved a mixed system of partial differential, ordinary differential, and algebraic equations describing the dynamics of multicomponent ion exchange in a fixed bed. A term can be added to the differential molar balance equation to account for axial dispersion of the fluid in the packed bed. For ionic species M the molar balance is

$$\frac{C_M}{\partial z} - \frac{1}{Pe_c}\frac{\partial^2 C_M}{\partial z^2} + \frac{\partial C_M}{\partial t} + D_{gM}\frac{q_M}{t} = 0 \qquad ...(15)$$

The sorption rate equation can be written as

$$\frac{q_M}{t} = Sh_M(q_M^* - q_M) \quad \ldots(16)$$

assuming a linear driving force for the sorption process and combined film and intraparticle mass transfer resistance: The equilibrium relationship for a metal species represented by equations (9) and (10) can be written schematically for a general counterion, either proton or metal, as

$$q_M^* = f(C_M, C_{CNT}) \quad \ldots(17)$$

and the electroneutrality equation for binary systems is

$$\frac{C_{M0} - C_M}{M_{WM}} = \nu \frac{C_{CNT0} - C_{CNT}}{M_{WCNT}} \quad \ldots(18)$$

To obtain equations 15–18, the original set of equations was transcribed with dimensionless variables using the following transformations of the variables:

$$C_M = \frac{C_{ML}}{C_0} q_M = \frac{q_{MS}}{Q} t = \frac{\tau\nu}{L_0} z = \frac{l}{L_0}$$

and yielding the following dimensionless groups:

$$D_{gM} = \frac{\rho_b Q}{C_0 \varepsilon} Sh_M = \frac{K_{fM} L_0}{\nu} Pe_c = \frac{L_0 \nu}{D_z}$$

The initial and boundary conditions pertaining to the situations under question are specified as follows:

$$t = 0 \quad l = (0, L_0) \quad C_M = 0$$

$$t > 0 \quad l = 0 \quad C_M = C_{M0} + \frac{1}{Pe_c}\frac{C_M}{z}$$

$$t > 0 \quad l = L_0 \quad \frac{C_M}{z} = 0$$

The system of equations 15–18 can be solved numerically using the Galerkin finite element method, combined with the implicit Euler method for time integration. Linear-basis functions would routinely be used for space discretization of both concentration C and uptake q, ensuring that the solution is mesh independent.

Calculation of ion-exchange isotherms

Equilibrium data provide the experimental basis for the fitting of the biosorption model. First, the concentration of binding sites C_t can be determined from the uptake experiments; equilibrium constants for the three ions considered here (K_H, K_{Cu}, and K_{Ca}) can then be found by the least square fitting. Once these parameters are known, the biosorption model can be used to calculate ion-exchange isotherms as follows: First, the normality of a binary mixture C_0 was fixed and the uptake Q, at the concentration of the sorbing species C_0, was calculated from equation 2 or 3. Second, a series of liquid-phase compositions was selected in such a way that the dimensionless concentration C/C_0 of one of the species covers the interval of interest. Finally, a series of dimensionless uptakes corresponding to the series of compositions was calculated using equations 2, 3, and 7.

Fixed-bed sorption column model

The dimensionless groups *Sh, Dg,* and *Pe,* together with the equilibrium parameters for a given binary system, represent the parameters of the fixed-bed model. However, because the equilibrium parameters can be determined from batch equilibrium studies, the values of C_0, v, L_0, and ρ_b are usually known, and Q can be calculated from equations 2 and 3, the only remaining unknown elements in the model are K_{fM}, D_z, and ε. The values of the three unknown elements K_{fM}, D_z, and ε must be determined simultaneously by fitting the fixed-bed model to an experimental breakthrough curve. The value of an overall mass-transfer coefficient for a given species generally depends on the flow rate, the axial dispersion coefficient, and the values of the diffusivities of the species in the liquid and solid phases. The diffusion coefficients of metal ions in diluted aqueous solutions can often be obtained from the literature.

CONCLUDING REMARKS

In preparation for the biosorption process design the sorption/desorption performance and characteristic properties (mass transfer, pressure and attrition resistance, chemical stability, etc.) of the biosorbent should be reasonably well known. Although equilibrium sorption tests yield some basic information, the sorption system behavior is invariably examined under continuous-flow conditions. The fixed-bed sorption column arrangement is usually the most effective process configuration. Its performance assessment and prediction based on both equilibrium and dynamic studies eventually leads to sizing of the equipment. Advanced methodologies used for this purpose are based on sophisticated computer-based modeling and optimization of the (bio)sorption process.

19

FLUIDIZED BED

Fluidized-bed bioreactors are directly linked to the use of biocatalysts (cells or enzymes) for transformations in an immobilized form. The solid particles of the immobilized biocatalyst are maintained in fluidization by means of the circulation of a fluid phase (either liquid, gas, or a mixture of both) that compensates their weight. In this way, good liquid mixing and mass transfer between the solid and the liquid phases can be obtained with low attrition. Also, fluidized-bed bioreactors can accommodate a gas phase and can be used to feed solids in suspension. High productivities can be achieved in these systems, but their hydrodynamic complexity and operational stability have to be well defined for a proper operation.

Expanded-bed adsorption is a technique based on fluidization for recovery of proteins directly from unclarified feedstocks such as fermentation broths, cell homogenates, and other solutions containing suspended particulates. An expanded bed is a chromatography bed with an increased void volume between the adsorbent particles, as compared to a packed chromatography bed. An expanded bed is formed when high-density adsorbent particles are lifted by an upward liquid flow through a column specially designed for expanded bed adsorption. When the bed has expanded and is stable, the particulate-containing feedstock is applied, and the protein of interest can bind to the adsorbent particles while the particulates in the feedstock can pass unhindered through the void of the bed.

Expanded-bed adsorption is a one-passage technique designed for large-scale recovery processes. Traditionally, when conventional packed-bed chromatography is used to recover proteins, the feedstock has to be clarified by centrifugation or filtration prior to application or the particulates can block the bed. Expanded-bed adsorption can handle particulate-containing feedstocks; as a result, clarification, concentration, and initial purification are accomplished in one unit operation. Reducing the number of unit operations in a downstream processing scheme decreases the risk of product loss, decreases the processing time, and lowers the overall production costs. The basic principles of expanded- bed adsorption and some applications are described in this article.

HISTORY

The biotechnology industry produces proteins for therapeutic and diagnostic use by using biological systems such as bacteria, yeast, and mammalian cells. A common bottleneck in the production of these proteins is the initial purification, during which the proteins are recovered from large volumes of cell- and debris-containing feedstocks, such as fermentation broths and homogenates. The purification processes are usually based on packed-bed chromatography techniques in which the protein is bound to an adsorbent packed in a column. A packed bed (fixed bed) cannot handle particulates very well and would eventually become totally blocked if an unclarified feedstock was passed through it. Another concern with these types of feedstocks is that because of the sometimes low production level of the protein of interest, the feedstock volume is large. A large volume takes a long time to pass through a chromatography column, and a long processing time increases the risk of proteolytic breakdown of the protein. Unit operations to clarify and reduce the volume of these feedstocks are therefore vital for a successful purification process. It is not uncommon for a purification process to consist of a sequence of unit operations just to achieve initial recovery.

The techniques traditionally used for feedstock clarification are centrifugation and filtration. Centrifugation is well suited to industrial processes because the throughput is high and it can be performed in a contained environment (a requirement when handling recombinant host organisms). The efficiency of the centrifugation process will depend on the properties of the feedstock, but it is usually difficult to remove all particulates by centrifugation, and the resulting fluid may therefore not be sufficiently free of particulates for direct application to a packed chromatography bed. Whole cells are relatively easy to remove by centrifugation but disrupted cells are not, and cells (especially mammalian) may even be damaged by the centrifugation process because of shear stress and centrifugal forces. New and improved centrifuges that are more gentle can be used for mammalian cells, but often the clarification problem remains because at harvest, culture fluids usually contain relatively large amounts of subcellular particulates, such as debris. When the feedstock is secreting bacteria or a bacterial homogenate, for example, clarification is even more difficult and centrifugation is not very efficient. When filtration is used for clarification of feedstocks, it is usually performed in cross-flow mode. The fluxes that can be used will depend on the properties of the feedstock, but it is often difficult to reach an acceptable economic limit (100 $L/m^2/h$) for a cross-flow filtration process because fluxes are typically less than 50 $L/m^2/h$. Cross-flow filtration can easily be operated as a completely contained system, and it yields a particulate-free solution.

The fouling of the filtration membranes is a major concern, and the resulting decrease in flux-per filter area often leads to long processing times. In addition, the cost of membranes must be taken into account in a process where membrane fouling is severe because the membranes have to be replaced frequently. Because of the drawbacks associated with both the centrifugation and the filtration techniques, they are often used in combination in a purification process to achieve the best clarification of the feedstock. The volume of the feedstock is not significantly reduced by either centrifugation or filtration, which is why product concentration does not usually increase until the protein has been adsorbed to a chromatography adsorbent in a packed-bed column. Because the feedstock volume is not reduced before application to the packed bed, processing time will be long. Thus, for a number of reasons, the best option for successful initial recovery of a protein from a large-volume cell- or debris-containing feedstock is direct adsorption of the protein. A variety of methods for direct recovery have been tried. One such method is to use a batch procedure in which an adsorbent is added directly to the unclarified feedstock in a stirred tank. The advantage of using a batch procedure is that the feedstock does not have to be clarified.

The disadvantage is that adsorption of protein to the chromatography adsorbent is inefficient, because the stirred tank acts as one single theoretical plate in a separation process. Several attempts, however, have been made to improve batch adsorption, for example, continuous-batch adsorption processes. Another alternative for direct recovery is fluidized bed adsorption. This technique uses high-density adsorbent particles in an up-flow fluidized bed column. As with a batch procedure, a fluidized bed behaves like a column with a low number of theoretical plates. The adsorbent particles and the feedstock are constantly "back-mixed," which usually results in poor adsorption of the protein by the adsorbent particles, making it necessary to recirculate the feedstock through the column, leading to long processing times. A number of solutions to improve the efficiency of fluidized beds have been suggested, such as columns fitted with perforated horizontal plates. This creates a column with a number of smaller, each completely back- mixed, fluidized beds. The number of fluidized beds depends on the number of plates inserted. The end result is a column with more than one theoretical plate and consequently a more efficient process.

Another way to improve the efficiency of a fluidized bed is to use magnetically stabilized fluidized beds. The adsorbent particles are magnetically susceptible and are fluidized while being stabilized by a magnetic field. Stabilizing the particles in this way prevents back-mixing and gives the column several theoretical plates. There are, however, drawbacks that have prevented this technique from being used at industrial scale, such as the heat generated by the magnetic field. The concept of expanded-bed adsorption (or expanded-bed chromatography) was introduced after another attempt to stabilize and prevent back-mixing in a fluidized bed. An expanded bed is a chromatography bed in which the bed voidage is larger than in a settled or packed bed. Furthermore, it displays no or very low back-mixing of adsorbent and liquid. An expanded bed can be also be described as a stable fluidized bed because the liquid flow shows a constant velocity profile over the whole cross-section of the bed (i.e., plug flow) (Worth mentioning is that the terms *fluidized bed* and *expanded bed* are used differently in the literature, and sometimes authors do not differ between the expressions.)

The early expanded bed experiments were performed using Sepharose Fast Flow as the adsorbent and a sintered glass filter as the liquid flow distributor at the bottom of the column. The results showed that if the bed expansion was limited to approximately twice the settled bed height, the liquid flow through the column was close to plug flow and the protein adsorption efficiency was similar to that of a packed bed. These promising results showed that in an expanded bed it is possible to combine the fluidized bed's ability to handle unclarified feedstocks with a packed bed's characteristic of efficient protein adsorption. The adsorbent used in these early experiments is intended for use in packed beds. The density of the particles is relatively low, which meant that the flow through the expanded bed column also had to be low, otherwise the bed would expand to an extent that adsorbent would start to pack at the top of the column. For this technique to be attractive at industrial scale, the flow velocities must be significantly increased. The sintered glass filter that was used as a flow distributor is not appropriate for large-scale operations for a number of reasons: particulates may get trapped in the filter, the filter is difficult to clean, and it is difficult to produce large homogenous sinters. Hence, it soon became apparent that for expanded-bed adsorption to be successful at industrial scale, new types of adsorbent particles and new column designs were needed. Many different types of adsorbents have been tested: agarose, zirconia, silica, glass, hydrophilized perfluorocarbon, and composites such as kieselguhr/agarose, cellulose/titanium dioxide, and dextran/silica. These adsorbents are suitable for certain applications, but they lack the combination of properties needed for an efficient expanded-bed adsorption process.

Agarose adsorbents are well suited for protein recovery but they are not dense enough to allow high flow velocities. Silica cannot withstand the harsh cleaning and sanitization conditions that are used in early recovery processes. Zirconia particles are the most dense, are stable at high pH, and are sterilizable. Porous glass and the described perfluorocarbons have promising sedimentation properties, but porous glass sometimes shows quite low protein-binding capacities, and the perfluorocarbons have low protein capacities because of the low particle porosity. With the recent development of STREAMLINE adsorbents (agarose/crystalline quartz composite, agarose-dextran/crystalline quartz composite, and agarose/metal alloy composite) and STREAMLINE columns, it was possible to secure a recovery system that has the properties needed for an efficient expanded-bed process. The columns have a flow distributor in the bottom that generates an even flow distribution over the whole cross-section of the bed. The density of the adsorbent permits process flow velocities (300-500 cm/h), and the size distribution creates a classified bed that enhances the stability of the expanded bed. The result is a bed with protein adsorption characteristics similar to those of a packed chromatography bed. Expanded-bed adsorption is a scalable, one-passage technique that is intended for industrial-scale operations. There have been a number of applications reported that use expanded bed adsorption with various feedstocks: secreting bacteria, periplasmic preparations of bacteria, bacterial homogenates, secreting yeast, yeast homogenates, mammalian cells, milk/whey, and inclusion body preparations.

BASIC CONCEPT

The term *fluidized-bed* is used to define those physical systems composed of a solid phase in the form of individual particles that move within a fluid phase and are not in continuous contact with each other. Fluidization of the solid particles is reached when the flow of fluid through the bed is high enough to compensate their weight. On the other hand, in order to be kept in the fluidized-bed reactor and not be washed out (elutriated), the superficial velocity of the fluid in the bed (that is, the ratio between the flow rate and the bed cross-sectional area) has to be lower than the settling velocity of the particles. When the flow rate of a fluid through a packed bed of solid particles steadily increases, the pressure drop increases proportionally to the flow rate, as long as the bed height remains constant. When the drag force of the fluid equilibrates the weight of the particles, the bed starts to expand and, after a transition period, reaches fully developed fluidization. At this point, further increments in the flow rate do not produce an increase in pressure drop, but instead lead to an increase of the height occupied by the solid particles in the reactor. If the flow rate is increased significantly, the elutriation of the solid particles occurs when the fluid's superficial velocity is higher than the solid's settling velocity. The fundamentals of the fluidization phenomena are discussed comprehensively in the chemical engineering literature. Although various configurations are possible, the most extensively used is the gas–liquid cocurrent up-flow reactor. In it, liquid usually comprises the continuous phase and is fed from the reactor bottom. Its flow upward in the reactor promotes fluidization of the solid particles.

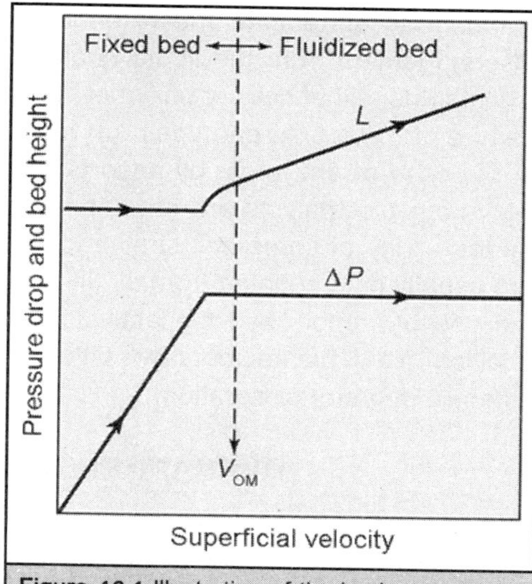

Figure 19.1 Illustration of the basic concept of fluidization.

Usually, the reactor will have two or three phases. In addition to the liquid and solid phases, the occurrence of a gas phase is quite common in those systems using cells as biocatalysts, either for aeration requirements (in which case, an air or oxygen stream is fed to the reactor) or because cell metabolism produces a gas product (for example, CO_2, CH_4). In systems using enzymes as biocatalysts, the most common situation is two-phase fluidization, without any gas phase. Very often, due to the low reaction rates of most biological transformations, long liquid residence times are needed for the completion of the reaction, and therefore the drag force created by the low liquid flow rate in a single pass reactor is not enough to promote fluidization of the solid particles. Fluidization is obtained either by external liquid recirculation or by the gas loaded to the reactor. In systems where a gas is produced by cell metabolism, the gas can also be an additional factor contributing to solid particle fluidization, although other effects are also observed in this case, such as internal liquid recirculation patterns. Fluidization at relatively low liquid flow rates is also favored in tapered fluidized-bed configurations; the liquid superficial velocity at the bottom of the reactor is higher due to the reduced cross-sectional area.

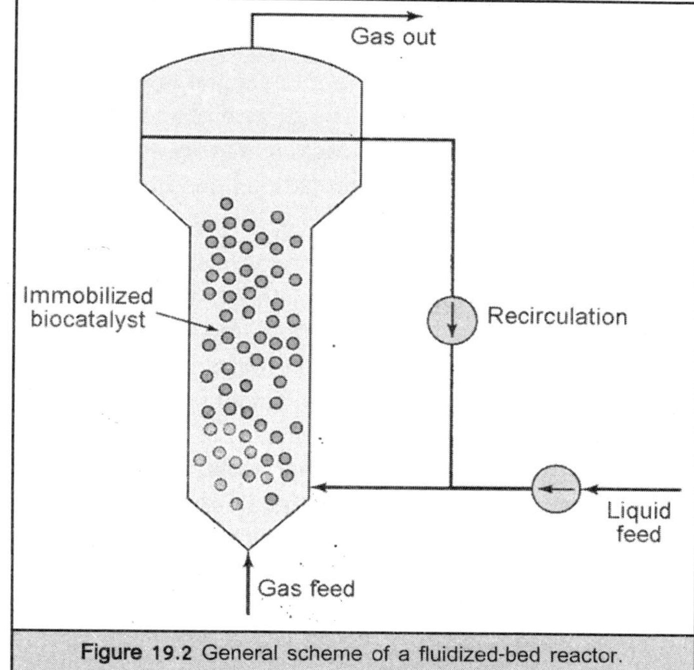

Figure 19.2 General scheme of a fluidized-bed reactor.

In general, one can distinguish three main sections in fluidized-bed bioreactors: (i) the bottom section, where feed (liquid, gas, or both) and recirculation are provided; (ii) the central main section, where most of the reaction takes place; (iii) and the top section, with a wider diameter that serves to decelerate the movement of the particles by decreasing the superficial velocity of the liquid, thus enhancing the retention of the solid phase and at the same time allowing gas disengagement from the liquid phase. It is a common trend for fluidized-bed bioreactors to use biocatalysts, either cells or enzymes, in the form of immobilized preparations. In general, the particles can be of three different types: (a) inert cores on which a biofilm is created by cell attachment, or in the case of enzymes, by adsorption or covalent binding immobilization; (b) porous particles in which the biocatalysts are entrapped; (c) cell aggregates obtained by self-immobilization caused by the ability of some cell strains to form flocs, pellets, or aggregates. Fluidized-bed bioreactors are usually differentiated from air-lift bioreactors by the fact that the latter do not specifically require the use of immobilized biocatalysts. Indeed, they were developed for free cell suspensions. In addition, air-lift bioreactors have different compartments, created by physical internal divisions, with different degrees of aeration.

OPERATION OF FLUIDIZED-BED BIOREACTORS

The use of fluidized-bed bioreactors can provide a number of advantages that makes them an interesting alternative for bioprocesses, especially for continuous operation. In comparison with conventional mechanically stirred bioreactors, fluidized-bed bioreactors provide a much lower attrition of solid particles, and almost any kind of immobilized biocatalyst preparation can be used without physical disruption. Biocatalyst concentration can be significantly higher because of immobilization, and the typical wash-out limitations of continuous bioreactors operating with free cells are overcome because the solid particles are physically retained in the reactor vessel; operation at flow rates higher than the maximum growth rate of the cells can be achieved. As a consequence, the final productivity of the bioreactor can be increased substantially. Compared with packed-bed reactors, fluidized-bed bioreactors can be operated with smaller-size particles and without the drawbacks of clogging, creation of preferential flow paths, or particle compression due to bed weight. Moreover, the smaller particle size minimizes the internal diffusional resistances, and the higher level of mixing enhances external mass and heat transfer from liquid to solid phase. The degree of internal mixing in a fluidized bed can vary to a great extent, and it depends on various factors: density and diameter of the solid particles, liquid and gas flow rates at the reactor inlet, endogenous gas production by cell metabolism, and recirculation rate.

For some operational conditions, high gas, liquid, or recirculation flow rates, the internal mixing will be very high, and the reactor will approach the behavior of a complete mixed tank. On the other hand, operation with low gas, liquid, or recirculation rates will provide a flow pattern close to plug flow, with some degree of axial mixing. This implies that reaction kinetics is an important factor to consider in the analysis and design of fluidized-bed bioreactors. Configurations favoring liquid mixing will be more appropriate for substrate-inhibited reactions, and configurations approaching plug flow will be indicated for product-inhibited reactions. Another advantage of fluidized-bed bioreactors is the ease of separation of the gas produced in most transformations involving cells (i.e., CO_2), or the feed of a gas stream to the reactor, for example, for aeration purposes. Also, fluidized-bed reactors make biocatalyst replacement easy, without disruption of the operation, enabling good control of the overall activity of the reactor. For example, one may replace particles with deactivated enzyme or removing an excess of biomass created by biofilms. On the other hand, solids attrition is higher in fluidized-bed than in packed-bed bioreactors. From the productivity point of view, the advantages of the fluidized bed, especially with respect to mass transfer rates, make it possible to obtain higher levels of overall productivity than in packed-bed reactors, in spite of the fact that the fraction of immobilized biocatalyst particles is lower for a fluidized bed. Fluidized-bed bioreactors are complex with regard to hydrodynamic aspects, especially taking into account that the properties of the biocatalyst particles may change considerably during the operation time and in the presence of three different phases (solid, liquid, gas) in many cases.

In fact, the nature of the particles (for example, their density and size, or their evolution with time, which are especially important with respect to certain kinds of immobilized cells), the liquid and gas flow rates employed, the type of reaction kinetics, and the kinetics of cell growth or enzyme deactivation influence each other and have a direct effect on the reactor design and performance. Using the biocatalyst in immobilized form also contributes to the complexity of a fluidized-bed bioreactor. The behavior of the immobilized biocatalyst, especially when cells are used, can be substantially different than that of free suspensions. The behavior must be determined at the kinetic level, the physiological level, and the genetic level, and the biocatalyst's relationship with the diffusional restrictions in the particles and the possible direct effects associated with the immobilization itself must be well understood and correctly described in order to build appropriate

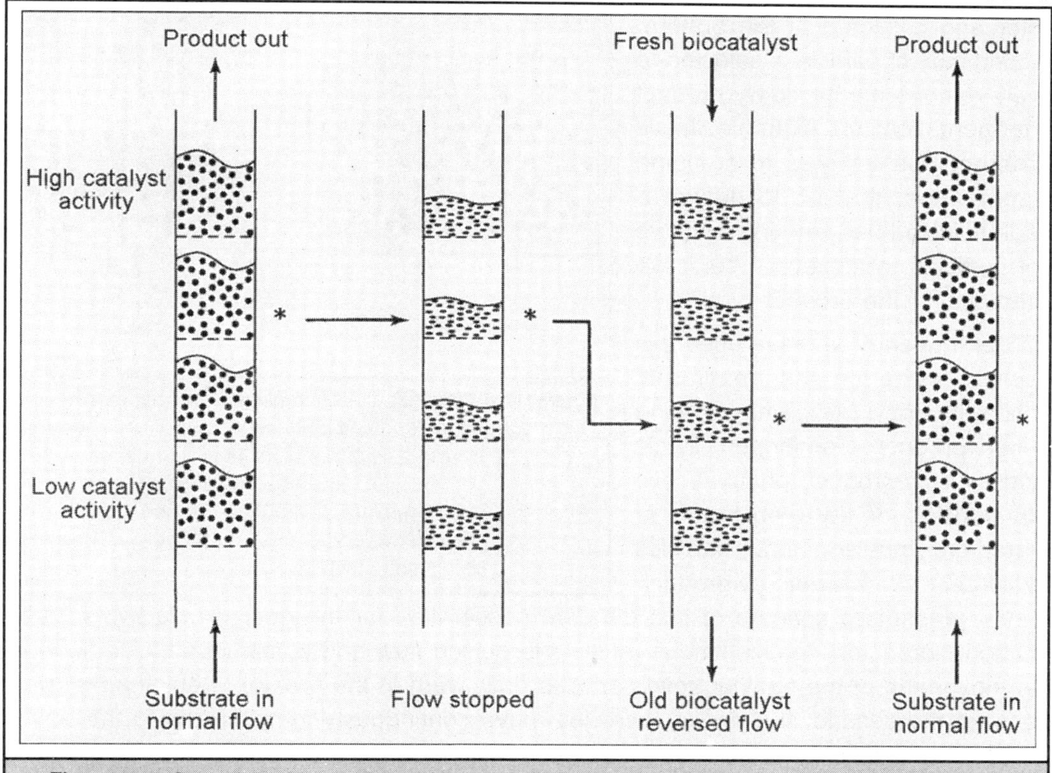

Figure 19.3 Operation phases of a multistage fluidized-bed reactor for a deactivating biocatalyst.

and reliable models of for reactor design, control, and scaling up. The potential of fluidized-bed bioreactors can be further exploited by considering multistage units and using two solid particles with different properties. Figure 19.3 gives an example of the concept of multistage operation, in particular, a countercurrent multistage fluidized bed working with immobilized enzymes.

The main characteristic of this bioreactor is the continuous transport of the solid particles of immobilized enzymes from one stage to another in a downward direction. The overall catalytic activity of the reactor remains constant as the exhausted enzyme is removed from the reactor bottom stage while fresh biocatalyst is added at the top stage. A second advantage of dividing the reactor into compartments is the very low degree of back-mixing of the biocatalyst, and the plug-flow regime attained in the liquid phase. The use of two solid particles with different properties, particularly with different densities, can be used in a fluidized bed to achieve the in situ separation of a product of the reaction; this is a clear advantage in systems with product inhibition or when unfavorable thermodynamic equilibria limit the conversion rate for a reaction.

For example, Davison and Scott have proposed a system based in two different types of particles with different densities. As one type of particle, containing the biocatalysts (in this particular example, cells of *Lactobacillus delbreuckii*), remains fluidized in the bioreactor, the second type, which is heavier and contains no cells, is introduced from the top of the bioreactor and collected at the bottom. This second type of particle is selected to selectively remove the inhibitory product of the fermentation, for example, lactic acid. Another possibility is to combine both approaches, that is, to design a multistage fluidized bed working with two types of particles to achieve a selective in situ removal of the product.

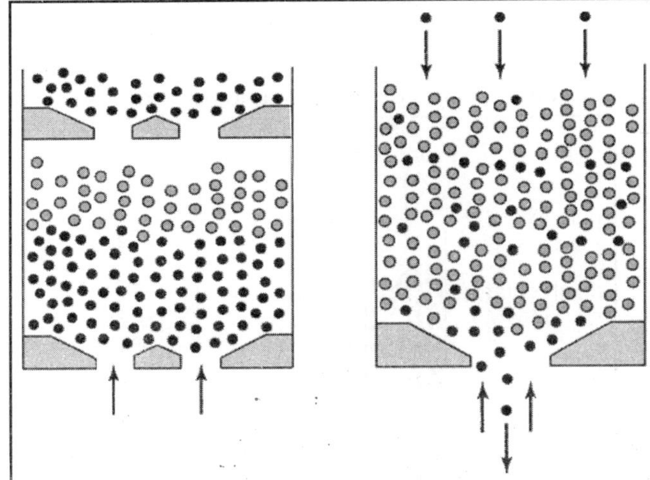

Figure 19.4 Schemes corresponding to two types of operation of a fluidized bed as countercurrent adsorptive reactors, with two types of solid particles: light/small immobilized biocatalyst (gray circle) and dense/large sorbent (black circle). Left: one compartment of the semicontinuous, multistage pulsed flow, Right: continuous trickle flow mode.

Van der Wielen et al. have used this approach to enhance the enzymatic deacylation of benzyl penicillin, providing the desired product, 6-aminopenicillanic acid, and as a by-product, phenyl-acetic acid, by means of light particles of immobilized enzymes and heavy particles of an adsorbent of the acidic by-product. Figure 19.4 presents a scheme of two different possibilities for the design of a two-solid-phase fluidized-bed bioreactor: semicontinous multistage pulsed flow and continuous trickle flow. In the first, the movement of the heavier solid particles downward in the reactor is obtained by periodic pulsations; in the second, the denser particles move continuously from the top to the bottom of the reactor.

THEORY OF FLUIDIZED BED

As mentioned earlier, an expanded bed is a fluidized bed that is devoid of or shows low back-mixing of liquid and adsorbent. The stability of the expanded bed makes it resemble a packed bed in that it shows an almost constant flow velocity profile and yields a bed with an increased number of theoretical plates (relative to a fully mixed fluidized bed). These characteristics are achieved by a combination of column and adsorbent design.

Packed Bed Chromatography

The liquid flow in a column used for expanded-bed adsorption must be uniform over the entire cross-section of the bed. A conventional packed-bed chromatography column with one or a few inlets cannot meet this criterion. In a packed bed, the bed itself creates a pressure drop that achieves the desired plug flow. In an expanded bed, there is a very low pressure drop over the column, thus requiring a specially designed flow distributor at the bottom of the column. This distributor must ensure that the flow is directed in a vertical direction only, because radial flow inside the column will cause disturbances in the expanded bed. It must also prevent the adsorbent particles from leaving the column while allowing cells and debris to pass through. Other requirements are that it must not create high shear stresses, or cells and shear sensitive molecules may break, and it should be designed to facilitate cleaning. There should be no stagnant zones in the column where cells and other contaminants can accumulate and, in the worst case, escape cleaning.

Additional Properties

The adsorbents used for expanded-bed adsorption must exhibit additional properties to those required of an adsorbent used in packed-bed chromatography. The density of the particles must be relatively high to enable high liquid flow velocities through the expanded bed without the adsorbent particles packing against the adaptor. The adsorbent particles in a fluidized/expanded bed are in constant motion; however, an efficient protein adsorption process requires that the solid dispersion (the movement of the adsorbent particles) is reduced to a minimum. This can be achieved if the adsorbent is polydispersed with respect to size, that is, the particles exhibit a size distribution. When such a polydispersed adsorbent with a small variation in particle density is expanded, each particle will find its equilibrium position in the upward flow. The result is called a classified bed, and at a closer look the larger particles will be found at the bottom of the bed and the smaller ones at the top. A distribution of density with size (a large particle exhibiting a larger density than a small particle) will further enhance the classification of the bed. One way of describing the behavior of such an adsorbent in an expanded bed is the Richardson-Zaki model for monodispersed particles:

$$u_0/u_t = \varepsilon^n$$

where u_0 is the superficial velocity of the fluid, u_t is the terminal falling velocity of a particle in infinite dilution, ε is bed voidage, and n is an index. The model describes expansion for a bed that has reached equilibrium (the expansion height is constant). The terminal falling velocity u_t is described by Stoke's law:

$$u_t = ((d_p)^2\ g(\rho_p - \rho))/18\mu$$

where d_p is particle diameter, g is acceleration due to gravity, ρ_p is particle density, ρ is fluid density, and μ is fluid viscosity. Stoke's law can only be used for particles with a Reynold number lower than 0.2. The Reynold number Re_p for a particle in a fluid is defined by:

$$Re_p = ud_p\rho/\varepsilon\mu$$

The Richardson-Zaki model is corrected for the particle size distribution of the adsorbent by using the perfectly classified bed model:

$$\Delta H(\varepsilon) = H_0(1 - \varepsilon_0)\Delta F_v/(1 - \varepsilon)$$

where H_0 is sedimented bed height, ε_0 is sedimented bed voidage, ΔF_v is the proportion of the total volume of particles having the diameter within the interval $d - \Delta d/2 < d < d + \Delta d/2$, and ε is bed voidage for the particles within that interval. This means that the particle size group between d and $(d - \Delta d)$ is found within the expanded bed between heights h and $(h + \Delta h)$. Thus, each segment of height is considered a bed with a specific voidage that can be calculated using the Richardson-Zaki equation. The total bed height is obtained by adding up the heights of each segment.

DESIGNING OF FLUIDIZED-BED BIOREACTORS

As already described, the design and operation of a fluidized-bed bioreactor has to take into account a number of aspects regarding both physical characteristics and reaction performance. These aspects include the hydrodynamics of the bioreactor (usually with three phases), characterization of the type of flow, degree of mixing for each phase and the heat transfer, mass transfer between phases, mass balancing of the species taking part in the reaction, intrinsic kinetics of the immobilized biocatalyst, diffusion transport of the species inside the solid particles, enzyme deactivation, and cell growth. Accurate knowledge concerning these points will allow complete characterization of a fluidized-bed reactor and the development of reliable mathematical models describing its performance. Next, we discuss the conceptual aspects intrinsically related to fluidized beds. The first step in hydrodynamic characterization requires a determination of the fraction of the total volume to be occupied by the different phases (gas, liquid, solid). This is known as *phase hold-up* and can be determined using different techniques. Strictly speaking, phase hold-up may not be uniform for the whole reactor, especially with low axial dispersion. However, in systems with appreciable mixing, phase hold-up can often be considered uniform.

With respect to mixing behavior, most attention is generally paid to characterizing the liquid phase because this is, in most cases, the continuous phase in bioreactors, that is, the one in which substrates are fed and products of the reaction accumulate. For a three-phase fluidized-bed bioreactor with cocurrent up-flow circulation of liquid and gas, the type of flow pattern is very much dictated by the value and the ratio of the liquid and gas superficial velocities in the reactor. Three main regimes are possible: (i) the dispersed flow regime occurs at high ratios of liquid velocity to the gas velocity and is characterized by the homogeneous dispersion of small gas bubbles in the liquid; (ii) the coalesced bubble flow regime occurs at increased gas flow rates and is characterized by the formation of bigger bubbles as a result of the coalescence of smaller ones; the coalesced bubbles have a nonuniform distribution in the liquid; (iii) the slug-flow regime is the consequence of a further increase of gas flow rates, occurs at high ratios of gas velocity to liquid velocity, and is characterized by the formation of large gas bubbles that, in small-diameter bioreactors, tend to completely occupy their cross-sectional area. Slug formation breaks the bed continuity and causes great instability. If the gas flow rate is increased dramatically, it will eventually become the continuous phase in the bioreactor. A very detailed study on the characteristics of these different regimes and the effect of the density and size of the solid particles on the regime transitions has been conducted by Zhang et al. The liquid flow pattern in the bioreactor is directly influenced by the degree of mixing associated with these different regimes, as well as by other factors such as internal gas generation, bead size distribution, and external liquid recirculation.

Basically, the flow will approach plug flow in systems with high velocity in the liquid phase. Quite often, bioreactors experience different degrees of mixing, and complete mixed flow can be observed in fluidized-bed bioreactors. The influence of the gas phase on the liquid and solid mixing in a fluidized-bed bioreactor has been studied by Gommers et al., considering two extreme situations: a reactor operating without gas and a reactor to which gas was introduced artificially from the bottom. The results showed that the gas phase greatly influences the degree of liquid mixing in the reactor, and that this effect increases sharply with the bed diameter. Another fact to consider is that in systems with gas generation associated with the reaction progress (for example, in most fermentations), the gas flow will change from the bottom to the top of the reactor, proportionally to the substrate consumption, and the axial dispersion will change with reactor height. An important aspect to consider in the study of the influence of the gas produced in a fluidized bed on the degree of liquid mixing is that the results will be affected by the experimental system used for gas injection, as discussed in detail by Buffiere et al. When liquid recirculation is used to promote fluidization because of the slow reaction rate and the long liquid-residence time required, plug flow is usually disrupted, and completely mixed flow is usually achieved. Other factors influencing the hydrodynamic behavior of fluidized-bed bioreactors are the properties of the solid and liquid phases.

In general terms, the weight and size of the solid particles will directly influence the liquid and gas flow rates required for bed fluidization. If the liquid residence time is fixed by criteria of substrate conversion, for example, then the heavier particles will require higher L/D ratios to increase liquid superficial velocity and enhance fluidization, or as an alternative, they will require high recirculation rates. In absence of recirculation, and when particles with a certain distribution in size and weight are used, solid-particle stratification is commonly found. Under stratification conditions, movement of the solid particles in the bed is very limited, and the particles are ordered by decreasing settling-velocities, from the bottom to the top of the bed. One of the consequences of this situation is that, in liquid plug-flow regimes, the particles at different reactor heights will experience different environments. For cells growing as biofilms around a solid particle, it is typical for the particles at the bottom of the bed to provide better conditions for cell growth (for example, substrate availability), and as a consequence, they will decrease their overall density and therefore migrate to the upper part of the reactor. Removal of excess biofilm may require external treatment of the particles. The operation of a stratified fluidized bed is illustrated in Figure 19.5, where the stratification of a bed with activated carbon as a solid support for cell growth is combined with the capacity to absorb the product of the reaction and, therefore, remove it selectively. The crucial effect of particle size and density distribution, as well as gas and liquid superficial velocities, on the phenomena of solid-particle stratification is discussed in detail in various recent studies. In general, solids mixing has been studied less comprehensively than liquid mixing, and it is less well understood.

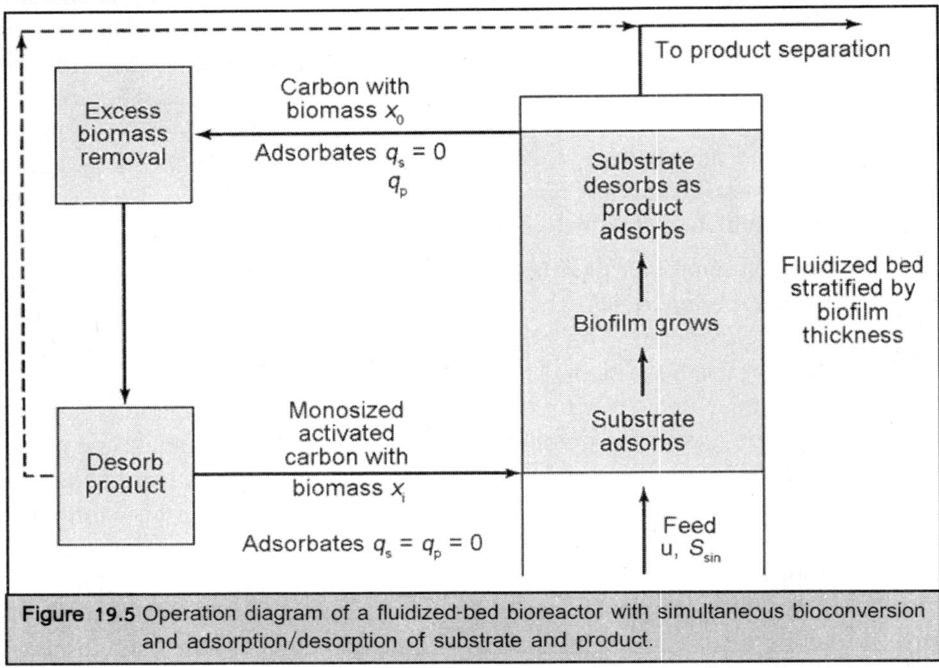

Figure 19.5 Operation diagram of a fluidized-bed bioreactor with simultaneous bioconversion and adsorption/desorption of substrate and product.

The possibility of using new experimental techniques has enabled researchers to propose new models for the interpretation of the solids mixing, for example, trajectory length distribution, in contrast to more classical models that assume that solids mixing can be defined by means of an axial dispersion model. Another important aspect to be considered from the hydrodynamic point of view is that fluidization properties depend on the difference in densities of the solid and liquid phases in the fluidized bed, and these values change during the operation time. This situation is of particular relevance for those systems where the absolute values for the liquid and solid densities are relatively close, such as for cells immobilized in natural origin polymers (alginate, agarose, etc.) or self-aggregated cells (pellets, flocs, etc.). In this case, relatively low variations in the absolute value of the solid density (usually associated with cell growth or the accumulation of CO_2 gas in the particles) and the liquid density (usually associated with substrate consumption) may cause an important percentile variation in the density difference between both phases, producing a relevant impact on the reactor hydrodynamics, as the regime changes from dispersed to coalescing, and on reactor stability.

This problem can be of particular relevance at the startup of a bioreactor operation, when the density values may change to a greater extent. A second aspect to be addressed in the characterization of a fluidized-bed reactor is the definition of the flux model. As mentioned previously, most of the attention is focused on the liquid phase, and the flux model is closely connected to the hydrodynamic conditions in the reactor that will generate a given degree of internal mixing, which is somewhere between the two extreme situations of perfectly mixed or plug flow. The determination of the real liquid flux model in the bioreactor is a necessary step for the application of the mass balance equations for the species taking part in the reaction. Stimulus–response techniques are commonly used for such a purpose and are based on the introduction of an inert tracer at the reactor inlet and the analysis of the response curve obtained at the outlet, which reflects the type of flux. The models that describe the liquid flux in a real reactor can be (i) an axial dispersion model, in which axial dispersion is superimposed on the liquid convective flux; (ii) a tank-in-series model, in which the bioreactor is considered as a series of CSTR reactors of the same volume; and (iii) a compartmented model, in which the flux model in the bioreactor is described as the combination of different ideal compartments.

For fluidized-bed bioreactors, especially when fermentation gas is produced, some interesting contributions have been proposed, such as the consideration of a variable dispersion coefficient, which increases its value in proportion to the fermentation gas generated in the reactor, up to a given point. In addition to hydrodynamic and mixing characteristics, another relevant aspect of a fluidized-bed bioreactor is its biocatalytic activity. Because immobilized biocatalysts are always used, the definition of the problem must simultaneously include the reaction characteristics (kinetics, stoichiometry, equilibrium) and transport characteristics, including both the transport between the liquid and the solid phase and the transport within the solid phase. When the general situation of a solid biocatalyst particle immersed in a liquid phase, is studied, different phenomena have to be considered. First, substrate is transported from the liquid phase to the solid external surface. This external mass transfer is usually modeled by means of a hypothetical liquid film that creates resistance to the transport, and it is characterized by the external mass transfer coefficient. The value of the coefficient, and thus of the external mass transfer velocity, depends on the physical properties of the liquid and the superficial velocity of the liquid around the solid; thus beds with more mixing will present better conditions for external mass transfer.

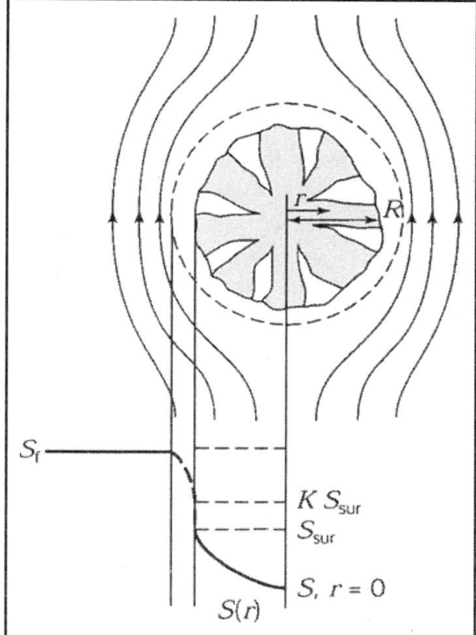

Figure 19.6 Inter and intraparticle mass transfer of a single porous spherical bead of radius R. Substrate concentration profiles across the stagnant liquid film and inside the solid particle, $S(r)$. S_f, concentration of the bulk liquid; S_{sur}, concentration on the solid surface; K, partition coefficient.

When external mass transfer resistance is negligible, the concentration at the surface of the particle is the same as in the fluid phase ($S_f = S_{sur}$). An additional possibility is the occurrence of partitioning phenomena in the substrate, between the particles and the liquid, as a result of the material properties. In that case, the ratio between the substrate concentration at both sides of the solid–liquid interface will be given by the partition coefficient, K. Equally, partitioning phenomena can be determined for the product. Once in the solid particle, the substrate will diffuse inside the solid, following Fick's law, and simultaneously the reaction will take place. Therefore the corresponding equations of kinetic reaction and diffusion in the solid have to be solved simultaneously in order to obtain the internal concentration profiles. To define this part correctly, the values for the intrinsic kinetic parameters of the reaction (that is, in absence of mass transfer limitations) and the effective diffusivities for the substrate and the product of the reaction in the solid have to be known. Moreover, the substrate conversion implies product generation, and this can affect the kinetics (for example, by product inhibition); usually the diffusion and reaction analysis is made simultaneously for both substrate and product. The overall activity of the biocatalytic particles will be dictated by the relative velocities of the two phenomena taking place inside them: reaction and diffusion.

Systems with low diffusion rates with respect to the reaction rates will be diffusion controlled; on the other hand, systems with high diffusion rates with respect to the reaction rate will be controlled by the reaction. In the first case, the low diffusion will limit the efficiency of the particle because the reaction potential of the immobilized biocatalyst will not be fully used. Usually, this is reflected in terms of the efficiency factor, which is defined as the ratio between the actual reaction rate occurring in the system, and the reaction rate that would occur if no diffusional limitations existed, that is, when all solid particles would present a uniform concentration, equal to that of their surface. The relationship between the effectiveness factor and some moduli, such as the observable modulus, being proportional to the ratio between reaction and diffusion rates.

One of the interesting properties of such graphs is that they are very similar for different types of geometries and kinetics. Therefore they enable a direct analysis of the degree of diffusional limitations in a given type of particle in a fluidized bed and also suggest quantitative modifications to be performed to avoid such limitations, for example, changes in the particle size or diffusion conditions. In general terms, fluidized beds are interesting with regard to these aspects because since they require comparatively small-diameter particles for better fluidization, the potential diffusion limitations are reduced. As mentioned at the beginning of this section, the combination of the different aspects we discussed allows the elaboration of mathematical models describing the behavior of fluidized-bed bioreactors. The reliability of these models depends greatly on the accuracy of the determination of the various parameters involved in the definition of the reactor and the reaction system (flux model, intrinsic kinetic parameters, external mass and heat transfer coefficients, internal effective diffusivities), and they will serve for various purposes such as conceptualization and understanding of the bioreactor itself (an important aspect required for building the model), design of similar bioreactors (especially in the case of scaling-up), simulation of the bioreactor operation at different conditions, and bioreactor control.

TECHNIQUES

The major difference between a standard packed-bed column and a column used for expanded-bed adsorption is the liquid distribution system. STREAMLINE columns are specially designed for expanded bed adsorption and have a fixed bottom distributor and a movable top distributor (adaptor). The bottom and top distributors have essentially the same design: a perforated plate with a single-weave screen or net facing the adsorbent. The distributor plate in the adaptor has a lower pressure drop than the bottom distributor. The purpose of the top distributor plate is to create a more even flow (compared with a column with one or a few inlets in the adaptor) when the column is run in downward mode during elution in settled bed mode. The steel screens help to stabilize the flow and confine the adsorbent particles to the column. These columns do not cause cell damage, and it has been shown that shear- sensitive cells pass through the column without breaking. Expanded-bed adsorption handles unclarified feedstocks, and for this reason, harsh cleaning procedures are necessary.

The columns are constructed of materials that withstand the chemicals used in such procedures and are also designed to have the minimum of stagnant zones where cells could accumulate and risk contaminating the next run. It has been shown that the columns can be efficiently cleaned and sanitized after passage of unclarified feedstocks such as whole cultures of bacteria and yeast. Columns range in size from small lab-scale columns with a 25-mm inner diameter to industrial-scale columns, presently the largest with a 1,200-mm inner diameter. They all have essentially the same design, but differ in the material used for the column tube: the smaller columns have tubes made of glass and industrial columns have tubes made of stainless steel. The larger columns also have a greater number of inlets and a higher pressure drop over the distributors than the smaller columns. One method of verifying successful scale-up and comparing the performance of different sizes of expanded-bed columns is to determine the axial dispersion of a tracer substance. Another method is to determine protein breakthrough capacity using a model protein in a buffer system. The results obtained from these methods give a good measure of bed stability, a factor that is crucial for a successful expanded- bed process.

The hydrodynamic properties of the columns are tested using the axial-dispersion method, and the "function"—the protein adsorption properties—is evaluated using the protein breakthrough method. The columns have been tested using both methods, and the results demonstrate that the performance of an industrial-sized column is very similar to that of a small lab-scale column. Because industrial-scale columns have to be automated, these columns are equipped with an adsorbent sensor that locates the position of the expanded bed surface. In response to the signal from the sensor, the adaptor is automatically adjusted to its correct position.

Table 19.1 Number of theoretical plates (N) and axial dispersion (D_a) for expanded-bed columns with inner diameters from 25 to 600 mm

Column	*N*	D_a (m^2/s)
STREAMLINE 25	50–70	3-5 10^{-6}
STREAMLINE 50	40–60	4-6 10^{-6}
STREAMLINE 200	35–50	5-7 10^{-6}
STREAMLINE 600	35–50	5-7 10^{-6}

Up Front Columns offer another approach to expanded-bed column design. The columns are without screens and without distribution plates, and the feedstock is introduced at the bottom of the column, right below a stirring device that distributes the feedstock over the cross-section of the column. The stirrer causes a mixed zone, but gradually above that zone plug flow is obtained. Columns of 20-, 50-, and 200-mm inner diameter are available. BioProcessing Ltd. uses modified standard chromatography columns for expanded bed adsorption. The standard column bottom support net or screen is replaced with a 300-μm nylon mesh, and then the column is packed with a layer (approximately 3 cm in height) of surface-modified solid glass beads (0.4-0.52 or 0.85-1.23 mm in diameter). This bed of beads serves as a distributor of fluid to the bed. The adsorbent particles are then added directly on top of the distributor beads. During operation of the column, the bed of the relatively heavy distributor beads remains static, and the bed of adsorbent particles expands.

Industrial Chromatography Flow

STREAMLINE adsorbents are specially designed for expanded bed adsorption. To ensure that the bed does not expand too much at industrial chromatography flow velocities, a dense core material has been incorporated into the agarose particles to increase the apparent density. The adsorbent has a defined Gaussian-like particle size distribution, which yields a stable classified bed when the adsorbent is expanded. The influence of particle size distribution on protein adsorption performance has been studied using three different particle fractions (120-160 μm, 250-300 μm, and 120-300 μm). The results showed that a wide particle size distribution gives an expanded bed with low axial dispersion and that small particles give high dynamic protein capacity (due to reduced diffusion length and large surface area). The results also showed that with increasing (settled) bed heights, the dynamic binding capacity increases over a wide range of flow velocities (due to reduced axial dispersion and increased numbers of transfer units for pore and film diffusion). The STREAMLINE DEAE, Q XL, SP, and SP XL, chelating and heparin adsorbents are spherical, macroporous, cross-linked 6% agarose particles with crystalline quartz as core material. The Q XL and SP XL adsorbents have dextran chains coupled to the agarose, and the Q and SP charged groups are attached to the dextran through chemically stable ether bonds. This increases the exposure of the Q and SP groups and results in very high protein binding capacities.

The adsorbents can be cleaned with 1 M NaOH. The matrix used for the protein A adsorbent contains highly cross-linked 4% agarose with a metal alloy core. The alloy is stable and composed of 77% Ni, 15.5% Cr, and 7.5% Fe. No leakage of metal was detected in 0.1 M glycine, pH 3.0, a solution commonly used in protein A chromatography. The metal leakage was below the detection limit of the analysis method (0.02 μg/mL eluent using induced coupled plasma atomic-emission spectroscopy). The protein A adsorbent is stable in all aqueous buffers commonly used in protein A chromatography and cleaning. No significant change was observed in chromatographic performance after either 1 week's storage or 100 cycles, normal use, at room temperature with 10 mM HCl (pH 2), 0.1 M sodium citrate (pH 3), 1 mM NaOH (pH 11), 6 M GuHCl, or 20% ethanol. Apart from the purpose-designed column and adsorbent, the equipment and the experimental set-up used for expanded bed adsorption is very similar to that used for packed bed chromatography.

Expanded Bed Process

The strategy for developing an expanded-bed process is in general analogous to that of a conventional packed-bed process. In order to save time, it is advisable to do the first experiments with clarified feedstock, using the expanded-bed adsorbent in a small, packed-bed column, for example, 15 mm in diameter and with 15-cm bed/height. (A bed height below 10 cm of these relatively large particles usually results in a decreased dynamic binding capacity. The increase in dynamic binding capacity with increased bed height is an effect of longer residence time for the sample.) From the packed-bed experiments, the most suitable adsorbent is selected (e.g., ion exchanger or chelating adsorbent), and the conditions for protein binding and elution are determined. It is important to establish that the conditions selected for protein binding are compatible with the properties of the unclarified, cell- or debris-containing feedstock (e.g., low pH might cause aggregation). Such problems are not apparent with clarified feedstock and should be investigated before the process has progressed too far.

Once the most promising method has been selected from the packed-bed experiments, it is then optimized in a lab-scale expanded-bed column using unclarified feedstock. At this stage, the effect of the now present cells or debris on protein binding is evaluated. In theory, if the net surface-charge of the cells and debris is negative, the protein and the cells or debris may compete for binding sites on an anion exchanger. A consequence of this may be that the feedstock load per milliliter adsorbent has to be decreased. When STREAMLINE DEAE was used to recover insulinase from whole yeast culture, the presence of cells did not cause any reduction in protein capacity. However, it has been reported that when the same adsorbent was used to recover glucose-6-phosphate dehydrogenase from a yeast homogenate, the presence of cell debris did cause a slight reduction in capacity of the adsorbent. Feedstock properties and their effects on the expanded-bed process are discussed further in "Feedstocks." Before the process is transferred into larger columns, appropriate cleaning procedures should be determined. The whole process is then verified at pilot scale before scaling up to the final production scale.

Flow Direction

Avoid getting air into the column; air trapped in the bottom distributor may disturb the bed. If air does enter the column, it can usually be removed with high-flow velocity pulses combined with back-flushes (reversal of flow direction). Before start-up always check that the column tube is positioned vertically; otherwise, flow will be turbulent when the liquid hits the inside walls of the column tube. The height of the settled bed (H_0) is noted before start-up.

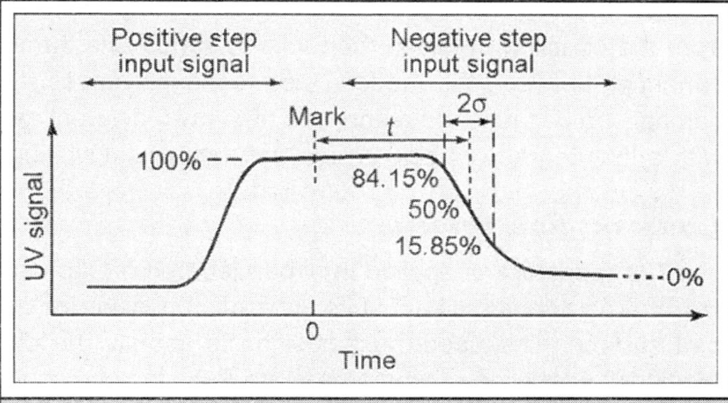

Figure 19.7 UV signal recording from the test produce for determination of the number of theoretical plates (*N*). $N = t^2/\sigma^2$, where *t* is the mean residence time and s is the standard deviation

Buffer practice

The bed is expanded and equilibrated with buffer. It usually takes 20–30 min at a flow velocity of 300 cm/h before the bed is stable and has stopped expanding. The height of the expanded bed (H) is then noted. At this stage, the bed can be checked empirically, before application of the feedstock, by determining the axial dispersion of a tracer substance. Before applying the tracer substance, the adaptor is lowered to a position where the adaptor screen is 0.5-1.0 cm above the expanded-bed surface. When the signal from the UV monitor is stable, the solution is changed to buffer-tracer substance mixture (e.g., 0.25% acetone v/v), that is, positive step input signal. The solution is changed tobuffer when the UV signal is stable at maximum absorbance (100%), that is, negative step input signal. The change is marked on the chart recorder paper, and the UV signal is allowed to stabilize at the baseline level (0%). The number of theoretical plates (N) is calculated from the negative step input signal:

$$N = t^2/\sigma^2$$

where t is the mean residence time, the distance from the mark on the chart recorder paper to 50% of the maximum absorbance; and σ is the standard deviation, half the distance between the points 15.85% and 84.15% of maximum absorbance.

The procedure gives the number of theoretical plates not only in the column but also the contribution from pumps, valves, tubing, and so forth. Because different experimental set-ups may vary in their contribution to N, the results should only be compared between runs performed using the same set-up. The negative input signal is recommended for evaluation (at least when acetone is used as tracer) because the reproducibility of the results is higher than that obtained from the positive step input signal (from 0% to 100%).

The relation between N and the axial dispersion coefficient (D_a) is

$$D_a = (u_0 \cdot H)/(2 \cdot N \cdot \varepsilon)$$

where u_0 is the superficial velocity, H is the expanded bed height, and ε is the bed voidage in expanded mode. ε is calculated from $H/H_0 = (1 - \varepsilon_0)/(1 - \varepsilon)$. In a settled bed, the bed voidage (ε_0) is approximately 0.4. The higher the value for N, the lower the value for D_a, indicating a low degree of back-mixing (a flow profile approaching plug flow). The relation between N and D_a is derived from the coupling between the tanks in series and the axial dispersion models by comparing the variances between the models' step-response curves. It is not completely accurate to make this comparison because the responses of the two different models are never identical. But for a system with a low degree of back-mixing, this way of calculating the axial dispersion is accurate.

Constant flow velocity

The feedstock is applied in expanded bed mode, usually at a constant flow velocity (e.g., 300 cm/h). The adaptor is kept at its uppermost position, or at a position 5–10 cm above the expanded-bed surface. The uppermost position is usually the choice when a new (unknown properties) feedstock is applied or when there are feedstock batch variations, making it difficult to predict how the feedstock will behave in the expanded bed from run to run. If feedstock properties, such as viscosity or biomass or others that affect expansion, are known to be consistent and do not vary, then the adaptor can be left at a position closer to the bed surface. The feedstock is always stirred during application to prevent cells and debris from settling. If there is a build-up of cells and debris on the adaptor, the column is intermittently back-flushed. The duration of a back-flush is about 5-

10 seconds, and the interval between back-flushes is usually 10 to 30 minutes, depending on the properties of the feedstock. Another method of applying the feedstock is to change the flow velocity (normally lowering it) during the run, so that the expansion height is kept constant. It has been reported that this approach, under certain conditions, may give a higher productivity than the constant flow velocity method does.

Washing

The wash is performed in expanded-bed mode at the highest possible flow velocity. This means that the flow velocity at the beginning of the wash is often the same as during feedstock application, but as the bulk of the cells and debris are washed out, the flow velocity is increased. At the beginning of the wash, the column is usually back-flushed as described in the previous section. During the wash, when the bulk of the particulates has been washed out, the adaptor is lowered closer to the bed surface. Lowering the adaptor significantly reduces the volume of wash buffer and shortens the time required for the wash step. If, instead, the wash is performed with a wide gap between the bed surface and the adaptor, mixing of the liquids is usually severe (a low density wash buffer being introduced after the high density/viscosity feedstock), resulting in a large wash volume. After the wash, when the UV curve has levelled out, the column is back-flushed to remove any trapped cells or debris from the distribution system. The bed is allowed to settle and the adaptor is lowered to the bed surface. The wash is continued in settled-bed mode, in a downward flow direction at 300 cm/h, for at least one settled bed volume or until the UV curve levels out. When the wash step has been optimized approximately, it then requires 15 settled bed volumes of wash buffer to flush out particulates and unbound proteins.

Mechanism of elution

Elution is usually performed in settled bed mode with downward flow. The flow velocity is lower than during the previous steps, usually 100 cm/h, in order to keep the elution volume as small as possible. A high flow velocity during elution will increase the elution volume. It has been shown that elution in expanded-bed mode may increase the volume by approximately 40%.

Process of cleaning

Cleaning is performed in expanded bed mode to allow any trapped particulates in the bed or column to be flushed out. The adaptor is typically placed at a position corresponding to twice the settled bed height, and the flow velocity is about 100 cm/h. Cleaning is facilitated if it is performed directly after each purification cycle. As with any protocol for cleaning, it will depend on the adsorbent and on the nature of the feedstock, etc. A general cleaning protocol suggested by the manufacturer for the different types of adsorbents (e.g., ion exchanger) is always a good starting point, as is a cleaning protocol used for a corresponding packed-bed process (if such exists). As previously mentioned, the cleaning is facilitated if as many of the particulates as possible have been washed out during the previous wash step. Back-flushes disturb the build-up of particulates, and it has been shown that back-flushes at a high flow velocity were effective in removing the build-up of sticky yeast from the distributor plate in an industrial-scale expanded-bed column. After cleaning, the adsorbent can be stored in the column in an appropriate storage solution.

Physical Properties

Particulate-containing feedstocks can be applied to an expanded bed without prior clarification steps such as centrifugation and filtration; however, there are certain physical properties a feedstock must have to make it suitable for an expanded-bed process. The size and shape of the particulates must be such that they can pass through the screen or net in the column, that is, the particulates must be smaller than approximately 60 μm in diameter. Feedstocks containing small, round single cells or fine precipitates (e.g., renatured inclusion body preparations) will easily pass through the screens, but large aggregates of cells will not. If the particulates are too large, they will build up under the bottom screen and eventually block the column. When the feedstock contains particulates that are too large, they can be removed by centrifugation or by filtering through a simple mesh screen. It is of course desirable for these steps to be avoided completely, but they are relatively easy to perform and should not be compared to the complete clarification procedures needed for a packed-bed process. It is important to stir all feedstocks during application, otherwise the particulates may settle and disturb the expanded bed.

Other parameters that will influence the performance of an expanded bed are the viscosity and biomass content of the feedstock. High values for these parameters will cause the bed to overexpand (relative to a buffer); if the bed expands up to the adaptor, the adsorbent particles will start to pack against the adaptor. This buildup will act as a filter and eventually cause the column to block. Furthermore, if the viscosity or the biomass content is too high, flow may become turbulent and channels may appear in the bed. If the turbulence is severe, the feedstock will pass through the channels in the bed, resulting in poor adsorption efficiency. The upper limits for feedstock viscosity and biomass content will vary from case to case, but for STREAMLINE with homogenates of *Escherichia coli*, there are a few guidelines. A dry weight of 3 to 4% gives the best result, 5% gives an acceptable result, and 7 to 8% is the upper limit. Viscosities up to 10 mPa s (measured at a shear rate of 1/s) give the best results; the upper limit is about 50 mPa s (at 1/s). In cases where the viscosity or biomass content of the feedstock does not cause the bed to expand too much, build-up on the adaptor can be prevented if the column is intermittently back-flushed. The only way to lower the biomass content of the feedstock is to dilute it. The disadvantage is the resulting increase in feedstock volume.

Dilution is only one of way of lowering viscosity. When the viscosity is caused by nucleic acids, the feedstock may be treated with a nuclease. When such a feedstock is a homogenate, further homogenization can be used to shear the nucleic acids and thereby lower the viscosity. The viscosity of a feedstock decreases with increased temperature, making it advantageous to apply the feedstock directly from the fermentor without prior cooling. Since temperature also has a positive effect on protein adsorption, expanded-bed adsorption should always be performed at room temperature (or without feedstock cooling), unless the protein of interest is, for example, sensitive to proteases. To keep aggregation of cells and debris to a minimum, any pH adjustment of the feedstock should be performed immediately before application to the expanded-bed column.

Aggregation of cells usually occurs at low pH and is therefore a problem associated with cation exchange, where pH has to be below the isoelectric point of the protein for it to bind to the adsorbent. If aggregation of cells and debris occurs within the expanded bed, when the feedstock is in contact with the adsorbent, and blocks the screen in the adaptor, the expanded bed may be operated without the adaptor screen (provided that the bed has not expanded all the way up to the adaptor). As mentioned previously, cells or debris may, under certain conditions, interact with the adsorbent particles. If this occurs, it is usually with anion exchangers because they are positively charged and may attract the negatively charged cells or debris. A slight change in the binding conditions can help, for example, the addition of some salt and/or lowering the pH. Alternatively, the feedstock load per milliliter adsorbent may have to be decreased, or it might be necessary to change to an adsorbent with another type of ligand, such as a cation exchanger.

COMMERCIAL VALUES

Expanded-bed adsorption is a technique with great potential for initial purification. The number of published industrial scale applications is limited because until recently columns and adsorbents suitable for use in expanded beds were not available. In addition to the three industrial-scale applications described below, an IND application has been filed with the Food and Drug Administration for a therapeutic recombinant protein in which expanded-bed adsorption is central to the purification process.

Conventional Packed Bed Method

The number of unit operations in the process for purification of recombinant human serum albumin (rHSA) from yeast was reduced by three when expanded-bed adsorption replaced the conventional packed-bed method. In the conventional packed-bed method, three filtration steps were necessary to prepare the yeast culture for the first cation exchange column; when using expanded-bed adsorption, the unclarified culture could be applied directly to the expanded cation exchanger. By reducing the number of unit operations, the total processing time was deceased and the overall yield was increased. In contrast to the conventional packed-bed process, the expanded bed process could be performed in a totally closed system, and the quality (determined as degree of coloring) of the rHSA was reported to be higher using the expanded-bed process. The process was successfully scaled up (determined by comparing yield of rHSA) from a 50-mm diameter column with 300 mL adsorbent to a 1,000-mm diameter process scale column with 150 L adsorbent. Approximately 2,000 L unclarified feedstock (1,000 L culture diluted with 1,000 L distilled water) was applied to the process scale column, and the average yield of rHSA from four different runs was 87% (ranging from 82 to 91%). Using a specially made column with a 4-mm inner diameter, the lifetime of the resin was investigated. The results showed that the adsorbent could be reused up to 1,000 times without compromising performance.

Recovery of Recombinant Monoclonal Antibodies

A process for recovery of recombinant monoclonal antibodies was directly scaled up from a small lab-scale expanded- bed column containing 70 mL cation exchange adsorbent to a process-scale expanded-bed column to which 12,000 L of cell culture was applied. In this single process step, all the cells were removed, the monoclonal antibody was concentrated fivefold, and the yield of antibody obtained was 95%.

Mimicked Purification Process

In a mimicked purification process, a 4.5% dry weight yeast culture was spiked with bovine serum albumin (2 mg/mL) and used as feedstock. The bovine serum albumin was then recovered from the feedstock using expanded-bed adsorption with an anion exchange adsorbent. The process was scaled up from a 25-mm diameter column to a pilot scale column with a 200-mm diameter and finally run in a fully automated system with a production-size column with a 600-mm diameter. The settled bed height was 15 cm throughout all scales, and the feedstock volume was approximately 750 mL in the 25-mm column and 430 L in the 600 mm column. The results from the different scales showed good agreement. The yield of bovine serum albumin was approximately 89% (ranging from 87 to 92%). The wash volume was approximately 11 settled bed volumes, and the volume of the elution peak was approximately 2 settled bed volumes.

Purification of Protease Inhibitor Aprotinin

The successful use of expanded beds at laboratory and pilot scales has increased significantly during the past few years, giving rise to a number of potential industrial-scale applications. One such example is the purification of the protease inhibitor aprotinin, which has good potential as a therapeutic and diagnostic compound. Aprotinin was expressed and secreted from the methylotrophic yeast *Hanensula polymorpha*, and the initial recovery was performed using 300 mL cation exchange adsorbent in a 50- mm diameter expanded-bed column. The only pretreatment of the yeast culture was a 1 + 1 dilution with water and an adjustment of pH. The final volume of the feedstock was 6.4 L. The dilution achieved two goals: it lowered the conductivity to favor binding to the ion exchanger, and it reduced the biomass and viscosity of the feedstock so that the bed did not expand too much. The yield of aprotinin was 76%, the purification factor was 4, and the concentration factor was 7. Another example is the recovery of a recombinant mitotoxin fusion protein, fibroblast growth factor–saporin (rFGF-2-SAP), which is a possible therapeutic agent for the treatment of diseases characterized by cellular proliferation (e.g., cancer).

The fusion protein was expressed in *E. coli*. After homogenization, it was recovered from the diluted homogenate using 300 mL cation exchanger in a 50-mm diameter expanded-bed column. Homogenate from 300 to 400 g (wet weight) cells was applied to the expanded-bed column. The yield of fusion protein was 65%, and the purification factor was 20. It was also reported that a somewhat modified process was scaled up to handle 8,000 g (wet weight) cells. This process was carried out under GLP (good laboratory practice) conditions, and one such process yielded an average of 1.6 g of protein. The yield and quality of the fusion protein were comparable between the different scales. Recombinant human nerve growth factor (rhuNGF) produced in Chinese hamster ovary cells is another example of a potential industrial scale application. Here, the rhuNGF was recovered using a 25-mm diameter expanded-bed column with cation exchange adsorbent using flow velocities up to 375 cm/h. The yield of rhuNGF was approximately 95% and concentrated approximately 45 to 50 times.

During method development, it was found that the dynamic binding capacity increased when the feedstock was applied at 37°C (normally at 25°C). The optimal conditions for recovery were determined in this small scale before the process was scaled up 65-fold. In another example, *E.*

coli was used to express modified exotoxin A from *Pseudomonas aeruginosa*. The exotoxin lacks the enzymatic activity but has retained binding activity. It accumulates in the periplasm of *E. coli*. After release of the exotoxin by osmotic shock, it was recovered using expanded-bed adsorption with an anion exchange adsorbent.

To reduce the viscosity of the extract, it was treated with a DNase prior to application to a pilot-scale expanded-bed column with a 200-mm diameter. Cell extract from 4.5 kg of *E. coli* was applied to 4.7 L adsorbent at 400 cm/h. Recovery by expanded-bed adsorption gave an exotoxin with a concentration three times greater than that obtained using the conventional packed-bed process and with a slightly higher yield. Furthermore, the entire process was completed in one third of the time. Humanized immunoglobulin G_4 was affinity purified from myeloma cell culture using expanded-bed adsorption at pilot scale. A 200-mm diameter expanded-bed column containing 4.7 L protein A adsorbent was used. The cell reactor was connected directly to the expanded bed, and approximately 100 L of cell culture was applied to the column at 37°C. The purification process (application, wash, and elution) took 2.5 h and 40 g of the antibody was quantitatively recovered. The purity of the antibody was comparable to that obtained by the corresponding packed-bed process. Other potential large-scale processes include the purification of proteins from milk or whey.

SPECTRUM OF APPLICATIONS OF FLUIDIZED-BED BIOREACTORS

Fluidized-bed bioreactors have been studied in a wide spectrum of applications as a consequence of the different advantages that they offer. Nevertheless, this interest has been translated with relatively low intensity into the industrial scale of operation. Possible reasons for this may be the greater complexity of the reactor compared with conventional types (especially regarding long-term operation, hydrodynamic behavior and evolution of the immobilized biocatalyst), and the time required in any industrial process to implement any change in technology. Some specific examples from different biotechnological processes illustrate the feasibility of the fluidized bed at a pilot plant and on the industrial scale. The area of waste-water treatment provides examples of high-volume operation and has made use of three-phase fluidized-bed bioreactors, as well as other related designs (such as the up-flow anaerobic sludge blanket reactor), to a significant extent. The wastewater from the baker's yeast industry, particularly from the Gist-brocades company in Delft, Netherlands, has been treated with two anaerobic fluidized-bed reactors in series, each one 21 m high and with 390 m^3 total volume, operating with bacterial cells immobilized by attachment onto 0.25- to 0.5-mm-diameter sand particles, a support that is typical for other fluidized- bed reactors applied to anaerobic and aerobic wastewater treatment.

The volume of the reactor employed for the disengagement of the CH_4 gas obtained was important, and the fluidized-bed volume was 215 m^3. The carbon–oxygen demand (COD) conversion of these reactors is 22 kg/(m^3 day), which is 65% of the COD in the load. A first set of these reactors started operation in 1984, and a second set in 1986. More recently, data from two full-scale anaerobic three-phase fluidized-bed plants have also been reported. A 700-m^3-total-volume reactor (500-m^3 working volume) has been installed for wastewater treatment at a sugar beet factory of United Northern Sugar Factories Co. in Clauen, Germany, using cells immobilized onto pumice and having COD conversion of 20 kg/(m^3 day). A 734-m^3-total-volume multistage reactor (650- to

700-m^3 working volume) has been installed for wastewater treatment at a yeast factory of Hamburg Co., in Hamburg, Germany, using free cells and a new principle that allows control, in each stage, of the ratio of the recirculated gas amount with respect to the rising gas flow. This enables optimization of the hydrodynamic conditions in each stage, according to the reactor loading.

Nitrification of the effluent from a municipal wastewater treatment plant in order to eliminate ammonium salts has also been carried out in large-scale fluidized-bed reactors with a cross-sectional area of 2.2 m2 and a height of 5 m, using 5 tons of sand particles 0.4–0.6 mm in diameter as support particles for cell biofilm development. The reported results show the feasibility of the system, from both the technical and economic points of view. Ethanol fermentation has been studied intensively in fluidized-bed reactors and brought up to the pilot-plant level in several cases. A pilot plant with a volume of 1 m^3, using *Saccharomyces cerevisiae* cells immobilized in carrageenan beads has been successfully operated for half a year. As an alternative to yeast, the bacteria *Zymomonas mobilis* can also be used. Very promising results have been reported on the operation of a series of two pilot reactors of 55 L each, under nonsterile conditions.

The support for the immobilization in this case was a macroporous glass particle, Siran, that allowed cell attachment to the internal surface of the pores. After an initial period of sterile colonization of the beads in a conventional stirred tank, the particles, with a high *Z. mobilis* cell concentration, can be transferred to the fluidized bed, which then can be operated at high dilution rates under nonsterile conditions because all the possible contaminants in the feed (a solution of hydrolyzed B-starch) are washed out from the reactor. Complete conversion of 120 g/L sugar concentration with a 4.25-h residence time can be consistently obtained, with ethanol productivities of 18 g/(L h), calculated with the total reactor volume as a basis. A complete technical and economic evaluation of the use of fluidized-bed reactors for the production of ethanol with *Z. mobilis* has shown promising features. Other examples of the use of fluidized-bed bioreactors on the pilot scale in fermentation processes include the production of alcohol-free beer with immobilized yeast in a 50-L pilot-plant fluidized-bed reactor and the production of penicillin using *Penicillium chrysogenum* immobilized in urethane particles in a 160-L pilot-plant reactor.

Penicillium chrysogenum cells have also been used in different studies with fluidized-bed reactors with celite as the support material. Also, the area of animal cell culture provides a number of examples of the application of fluidized-bed technology, taking into account that smaller production volumes are usually required in this case. The use of immobilized cells is particularly required for those cells that are anchorage dependent, and a variety of porous carriers and beads have been developed for this purpose. The production of a human anti-HIV-1 antibody using recombinant Chinese hamster ovary cells grown onto a macroporous polyethylene carrier has been carried out in an 84-L-volume pilot-scale fluidized-bed novel reactor equipped with a low shear stress internal impeller for the recirculation of the liquid.

20

MICROBIAL PHOTOBIOREACTOR

Photobioreactors are bioreactors in which phototrophic microorganisms or plant cells are grown or used to carry out a photobiological reaction. In a broad sense, the open shallow basins used for microalgae cultivation could also be viewed as photobioreactors. In this article, however, the term *photobioreactor* is used to indicate only closed systems, that is, reactors that do not allow direct exchange of gases (e.g., CO_2, O_2) or contaminants (dust, microorganisms, etc.) between the culture and the atmosphere. At present, commercial production of phototrophic microbial biomass is limited to two microalgae (*Dunaliella* and *Chlorella*) and one cyanobacterium (*Arthrospira*). Mass cultivation oı these species is carried out in open ponds by means of a selective environment (high bicarbonate and high pH for *Arthrospira*, high salinity for *Dunaliella*), or by making use of large amounts of inoculum produced under strictly controlled conditions and taking advantage of a high growth rate (*Chlorella*). Most microalgal species cannot be maintained long enough in outdoor open-culture systems because of contamination by heterotrophic microorganisms (fungi, bacteria, protozoa) and competition by faster-growing microalgae that tend to dominate, regardless of the species used as inoculum.

With the few exceptions mentioned earlier, virtually all attempts at mass cultivation of specific species of phototrophs in outdoor open ponds for prolonged periods have failed. An alternative choice is to use photobioreactors, which can provide an environment that is protected from direct fallout, are relatively safe from invasion by competing microorganisms, and in which conditions (pH, pO_2, pCO_2, temperature) can be better controlled so that dominance of the desired species is established for a sufficient period of time. The use of photobioreactors could extend the exploitation of microbial phototrophs from the few strains commercially grown today to the more than 50,000 known species, many of which are potential sources of high-value products. Examples of such products are specialty lipids (sterols and polyunsaturated fatty acids); antifungal, antibacterial, or antiviral compounds; radioactively or isotopically labeled molecules (such as ^{13}C-labeled glucose and amino acids); vitamins (e.g., vitamin E); pigments (such as β-carotene, astaxanthin, and phycobiliproteins); pharmaceuticals (e.g., antineoplastic agents); and aquaculture feeds. The latter represent a good example of a case that could benefit in a decisive manner from the introduction of new photobioreactor concepts. Understanding photobioreactors requires background knowledge on the fundamentals of the growth of phototrophs. The first part of this article therefore describes the principal characteristics of phototrophic microorganisms and their growth kinetics. The second part deals with photobioreactor designs, advantages, limitations, and commercial applications.

PHOTOTROPHIC MICROORGANISMS

All organisms require a source of energy for growth to take place. Phototrophic organisms, or phototrophs, use light as an energy source. Light, or visible radiation, is the electromagnetic radiation in the 400- to 700-nm range, to which the human eye is sensitive. Because it is used by algae and plants for photosynthesis, visible radiation is also called photosynthetic active radiation (PAR). Photosynthesis is the process by which phototrophs capture and convert the energy of photons (quanta of electromagnetic radiation, i.e., the smallest "packets" of light energy) to biochemically usable energy. It is through photosynthesis that phototrophs obtain the energy (ATP) and the reducing power (NADH or NADPH) necessary to incorporate CO_2 and to manufacture the organic molecules they require for growth. Photosynthesis is generally associated with higher plants, yet about half the photosynthesis on earth is carried out by microorganisms.

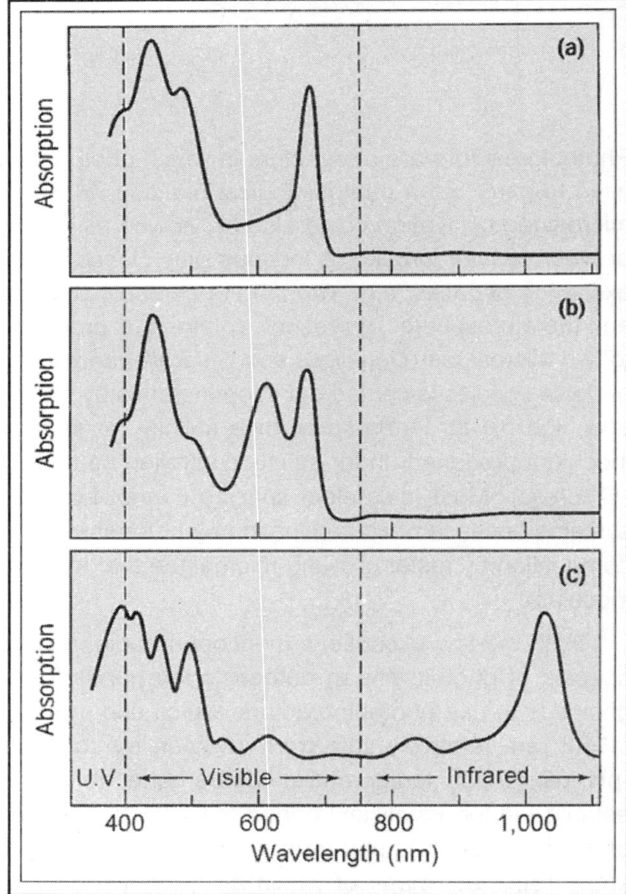

Figure 20.1 In vivo absorption spectra of (a) a microalga (*Scendesmus acutus*), (b) a cyanobacterium (*Synechocystis sp.*), and (c) a purple bacterium (*Rhodopseudomonas viridis*).

There are three major groups of phototrophic microorganisms: the photosynthetic purple and green bacteria, the cyanobacteria (also referred to by botanists as blue–green algae), and the microalgae. Photosynthetic bacteria, and cyanobacteria are prokaryotes; microalgae have a eukaryotic-cell-type organization. Photosynthetic bacteria differ from cyanobacteria and microalgae because of their photopigments and photosynthetic processes. The ability to perform photosynthesis depends on the presence of light-absorbing pigments, the chlorophylls: large planar molecules composed of four substituted pyrrole rings with a magnesium atom coordinated to the four central nitrogen atoms. Chlorophylls absorb light primarily in the blue–violet and red regions of the spectrum and absorb weakly in the green region; this is why many phototrophs are green in color.

Photosynthetic bacteria differ from cyanobacteria and microalgae in that their chlorophylls, called bacterio-chlorophylls, have different molecular structure and absorption characteristics. For example, bacterio-chlorophyll *b* of some purple bacteria has an in vivo maximum absorption in the infrared region between 1,020 and 1,040 nm. For purple bacteria, photosynthetic active radiation covers a much wider wavelength range than it does for microalgae and cyanobacteria, from the near ultraviolet to the long-wave infrared region. Besides the chlorophylls, other photosynthetic pigments, called accessory pigments, are also involved in the capture of light. Accessory pigments absorb light in the blue–green through yellow–orange range where chlorophylls absorb poorly and transfer the trapped energy to the chlorophylls, making photosynthesis more efficient over a broader range of wavelengths. The most widespread accessory pigments are the carotenoids, fat-soluble molecules that absorb light in the blue region of the spectrum, and the water-soluble phycobiliproteins of red algae and cyanobacteria that absorb in the green–yellow region.

The accessory pigments also protect phototrophs from sunlight, which can damage the photosynthetic apparatus when it is too intense. Cyanobacteria and microalgae also differ greatly from photosynthetic bacteria because of their photosynthetic process. Like higher plants, cyanobacteria and microalgae are oxygenic phototrophs; that is, they exhibit a type of photosynthesis that involves two photosystems (PSI and P SII), uses H_2O as electron donor, and evolves oxygen as a result of water photolysis. Green and purple photosynthetic bacteria lack photosystem II and cannot use water as an electron source; hence, they do not produce oxygen photosynthetically. For this reason they are called anoxygenic phototrophs. These definitions apply irrespectively of whether the organism obtains its cellular carbon principally from inorganic (autotrophically) or organic (heterotrophically) sources. Microalgae and cyanobacteria are mostly aerobic (they use oxygen as electron acceptor for respiratory metabolism) and autotrophic; purple bacteria are mainly anaerobic and heterotrophic; green bacteria are prevalently anaerobic and autotrophic. Although this article deals exclusively with photobioreactors for the cultivation of oxygenic phototrophs (microalgae and cyanobacteria), some of the aspects covered and most of the conclusions can be extended to other phototrophic groups. For brevity, the term microalgae is used here in its botanical meaning, which includes cyanobacteria.

Role of Light in Photosynthesis

The rate of photosynthesis achieved by a phototrophic cell depends on the rate of photon absorption and on the efficiency with which the absorbed photons are used. The rate of absorption and the efficiency of conversion of absorbed photons are in turn determined by the characteristics of the photosynthetic apparatus and by the spectral quality and intensity of the incident light.

Quality of light and photosynthesis

The wavelengths absorbed by a phototroph depend on its primary and accessory pigment content as shown by its absorption spectrum. Only absorbed wavelengths are used, but not all the wavelengths absorbed are equally efficient for photosynthesis. Therefore, the spectral suitability of a light source for a phototroph is better represented by the *action spectrum* of photosynthesis (i.e., the curve obtained by plotting the *specific rate of photosynthesis* [*P*]† against wavelength across the photosynthetic range for the cultured organism). Although action spectra give more information than absorption spectra, the latter are more commonly reported because they are easier to determine and interpret. The curve obtained when growth, rather than photosynthesis, is plotted against wavelength is called a *growth action spectrum*.

Role of photosynthesis

The rate of photosynthesis is not simply proportional to the rate of photon absorption, since photons may be captured by the pigments much faster than the photosynthetic apparatus can make use of them. If photosynthesis is measured at different light intensities, the specific photosynthesis rate *(P)* can be plotted as a function of light intensity *(I)* in the well-known light response curve of photosynthesis, or the *P–I* curve. In the dark, where there is, of course, no photosynthetic activity, O_2 is consumed and CO_2 is released due to cellular respiration. As the light intensity gradually increases, photosynthetic O_2 evolution (and CO_2 consumption) takes place and a light intensity value, called the light compensation point or compensation irradiance (I_c), is eventually reached, at which value photosynthetic O_2 evolution balances respiratory O_2 consumption. Beyond this value, production exceeds consumption and net photosynthesis is achieved. From here on, *P* increases linearly with *I* up to a certain point, beyond which the rate of increase of *P* per increment in light intensity declines and eventually levels off. The slope of the linear portion of the curve, called alpha (α), represents the rate of photosynthesis per unit incident light or relative quantum yield (also called maximum photosynthetic efficiency or maximum quantum yield), and is a measure of the efficiency with which the organism uses low light intensities.

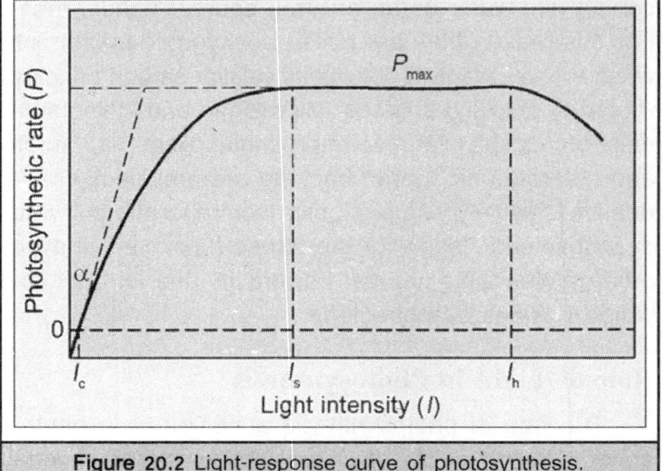

Figure 20.2 Light-response curve of photosynthesis.

The relative quantum yield is expressed as moles of O_2 evolved per mole of photons of incident PAR. Plotting the absorbed light (incident light minus reflected and transmitted light) instead of incident light will give the absolute quantum yield. The reciprocal of the quantum yield (i.e., the number of photons required to evolve one molecule of oxygen) is called the quantum requirement of photosynthesis. In the range of intensities where *P* does not vary with *I*, photosynthesis is said to be light saturated and the light intensity is said to be saturating (I_s); *P* reaches its maximum value (P_{max}) in this range. If light intensity further increases, *P* may decrease again. This phenomenon involves damage to some components of the photosynthetic apparatus and is referred to as photoinhibition. At low light intensities, *P* is limited by the light reactions of photosynthesis (thermodynamic constraint). At saturating light intensities, P (P_{max}) is limited by the so-called dark reactions, in which the products of light reactions are used to fix CO_2 (metabolic constraint). $P–I$ curves are species specific and must be determined under conditions such that all the cells receive the same photon flux and are not mutually shaded.

Hill and Bendall's widely accepted Z-scheme of oxygenic photosynthesis incorporates two photosystems (PSII and PSI), and establishes that it is necessary to move four electrons through each photosystem to evolve one molecule of O_2 and fix one molecule of CO_2. Since the transfer of each electron requires two photons (one in PSII and one in PSI), it follows that the theoretical minimal quantum requirement cannot be lower than 8 (i.e., not less than eight photons are needed to move four electrons from water to NADP through the two photosystems) and the maximal quantum efficiency cannot be higher than 0.125 (1/8). The curve relating growth to light intensity has a pattern similar to the $P–I$ curve (with growth rate increasing proportionally to light intensity when intensity is limiting and being independent of intensity when saturating values are reached), since phototrophic growth is a direct consequence of the photosynthetic rate.

Culture of Phototrophic Microorganisms

Phototrophic microorganisms can be grown in batch or continuous culture. A batch culture is initiated by the transfer of a small amount of cells (inoculum) into fresh growth medium; after the inoculum, nothing is added to (except light and carbon dioxide), or removed from (except oxygen), the system.

Growth rate

Growth can be defined as the orderly increase of all the cell constituents. During this process, unicellular organisms grow to a maximum size, then divide to form two daughter cells. The daughter cells in turn repeat the process, and the number of cells doubles at regular intervals of time. The time required for a growing population to double is called the generation time, g. The rate of increase in biomass concentration is generally expressed as the specific growth rate, μ, which is related to the generation time by $\mu = \ln 2/g$. The specific growth rate, which can be defined as a proportionality constant for the auto- catalytic growth reaction, is described by the following equation:

$$\mu = 1/X(dX/dt) \quad \ldots(1)$$

where μ is the specific growth rate (h^{-1}), X is the biomass concentration ($g\ L^{-1}$), t is time (h), and dX/dt is the variation in biomass concentration with respect to time ($g\ L^{-1}\ h^{-1}$), that is, the culture productivity per unit volume. The volumetric productivity ($dX/dt = \mathrm{I}X$) of a culture in a pond or a flat reactor can be easily normalized on a per-unitarea basis by considering the culture volume (V) corresponding to the unit of illuminated surface area (A). Therefore, the areal productivity can be calculated as $\mu XV/A$, or, since V/A is the depth d of the culture, as μXd.

Growth curve

When a phototroph grows under suitable conditions in batch culture, its growth curve follows a sigmoidal pattern showing six principal phases. Each phase is characterized by a particular value of the specific growth rate and reflects a particular metabolic state of the cell population. After the inoculum, growth does not necessarily start right away since most cells may be viable, but not in condition to divide, especially when the parent culture is old. The interval necessary for the transferred cells to metabolically adapt to the new situation and start to grow is called the lag phase. After the lag phase, the culture enters into the acceleration phase, during which μ increases continuously. When a constant growth rate is reached, the culture is said to be in the exponential growth phase.

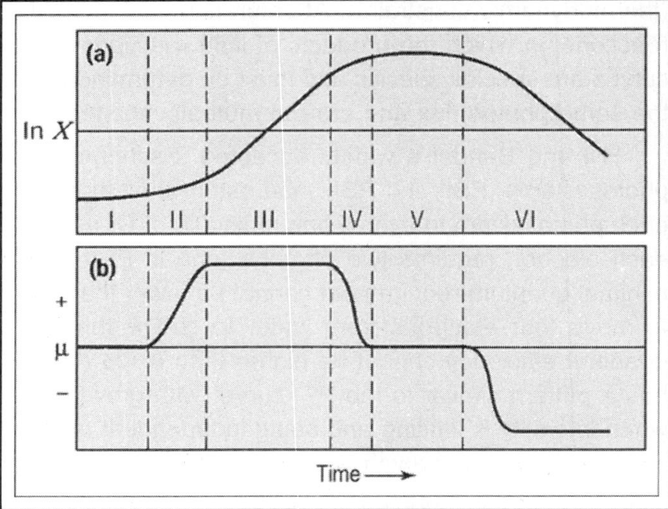

Figure 20.3 (a) Growth curve of a phototrophic microorganism in batch culture. (b) Changes in growth rate (μ) during the six typical growth phases.

During exponential growth, all the cells are light saturated (if a saturating light intensity is provided at the culture surface) and growth proceeds according to equation 1. The maximum μ is achieved in this phase, and in this respect phototrophs do not behave differently from chemotrophs. (*Note*: Organisms that obtain energy for growth from oxyreduction reactions involving either organic or inorganic electron donors.)

The exponential growth phase of phototrophs in batch culture normally lasts for a relatively short period because the cells start to shade each other as their number increases; growth becomes light limited and μ decreases (deceleration phase). Self-shading and the fact that the energy source is instantaneously provided at the culture surface and cannot be stored in the reactor are the two fundamental factors that differentiate photobioreactors from reactors for chemotrophs. During the deceleration phase, although μ decreases continuously, the culture productivity may remain stable at maximum value because of the continuous proportional increase in cell concentration *(X)*. Since the increase in cell biomass becomes almost linear, sometimes the initial part of this phase is called linear phase.

Conditions of linear growth are typically sought in mass cultures of phototrophs, since they lead to the highest productivity. Following the linear growth phase, the cell population continues to increase, but μ decreases until it reaches zero, at which point the culture enters the stationary phase, during which the cell concentration remains constant at its maximum value. The stationary phase is followed by the death phase, in which the population of viable cells decreases.

To achieve continuous exponential light-saturated growth, the culture process has to be changed from batch to continuous; this is accomplished by continuously removing the newly formed cells. This condition is not adopted in photobiological process targeting production of biomass or related products, since it leads to low efficiency of conversion of the irradiance impinging on the culture surface and to low productivity.

Volumetric Productivity

If nutrients are provided in suitable amounts and temperature is maintained at the optimum value, the productivity of phototrophic cultures becomes a function of light availability and cell concentration. In dilute cultures, self-shading is minimal and all the cells receive the same amount of light, independent of their position. It is self-evident that, at these low densities, even if the light intensity provided would permit the maximum growth rate, the culture will achieve low efficiency of light conversion and low productivity, since most of the impinging photon flux will pass through the culture unabsorbed. Mass phototrophic cultures, in order to obtain maximum productivity, must be kept dense enough to absorb all the light impinging on their surface. Under these conditions, self-shading, due to light absorption by cell pigments and scattering by the cells causes light intensity to fall off exponentially with the culture depth. In dense cultures, the light availability for each single cell is much reduced; this brings about a significant diminution of the specific growth rate, but productivity is, in any case, higher than in dilute cultures because the decrease in growth rate is more than compensated by the increase in cell concentration. For achieving maximal productivity in mass cultures it is mandatory to operate at a proper combination of culture depth and cell concentration.

This concept, known as operating at the optimal population density (OPD), has been investigated by many researchers, most recently by Richmond and coworkers. In general, maximal productivity is obtained at relatively high population densities and at specific growth rates that are about one-half of the maximum. Increasing the cell concentration above the OPD is generally associated with a reduction in productivity because of the presence of a large dark zone inside the reactor, which leads to high maintenance requirements in relation to the overall photosynthetic activity of the culture. Given the nonuniformity of the light environment that dominates in dense cultures, one requirement for achieving high photosynthetic efficiencies is that the individual cells be intermittently moved into and out of the illuminated zone. Although it has been shown that conventional mixing can do little to improve photosynthetic efficiency through what is known as the "flashing light effect," mixing is, in any case, required to provide intermittent exposure of the cells to light and darkness. A means to increase volumetric productivity in phototrophic cultures without increasing the light intensity at the culture surface is to shorten the light path by increasing the illuminated surface-to-volume ratio (S/V ratio) of the reactor.

Mass Phototrophic Culture

Mass phototrophic cultures are normally light limited, even though the light intensity striking their surface may be several times above saturation. For example, in outdoor algal cultures at midday, cells in the top layers will receive light at intensities above I_s (they will be photosaturated), or may even be photoinhibited, but a considerable portion of the culture will be photolimited or even below the compensation point (bottom layers). The rate of photosynthesis will thus be different in different regions of the culture, since it depends on the local environment experienced by the single cells, but the photosynthetic rate of the culture as a whole will be an average of all the individual rates. In the same way, while for a single cell the instantaneous specific growth rate, which depends on the instantaneous rate of light absorption and of photosynthesis, varies depending on the position of the cell, the average specific growth rate of the culture depends on the average amount of light received by the cells. It is hence a function of the culture depth (or light path) and of the cell concentration, as well as of the intensity of the incident light. The exponential decay of the intensity of light as it penetrates into dense cultures is generally described by Lambert-Beer's law. This equation, however, accounts for the sole absorption by pigments and neglects light scattering by cells.

More sophisticated models, (e.g., the radiative transfer theory developed in astrophysics), have been proposed to obtain radiant energy profiles inside reactors with a reasonable accuracy. Models of photosynthetic growth are based on the assumption that growth rate is closely related to the light availability inside the culture. It is thus necessary to consider both the effect of the light intensity available at each point of the reactor on growth rate and how the light environment is modified by absorption and scattering. This creates a high degree of complexity for the modeling of phototrophic cultures. A complete treatment of the kinetics and energetics of photosynthetic growth in photobioreactors has been presented by Cornet et al. Accurate models for calculating local and average irradiance inside photobioreactors as functions of the design and of the impinging irradiance, and for relating growth rate and productivity to average irradiance have been developed by Molina Grima and coworkers.

Productivity in Phototrophic Cultures

The photosynthetic efficiency of phototrophic growth is defined as the energy stored in biomass per unit of light energy absorbed. This parameter is of fundamental importance in mass cultures, since high photosynthetic efficiency leads to high biomass production per unit of illuminated area. As mentioned earlier, according to the current model of photosynthesis, eight photons are required to transfer four electrons from water to NADP and bring about the reduction of one molecule of CO_2 to carbohydrate in the dark reactions of photosynthesis. Since one mole of photons in the visible range contains on average about 50 kcal of free energy, and since the energy stored in carbohydrate during photosynthesis is about 114 kcal/mol of reduced CO_2, the theoretical maximum photosynthetic efficiency is approximately 28.5% [114/(8 x 50)]. The biosynthesis of the other macromolecules making up the microbial cell (lipids, protein, nucleic acids) requires additional reducing power and ATP and hence additional photons per CO_2 incorporated; this brings the quantum requirement for growing cells to 10-12, rather than 8, and reduces the maximum efficiency to about 20%.

In practice, this efficiency cannot be reached with sunlight because of several losses and limitations, which include the fact that PAR represents only 43% of the total solar radiation at the earth's surface, the loss of part of the incident radiation by reflection at the culture surface, and the fact that part of the absorbed energy is invested in cell maintenance. Another major limitation, often ignored, is represented by the light saturation effect, that is, the fact that the photosynthetic apparatus of most phototrophs saturates at only 10–20% of full sunlight (about 2,000 μmol photons m^{-2} s^{-1}). According to the equation derived by Bush ($f = I_s/I[\ln I/I_s + 1]$), which expresses the fraction (f) of the thermodynamic efficiency that can be attained as a function of the incident light intensity (I) and the saturation intensity (I_s), the quantum efficiency comes very close to the theoretical at low light intensities (<100 μmol photons m^{-2} s^{-1}). When the light intensity approaches that of full sunlight, utilization drops to less than 20% and most of the light energy absorbed by the culture is dissipated as fluorescence and heat.

There are not many options available to overcome the limitation imposed by the light saturation effect. In dense phototrophic cultures, mixing partially compensates for the limitation imposed by the light saturation effect, but does not overcome the substantial inability of phototrophs to use effectively high irradiances. Benemann discussed the alternative concept of increasing the light intensity at which photosynthesis saturates by decreasing the amount of accessory pigments in the cells. This however, is not easy to achieve and may lead to another problem: Contaminating algae with higher pigment content, because of their ability to absorb more light in the light-limited zones of the culture, will overgrow the cultured alga.

Based on studies that took this direction in the 1950s, Tredici and Chini Zittelli have recently proposed spatial light dilution as a means to increase the photosynthetic efficiency of outdoor algal cultures. Considering all of these limitations, it is reasonable to assume that only about 4–5% of solar energy can be converted to the chemical energy of biomass, under even the best conditions. For locations characterized by an annual average insolation of 4,500 kcal m^{-2} d^{-1}, and assuming a caloric content of the biomass of 5.4 kcal g dry wt^{-1}, this efficiency will result in a calculated average annual productivity between 33 and 42 g m^{-2} d^{-1}.

Outdoor cultures

The productivity of outdoor cultures is one of most debated issues connected with mass cultivation of phototrophs. Extrapolations from efficiencies of light conversion obtained in the laboratory had led to predict annual productivities as high as 140 g m^{-2} d^{-1}, but actual productivities are only one-tenth or less these figures, with about 40 g m^{-2} d^{-1} being considered the highest value obtainable by outdoor algal cultures under favorable conditions. A common mistake made (more or less consciously) in this respect is that the results obtained on a small scale over short periods and under optimal conditions are extrapolated to all-year-round, full commercial scale. As a matter of fact, despite statements in the early years of algal biotechnology, large algal cultures are not much more productive than terrestrial systems, and generally are much less reliable. In the author's opinion, however, productivity should not be considered as being the decisive factor in determining the viability of algal biotechnology. It is true that productivity greatly influences the economy of the process, but other factors, like sustainability and reliability, are just as, or even more, important. Several models have been proposed to predict the productivity of outdoor cultures of phototrophs both in open ponds and in photobioreactors.

Expression of productivity

In the case of algae cultivation in open ponds, it is common to evaluate productivity as areal yield (i.e., productivity per unit of occupied land area per unit time, expressed either as g m^{-2} d^{-1} or tons ha^{-1} $year^{-1}$). This is both practical, since it makes comparison with terrestrial systems easy and immediate, and correct, since areal productivity in ponds is proportional to the amount of light entering the system per unit illuminated area. In a pond this corresponds substantially to the land surface area occupied by the system. With photobioreactors the situation is much more complex. Except in the cases of flat horizontal systems and tubular horizontal photobioreactors with no empty space between contiguous tubes (in both these cases the areal productivity referred to the occupied land is correct and easy to interpret), expressing productivity on a per land surface area basis is misleading. For example, in some horizontal serpentine bioreactors it is difficult to decide whether and how to compute the empty space between contiguous tubes and how to account for the fact that the reactor intercepts a different proportion of the radiation impinging on the horizontal, depending on the elevation of the sun and, hence, on the hour of the day.

With elevated systems, the inadequacy of the areal yield is even more evident, since the amount of radiation intercepted by a vertical or sharply inclined reactor is not related to the land surface area it occupies. Expressing the productivity of elevated systems on a per unit occupied land surface area basis (i.e., with reference to the horizontal projection of the reactor on the ground) may lead to high unrealistic figures.

Productivity can also be referred to the illuminated surface area of the reactor (biomass produced per unit illuminated surface area per unit time, in g m^{-2} d^{-1}), but this parameter is also not adequate for all situations, since it can be difficult to define the correct illuminated surface area to which to refer productivity. In tubular reactors, for example, the illuminated surface area generally corresponds to half the total surface area of the reactor; using it as basis for calculation of productivity in most cases underestimates reactor performance. In some vertical reactors (e.g., large-diameter coils that receive a significant amount of radiation on their inner central part at midday, but not in the early morning nor in the evening), the illuminated surface area changes during the day with sun elevation.

Another complication with vertical reactors is determined by the fact that they intercept a significant amount of diffuse radiation on the surface not directly illuminated by the sun, and this is not always taken into account. For these reasons, photobioreactor performance, like that of fermenters, is frequently reported in terms of volumetric productivity (amount of biomass produced per unit reactor volume per unit time, commonly expressed in gL^{-1} d^{-1}). But volumetric productivity is a function of the number of photons that enter the unit reactor volume in the unit time and, as such, is dependent on the illuminated S/V ratio of the reactor. The higher this ratio, the higher the volumetric productivity. It is well known, for example, that small-diameter tubular reactors or thin flat systems achieve high volumetric productivities, even if they perform poorly. We should be aware of the fact that a volumetric productivity of 1 gL^{-1} d^{-1} assumes a completely different significance if obtained in a tubular reactor of 1 cm or of 5 cm diameter.

To overcome these limitations, Tredici and Chini Zittelli suggested evaluating the performance of vertical photobioreactors on the basis of the photosynthetic efficiency achieved by the culture. Unfortunately this parameter is not always easy to calculate (e.g., in coiled reactors) and since it depends heavily on the irradiance (the lower the irradiance, the higher the efficiency) favors experiments carried out under low light intensities, such as in winter. The productivity of a photobioreactor could be expressed on a per shaded surface area basis (the productivity of the reactor divided by the average ground surface area shaded by the reactor). This parameter, never suggested before, substantially reflects the efficiency of light utilization, but also takes the actual irradiance conditions into account and offers the possibility to compare very different culture systems. For example, the performance of an open pond and of a vertical coiled reactor could be compared on this basis, whereas it is rather difficult to envisage any other suitable comparison parameter for these two very different systems. The per shaded surface area parameter cannot be used alone, however, since in some cases (e.g., serpentine reactors) it does not give a correct evaluation of the efficiency of land utilization.

In conclusion, each method for expressing productivity has its own limitations, and complete evaluation of a reactor performance in terms of productivity will be achieved only by a combination of the different available methods. For example, expressing the productivity on the basis of unit shaded area, unit occupied area (including eventual empty spaces), and unit volume will allow evaluation (and comparison) of photobioreactors in terms of efficiency of light conversion, efficiency of land use, and efficiency with which the unit volume of reactor is used. This combined approach is highly recommended.

Photosynthetic efficiency in artificial light

Solar energy is a relatively cheap energy source, but given the low photosynthetic efficiency attainable under full sunlight, and the fluctuations in supply determined by the weather, the time of the year, and the natural day/night cycle, the use of artificial light sources to grow phototrophs has been taken into serious consideration, at least for production of high-value compounds. The use of artificial illumination offers the possibility of selecting light quality and intensity, as well as the length of the illuminated period, and thus of significantly enhancing the reliability of the production process. This, together with the higher photosynthetic efficiency attainable (a photosynthetic efficiency of 15% is not uncommon with artificial light of relatively low intensity), may compensate in many cases for the higher cost.

One of the fundamental factors guiding choice of the light source should be the efficiency of conversion of electric energy into PAR energy. This efficiency varies considerably with the type of lamp; for example, it is only about 5% in tungsten–halogen (incandescence) lamps, 10–15% for fluorescent tubes, and 20–25% for high-pressure sodium lamps. Only with xenon lamps and light-emitting diodes are very high efficiencies attained (50% and up to 80%, respectively). Considering both the efficiency of conversion of electrical energy into PAR and the efficiency of conversion of light energy into biomass, an overall efficiency of conversion of electrical energy into the energy of biomass between 0.7% and 2% has been attained by *Arthrospira* cultures grown under fluorescent lamps.

Another aspect to consider is the spectral quality of the emitted light, which ought to match the action spectrum of the cultivated organism as closely as possible. For example, tungsten lamps have a low output in the blue region and are therefore not appropriate for oxygenic phototrophs. High-pressure sodium lamps are relatively efficient as regards electrical energy conversion but emit in a limited yellow–orange region. Practicality often prevails over efficiency considerations in the choice of an artificial light source. For example, fluorescent lamps are widely adopted, independently of the spectral quality of their emission, because they provide a relatively uniform irradiance over a large surface.

DETAILED STUDY OF PHOTOBIOREACTORS

Categories of Photobioreactors

Photobioreactors can be classified on the basis of both design and mode of operation. In design terms, the main categories of reactors that can be envisaged are (a) flat or tubular; (b) horizontal, inclined, vertical, or spiral; and (c) manifold or serpentine. An operational classification of photobioreactors would include (d) air or pump mixed and (e) single-phase reactors (filled with media, with gas exchange taking place in a separate gas exchanger), or two-phase reactors (in which both gas and liquid are present and continuous gas mass transfer takes place in the reactor itself). Construction materials provide additional variation and subcategories—for example, (f) glass or plastic and (g) rigid or flexible photobioreactors. Axenic photobioreactors are reactors operated under sterile conditions. Although a major characteristic of photobioreactors is their ability to limit contamination, it must be made clear that an effective barrier, and thus operation under truly sterile conditions, is not achieved, except in the few special designs developed expressly for that purpose. This is a major difference between photobioreactors and bioreactors for chemoheterotrophs, which typically operate under sterile conditions.

Like fermenters, photobioreactors can be divided into laboratory photobioreactors used for research, pilot-plant photobioreactors used to scale-up and optimize the process, and production units. Unlike fermenters, there are only a few examples of commercial photobioreactors in operation. Some companies sell photobioreactors in a variety of sizes for use in teaching, research, and commercial applications. The emphasis in this article is on pilot-culture systems and the two commercial-scale plants realized to date. Laboratory systems, such as illuminated carboys and fermenters, immobilized cell systems, and optical-fiber reactors are not considered.

Earlier Work Done

The use of photobioreactors for microalgae mass culture dates back to the late 1940s and was a logical consequence of the investigations carried out with *Chlorella* on the fundamentals of photosynthesis. In those early days of algae culture, the emphasis was on continuous controlled cultivation; hence, closed reactors were given priority. Open systems were considered inappropriate to guarantee the necessary degree of control and optimization of the process. One of the first closed systems was set up at Stanford University in California in the late 1940s and involved use of vertical tubular reactors. The units consisted of glass columns 1.8 m in height and 10 cm in diameter, constricted at the bottom to prevent algal settling.

The system was operated turbidostatically, with medium addition being regulated by a photoelectric cell. Culture temperature was maintained constant by means of a cooling coil inserted into the reactor. Carbon dioxide–enriched air was injected at the bottom of the column to obtain mixing and pH/pCO_2 control. The system was experimented with in the cultivation of *Chlorella* under both artificial light and sunlight. Maximum volumetric productivity indoors was 0.48 g L^{-1} d^{-1}. Outdoors, productivity averaged 0.28 g L^{-1} d^{-1}, with a maximum of 0.35 g L^{-1} d^{-1}. The reason for the lower outdoor productivity was not clear. In the early 1950s, the Arthur D. Little Company at the Massachusetts Institute of Technology, under contract with the Carnegie Institution of

Washington, developed two basic types of photobioreactors: an array of 10 vertical rigid plastic tubes and a 50-m^2 horizontal flexible tubular unit. The vertical units, which were essentially the same as those operated at Stanford University, provided inoculum to the larger horizontal unit and were reported to operate satisfactorily, although no productivity data were actually presented. The horizontal tubular unit may be considered the first pilot plant for microalgae production. It consisted of 4-mm-thick polyethylene tubes that, once filled with algal culture up to a depth of about 8 cm, assumed a more or less elliptic shape of 1.2-m width. The tubes were joined at the ends to form a raceway 21 m in length, for a total surface area of about 56 m^2.

The culture (*Chlorella*) was circulated by means of a centrifugal pump. The pumping system was not capable of generating circulation velocities over 7.5 cm s^{-1} and there was considerable settling of the culture. With the same pump and a tube width of 0.43 m, flow rates of 25 cm s^{-1} were achieved and problems of settling were eliminated. Leakage and contamination by *Chlorococcum* and protozoa were the major problems encountered in the operation of the system. Spray cooling proved inadequate for temperature control, and it was necessary to install a heat exchanger. Despite these difficulties, a 40-day run attained an average areal productivity of 9 g m^{-2} d^{-1}. The data collected with this pilot plant provided the material for an economic analysis of algae production in photobioreactors, which concluded that the cost of products obtained in photobioreactors was excessive in comparison to that of equivalent materials obtained using conventional methods. Cooling of the culture was found to be a major cost factor.

At about the same time in Japan, Tamiya and coworkers developed a system consisting of a serpentine tubular reactor immersed in a water bath for temperature control and an external column for gas exchange. This system provided the basic design criteria for all subsequent, essentially identical, serpentine reactors (e.g., those adopted by Gudin in France and Molina Grima and coworkers in Spain) (see later section). The reactor was made up of 3-cm-diameter glass tubes joined to form a serpentine and had a total length of 33 m. The total culture volume was 40 L. The culture (*Chlorella ellipsoidea*) was circulated at a linear velocity of 15–30 cm s^{-1}. A separate column, in which a mixture of 25% CO_2 in air was injected at a rate of 1–1.5 L min^{-1}, provided mixing and gas exchange. Productivity was directly related to solar irradiance with the highest value, achieved at the maximum irradiance, of about 0.6 g L^{-1} d^{-1}. After several months of experimentation, the authors concluded that one of the main limitations was temperature control and that a significant increase of productivity could be attained through cultivation of algae exhibiting higher temperature optima and through vigorous mixing aimed at increasing growth by providing intermittent illumination.

Following these initial studies in the 1950s, little work was carried out in this field for several decades. Two exceptions were the work of Setlik in Czechoslovakia and Juttner in Germany. The reasons for this almost total abandonment of research activity on photobioreactors are not clear, although the successful production of *Chlorella* in open ponds in Japan, the development of microalgae-based wastewater treatment in California, and later success with open-raceway systems for outdoor production of *Spirulina* (*Arthrospira*) and *Dunaliella* perhaps explain the relative lack of interest in closed systems. The difficulties experienced by the A.D. Little Co. and Fisher's negative conclusion in his economic analysis also contributed to discouraging further studies on the application of photobioreactors.

Modern Photobioreactors

Several types of photobioreactors have been designed and experimented with since the work of Pirt and coworkers sparked renewed interest in closed systems in the early 1980s. Most of these are small-scale systems; some are at the pilot or prepilot stage of development. Only in a very few cases has scale-up to commercial size been attempted. For convenience, these systems are described herein by categories: (i) axenic photobioreactors, (ii) vertical tubular photobioreactors, (iii) serpentine photo-bioreactors, (iv) manifold photobioreactors, (v) helical photobioreactors, and (vi) flat photo-bioreactors. Small systems and reactors for which experimentation has been conducted prevalently indoors with artificial light are not considered.

Glass tubular reactors

Setlik and later Juttner and Kruger and Eloff developed glass tubular reactors for the axenic cultivation of microalgae. Juttner's system, which substantially reproduced Setlik's design, was an all-glass, 110-L vertical flat coiled photobioreactor that could be sterilized by steam. The reactor, made of 4-cm-ID tubes, had a total length of 79 m. The algal suspension was circulated by a glass centrifugal pump at a flow rate of 20 to 60 cm s^{-1}. The temperature was regulated through a 2-m water-jacketed section. Gas exchange took place in an external exchanger in which a CO_2–air mixture was injected at a rate of 300 L h^{-1}. The system attained a maximum productivity of about 0.5 g L^{-1} d^{-1} under artificial illumination. Cell damage caused by the centrifugal pump was a problem in the case of fragile cyanobacteria; other problems encountered were foam formation and cell adhesion to the glass walls.

Juttner concluded that cell adhesion was the most discouraging feature and suggested that cultivation ought to be limited to "nonsticky" algae. Wall growth and fouling are still today major limitations in photobioreactors, although the problem is not always recognized as such because experiments generally do not last long enough for these effects to become evident. A similar 60-L system, composed of 12 5-cm-ID glass tubes each 2 m in length, coupled by U-bends to form a vertical flattened spiral, was operated by Kruger and Eloff, who reported a productivity of about 0.13 g L^{-1} d^{-1} with a nontoxic species of *Mycrocystis*.

Soeder et al. devised a rocking tray for the sterile cultivation of *Scenedesmus*. It consisted of a flat horizontal unit (inner dimensions 59.5 x 36 x 9 cm) made of stainless steel and covered with a glass lid. The tray was placed on a support connected to a rocking device. Temperature was controlled by circulating thermostated water through the double bottom of the unit. Although these systems appear difficult to scale up and are rather expensive, they are mentioned here since with the exclusion of laboratory reactors they are the only systems that allow sterile cultivation of phototrophic microorganisms.

Column reactors

Vertical tubular reactors (or column reactors) are simple systems in which mixing is achieved by injecting compressed air at the bottom of the column. Several vertical tubular photobioreactors were developed following Cook's first design. An example is the 30-L (2-m height, 15-cm ID) tower-type glass reactor devised by Juttner. The main characteristic of this system was a wide, hollow glass finger inserted from above into the column to reduce dark space inside the reactor. Since the finger was open at the upper end, it could also be used to regulate the culture temperature. A thickness of the annular culture chamber of about 3.5 cm was found to be optimal for the growth of the light- and shear-sensitive *Microcystis aeruginosa*. The rigid vertical columns, typically 2–2.5 m in height and 30–50 cm in diameter, that are extensively used in hatcheries to produce algal biomass for feeding the larval stages of marine bivalves and fishes must be included in this category.

Most commonly, these systems are made of translucent fiberglass sheets formed into cylinders. Air is bubbled at the bottom for mixing, and either artificial illumination or natural light (or a combination of both) is provided. James and Al-Khars cultivated *Chlorella* and *Nannochloropsis* in 200-L, 30-cm-diameter fiberglass cylinders of this type. Five such units were connected in series and operated continuously. Illumination was provided externally by a vertical bank of daylight fluorescent lamps. Although as a consequence of the low *S/V* ratio of the cylinders, mean volumetric productivities (0.17 g L^{-1} d^{-1} with *Chlorella* and 0.05 g L^{-1} d^{-1} with *Nannochloropsis*) were not high, exceptionally high productivities per unit occupied area were reported (up to 200 g m^{-2} land area d^{-1} for *Chlorella*). However, these values must not be taken as constituting a valid parameter for evaluation and comparison of this system, since they suffer from the limitation inherent in expressing the productivity of elevated reactors on a per occupied land area basis, as discussed earlier. Cylinders illuminated from inside, which may be considered a variation of the annular reactor devised by Juttner, have been proposed for the production of marine microalgae biomass in hatcheries. Although internally lit cylinders have high construction costs, they are less expensive to operate than completely filled columns. They also typically attain higher volumetric productivity and greater efficiency of light utilization since the photon flux provided to the culture is completely trapped by it.

A patented internally illuminated annular airlift column is used by Martek Co. for specialty chemicals production from microalgae. Miyamoto et al. experimented with vertical tubular reactors (2.35-m height, 5-cm ID) made from the low-cost (about $2.50 per tube), mass-produced glass tubes used in the fluorescent lighting industry. Volumetric productivities of about 0.6 g L^{-1} d^{-1} were obtained with *Nostoc* and *Anacystis* in outdoor cultivation. Although these systems do not represent an innovation, they are materially useful because of their low cost and relatively high *S/V* ratio in comparison to previous reactors of this type. Using a similar system consisting of 32 glass tubes 1.5 m in height and with a 2.6-cm ID, Hu and Richmond obtained a volumetric productivity of 1.6 g L^{-1} d^{-1} with *Isochrysis galbana* cultivated outdoors. Transparent polyethylene tubing of various diameters and thicknesses is available in long rolls at very low cost. A disposable vertical reactor can be easily built by cutting a suitable length of tubing and heat-sealing one end. This bag or sleeve reactor can be suspended from a framework or supported within a mesh frame. Such reactors are used indoors with artificial illumination (generally vertically mounted fluorescent lamps) or outdoors under sunlight. The temperature reached in the manufacturing of the polyethylene tubing is high enough to sterilize the material.

These systems have a relatively short life because of biofouling but are easy and inexpensive to replace at the end of the culture run. Two-meter-long transparent polyethylene sleeves sealed at the bottom and hung on an iron structure have been used to grow *Porphyridium* and *Dunaliella* outdoors (in Israel), obtaining a significantly higher productivity compared to open ponds. The main drawbacks of sleeve reactors are the relatively low *S/V* ratio and the low efficiency of utilization of artificial light when it is provided from outside of the reactor.

Snake shaped photobioreactors

Serpentine photobioreactors are systems in which several straight culture tubes are connected in series by U-bends to form a flat loop that can be arranged either vertically or horizontally. Gas exchange and nutrient addition take place in a separate vessel. Circulation between the flat loop and the gas exchanger is achieved by the use of a pump or of an airlift. Several reactors of this type have been developed in the last two decades following the original design by Tamiya. In the early 1980s, Gudin and coworkers, working at the Centre d'Etudes Nucl□aires de Cadarache, France, devised (and since that time have further developed) a horizontal serpentine reactor the main characteristic of which is the use of a water pond to obtain temperature control (a method anticipated by Tamiya). After a first 1-m^2 version consisting of glass tubes of 3-cm diameter, a 10-m^2 and then a 100-m^2 plant were built using larger diameter flexible plastic tubes. The 100-m^2 system, experimented with from 1986 to 1989, was composed of five identical 20-m^2 units, each of which consisted of 20 tubes, 20 m in length and 6 cm in diameter, made of 0.25-mm-thickness low- density polyethylene. The total culture volume was about 7 m^3.

Temperature control was attained by floating or immersing the reactor in a water bath by inflating or deflating a second layer of tubes placed below the reactor. Initially, the culture was circulated by means of a pump, but later airlift systems were adopted to limit damage to shear-sensitive cells and at the same time provide CO_2 supply and O_2 degassing. Productivities from 20 to 25 g m^{-2} d^{-1} were obtained with *Porphyridium cruentum* in a 2-month continuous run. The use of rigid rather than flexible tubes in the more recent work by Gudin's group has permitted operating a self-cleaning system consisting of two plastic balls, one with higher density than the culture medium and one lighter, which are hydraulically pushed through the system. Although flotation immersion in a water basin can provide efficient thermoregulation, the cost of such a system is prohibitive for most applications. In 1992, Dr. Gudin founded the Heliosynthese S.A. company for development of tubular photobioreactor technology in France. In 1996, Heliosynthese changed its name to Thallia Pharmaceuticals S.A. Thallia presently makes use of serpentine reactors to produce high-value algal compounds for the nutriceutical and pharmaceutical markets.

The major products under development are polyunsaturated fatty acids, carotenoids (zeaxanthin from *Porphyridium cruentum*, astaxanthin from *Haematococcus pluvialis*) and antioxidants (superoxide dismutase). Reactors of the type developed by Gudin are also being operated by Molina Grima and coworkers in Spain. These systems consist of a horizontal loop, made of Plexiglas tubes (about 100 m in length and 2.6 cm in diameter), which is connected to a 3- to 3.5-m- high airlift. The tubular solar receiver is immersed in a shallow pond for temperature regulation. Dissolved oxygen, pH, and temperature probes are connected to an online control unit and a computer for data acquisition. The Spanish group has made important improvements to the reactor originally purchased from Heliosynthese, and with it has carried out, and is still conducting, extensive research activity with the dual aim of shedding light on basic aspects of tubular photobioreactor technology (e.g., mass transfer and light distribution), and of defining guidelines for application of closed systems to production of specialty chemicals from microalgae (e.g., polyunsaturated fatty acids).

In the early 1980s, Pirt and coworkers developed a serpentine tubular reactor at the University of London's Queen Elizabeth College. The photobioreactor consisted of a photostage formed of 52 Pyrex glass tubes (each approximately 1 m long and with a 1-cm ID) horizontally stacked. The tubes were connected to Pyrex glass U-bends by silicone-rubber tubing to form a vertical loop. The loop outlet was connected through a vertical riser to a degasser. A second tube connected the degasser to the inlet of the loop. The reactor covered an area of about 0.5 m^2. The culture suspension (about 4.6 L) was circulated by either a peristaltic pump, a rotary positive-displacement pump, or an airlift; the latter method was preferred because of the adverse effects observed with the two pumps.

With a defined consortium consisting of a chlorella-type green alga and three chemotrophic bacteria, cell concentrations in excess of 20 g L^{-1} and a very high productivity of 2.2 g m^{-2} h^{-1} were achieved at a light intensity of only 38 W m^{-2} (PAR). The work carried out by Pirt and coworkers was remarkable for three reasons: It provided the first detailed analysis of the fundamental engineering parameters of a closed photobioreactor, introduced the concept of high *S/V* ratio reactors (127 m^{-1} in this case), and attained extremely high light-conversion efficiencies. It is worth noting that the about 40% reported efficiency for *Chlorella* growth corresponds to a quantum yield of about 6, as compared to the theoretical value of 8. Pirt's analysis, however, did not consider the issue of the accumulation of photosynthetically generated oxygen, which represents a key limiting factor in large-scale tubular reactors, and ignored the light saturation effect, assuming that the efficiencies observed at low light intensity could be achieved in full sunlight as well. Although Pirt's serpentine reactor did deal efficiently with, and solve, many of the problems commonly encountered in closed systems, the concept of very high *S/V* ratio reactors revealed all its drawbacks when the experimental setup was scaled up to industrial level.

The vertical serpentine reactor designed by Pirt was patented in 1983. The patent also covers the design of a full-scale, near-horizontal manifold system according to which the culture flows from a tank through a descending conduit to the higher ends of a set of 50-m-long transparent tubes supported by a slightly inclined framework. Each tube communicates at its lowest point with its own riser, which leads up to the common reservoir. The tubes are arranged close together so as to intercept a high portion of the solar radiation impinging on the area occupied by the reactor. Both this manifold and the vertical serpentine design were scaled up to commercial size in the late 1980s by PBL. In the mid-1980s at the Centro di Studio dei Microrganismi Autotrofi, Florenzano and coworkers started their experiments with horizontal serpentine tubular photobioreactors for the cultivation of *Arthrospira*. One of the first pilot reactors they set up was made of 14-cm-diameter, 0.3-mm-thick flexible polyethylene tubes. Because of problems of leakage, these collapsible sleeves were later replaced with rigid Plexiglas tubes.

The standard unit consisted of straight tubes, with a 13 cm ID and 4 mm thick, laid parallel on a white sheet and joined by PVC U-bends to form a loop 500 m in length. As it exited the reactor, the culture suspension flowed into a receiving tank, from which a diaphragm pump raised the suspension to a feeding tank placed at 3 m above ground level. The feeding tank was provided with a siphon that allowed discharge of about 340 L of culture into the photobioreactor every 4 minutes, thus sustaining a culture flow of 26 cm s^{-1}. The photobioreactor had a total volume of 8 m^3 and covered an area of 80 m^2. The productivity of *Arthrospira* in this system was 50% higher than in open ponds, due both to the better diurnal temperature profile and to the extended cultivation period. More recently, the same group developed and experimented with a 145-L two-plane serpentine reactor.

The system consisted of a 245-m-long loop made of 2.6-cm ID Plexiglas tubes arranged in two planes to maximize capture of the radiation impinging on the horizontal, and thus increase areal yield. The two planes were connected to form a single loop and the culture, circulated by means of two airlifts, flowed from one plane to the other. A maximum areal productivity of 27.8 g m^{-2} d^{-1} was achieved in July. It would appear that for such a high-cost system, the loss of volumetric productivity due to shading by the upper plane would exceed the benefits deriving from improved areal yield and land saving. It must be noted, however, that the volumetric productivity achieved with this two-plane photobioreactor (1.5 g L d^{-1}) was relatively high. The main merit of this study was that it addressed fundamental engineering parameters such as the effect of the rheological behavior of the algal culture on the mixing characteristics of the system, and the energy requirement for turbulent flow, which are often ignored. Aquaserch (Kona, Hawaii) has recently devised and patented a serpentine tubular reactor made of low-density polyethylene tubes. As in several previous serpentine designs, cooling is obtained by immersion of the reactor in a pond. The system, called the Aquaserch Growth Module (AGM), has been tested with different microalgae and is presently used to grow *Haematococcus* for astaxanthin production. The AGM is claimed to be the largest commercial-scale photobioreactor ever devised and one of the least expensive.

Manifolds photobioreactors

In manifold photobioreactors, a series of parallel tubes is connected at the ends by two manifolds, one for distribution and one for collection of the culture suspension. This type of photobioreactor represents one of the simplest configurations possible. As shown by Pirt et al., about 15% of the energy consumed for recycling the culture in serpentine reactors is spent in moving the culture suspension around the bends; thus, in comparison to serpentine reactors, in which the culture inverts the direction of motion at each turn, manifold systems allow a significant saving of energy. In 1991 in Javan, Sagdiana PGT built and used a large-scale glass reactor of the manifold type for cultivation of *Chlorella*. The system, very likely the largest glass reactor ever built, comprised 10 units consisting of two banks of 28 5.7-cm-ID tubes, each 73.5 m long, for a total of about 41,000 m of tubing. The tubes were connected in parallel to two metal headers each 20 cm in diameter.

The culture suspension (13 m^3 in each unit) was circulated between the two banks of tubes at a speed of 1.1 m s^{-1} by a 290-m3 h^{-1} capacity pump. Each unit was provided with one 3.6-m-high, 2.4-m-diameter gas exchanger. Temperature control was obtained through circulation of water from a single cooling tower serving all 10 units. The reactor was operated from August to October 1991, during which time it produced 0.5 ton of *Chlorella* biomass. Instability of the production process and contamination by *Scenedesmus* were reported to be the main problems. Tredici and coworkers, working at the University of Florence (Italy), have developed several types of near-horizontal manifold-type photobioreactors (NHTR). Units made of rigid or flexible tubes ranging from 6 to 85 m in length and from 4 to 5 cm in diameter have been built and used to grow *Arthrospira platensis, Anabaena siamensis,* and *Nannochloropsis* sp. under the climatic conditions of central Italy. Two eight-tube reactors of this type, 20 m in length, have been installed at the University of Hawaii as part of a joint project.

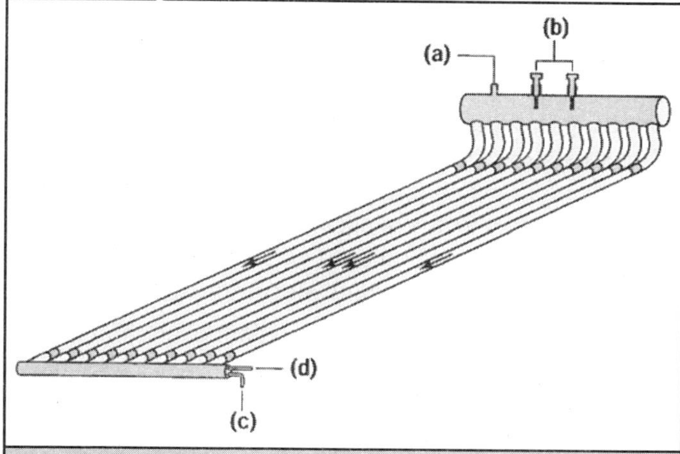

Figure 20.4 Scheme of the NHTR photobioreactor: (a) gas outlet; (b) electrodes and sensors; (c) culture outlet; (d) air and CO_2 injection point.

Typically, a NHTR consists of some 10–15 flexible tubes, each 45 m long and 4 cm in diameter, connected by two manifolds; the upper manifold, about 30 cm in diameter, acts as the degasser. The tubes are placed side by side on white corrugated plastic sheeting, facing south and inclined at a slight angle to the horizontal (4–6°). The corrugated sheeting keeps the tubes well aligned at a uniform inclination, reflects light onto the culture, and also provides drainage for rain and cooling water. Temperature control during the day is obtained by water spraying; the cooling water is collected and reused. The largest unit experimented with thus far occupies an area of 30 m^2 and contains about 600 L of culture. The main characteristic of the NHTR reactors developed in Florence is the use of air injection, at rates between 0.02 and 0.1 L L^{-1} min^{-1}, to attain mixing and gas exchange. The air is injected at the lower end of each tube by means of a perforated pipe inserted into the bottom manifold. In a different configuration used more recently, a number of tubes (typically one out of four) are not gassed, and are used as return flow tubes to increase the circulation speed of the culture and to obtain better mass transfer.

Carbon dioxide addition to the air stream or in the return tubes is regulated through a pH-stat system. Volumetric productivities of up to 1.4 g L^{-1} d^{-1} and a real productivities of more than 28 g m^{-2} d^{-1} have been obtained with *Arthrospira platensis* grown in these manifold systems. Besides simplicity of operation and low maintenance cost (due to the absence of moving parts), this internal gas-exchange reactor has other advantages: low shear stress compared to pump mixed reactors; no need for pumping the culture to a gas exchanger; and reduced fouling and wall growth due to the scouring effect of the gas bubbles. Moreover, this system appears to have potential for unit scale-up to about 100 m^2, and is relatively low cost in comparison to other designs. A preliminary estimate for a unit of 30 m^2, built and tested at Montepaldi, sets the cost of this type of reactor at about \$60/m^2, with the following cost per square meter: land, \$1.50; site clearing, grading, and leveling, \$6; liners, \$3.50; support framework, \$14.50; corrugated sheeting, \$5.50; tubing, \$13; manifolds, connectors, and piping, \$11.50; cooling system, \$1.50. A cost analysis for a large NHTR system for biological hydrogen production was recently presented.

A parallel-flow tubular photobioreactor was devised and experimented with in outdoor cultivation of cyanobacteria at the Microalgal Biotechnology Laboratory of Ben Gurion University. The system consisted of eight 20- m-long, 3-cm-ID polycarbonate tubes laid parallel on the ground or on a horizontal platform and connected by manifolds made of 6.3-cm-diameter black polyethylene tubing. Circulation and gas exchange were obtained by means of a 2.2-m-high, 4.6-cm-ID airlift connected to a 20-L gas separator. A manifold received the culture from the separator and distributed it among a first series of culture tubes. At the far end of these tubes, the culture suspension was collected by a second manifold and distributed among a second series of returning tubes through which the culture ran back to the airlift riser and the gas separator. Small differences in flow rates were measured between the tubes. Temperature control was provided by water spraying. A productivity of 0.55 g L^{-1} d^{-1} was obtained with *Arthrospira platensis* and *Anabaena siamensis*.

The main advantages of the system over serpentine reactors are the reduction of head losses and lower oxygen concentrations; these two factors facilitate scale up to industrial size. A manifold elevated system called the *a*-type tubular photobioreactor was devised and experimented with in Singapore. The reactor consisted of two sets of 10 parallel transparent PVC tubes (25 m long with a 2.5 cm ID), both placed at an angle of 25° to the horizontal but inclined in opposite directions. The algal suspension was lifted by air injection through a series of vertical PVC air-riser tubes (3.5 cm in diameter, 5 m in height) to a receiver tank from where it flowed down the first set of 25°-inclined tubes. These were connected to the base of a second series of air risers that lifted the culture to another receiver tank, from where the culture descended again through the second set of 25°-inclined tubes to the base of the first series of risers. The reactor thus looked like the Greek letter α. This elevated reactor, containing about 300 L of culture and covering a land area of about 12 m^2, was shown to be particularly well-suited to operation at low latitudes. *Chlorella pyrenoidosa* was grown for several months and reached very high cell densities (>10 g L^{-1}).

The major problems were biofouling, solved by injecting air from the upper end of the inclined tubes, and foam formation at the higher cell densities. An areal productivity of 72 g m^{-2} occupied land d^{1} was reported; as mentioned previously, the latter parameter can be misleading when referred to elevated systems. At the University of Manchester, Applied Photosynthetics Limited (APL) has recently developed a photobioreactor known as the Bio-Fence. The company offers several sizes of modular reactors of this type, designed specifically for the cultivation of the marine algal species used in aquaculture or for waste treatment with either microalgae or photosynthetic bacteria. The BioFence consists of rigid transparent tubes racked together in banks and connected by manifolds in a fence-like structure. The culture suspension is circulated continuously between the photostage and a holding tank by a centrifugal pump or by an airlift. A patented scouring mechanism provides continuous cleaning of the inner walls of the tubes. The pH of the culture is controlled automatically by injection of CO_2 in the photostage.

The most distinctive feature of the Bio-Fence is the stacked manifold configuration, which provides optimal gas exchange with elimination of pressure drops within the reactor. A 100-L unit requires a tube stack 5 m long by 0.75 m high. Units with a total volume of 500 L, with a tube length of 5 m, or 1,000 L volume, with a tube length of 10 m, can be assembled. The Bio-Fence is produced and marketed as units of from 50- to 1,000-L capacity that can be connected in parallel to any scale. Small steam-sterilizable systems for axenic operation are also available. In the case of systems connected in parallel, the units can be placed as close as 30 cm apart, as in the vertical system built by PBL in Spain and Pulz's alveolar plates (see later section). This arrangement will reduce volumetric productivity in case of outdoor utilization because of shading among the units but will attain two important results: high efficiency of land utilization (due to the very compact structure) and high photosynthetic efficiency (due to dilution of sunlight). The possibility of installing fluorescent tubes between the parallel units for additional lighting has also been considered. Typical cell concentrations achieved in the Bio-Fence are about 1 g L^{-1}. Productivity data are not available. The Bio-Fence is an accurately designed, well-engineered reactor. The only limitation seems to be the presence of a large dark tank that might reduce productivity in case of an excessively high ratio between dark-stage and photostage volume.

Flexible tubing photobioreactors

Helical photobioreactors consist of small-diameter, generally flexible tubing wound around an upright structure. This design is not new; it was used in the 1950s by Davis to grow *Chlorella* and later adopted, in a flattened version made of glass tubes, by Setlik, Juttner, and Kruger and Eloff. This coiled arrangement allows deployment of relatively long tubes on a small land area. Helical photobioreactors called Biocoils were devised and patented in the late 1980s by Biotechna Ltd. The Biotechna reactor consists of low-density polyethylene or clear food-grade PVC tubing, 2–3 cm in diameter, wound in a helix around an upright cylindrical core structure, typically 8 m in height with a core diameter of 2 m. Compressed air or peristaltic or centrifugal pumps may be used to induce culture circulation, with the culture being pumped downward through the coiled tube or upward through the tube to a header tank. Temperature control is achieved by a heat exchanger placed between the photostage and the pump. Pilot "biocoil" facilities have been installed at Luton (UK) and in Australia. Biotechna-Graesser A.P. Ltd. marketed biocoils of different sizes (from a few liters capacity up to 10,000 L) for microalgae cultivation or for use with algae and heterotrophic bacteria in the treatment of wastewaters or toxic wastes.

Small biocoil units are being operated for experimental and teaching purposes at King's College and at Murdoch University. A 120-L helical bubble reactor was devised and operated by Tredici and Chini Zittelli to grow *Anabaena siamensis* and *Arthrospira platensis* outdoors. The reactor consisted of three 49-m-long transparent PVC tubes (3-cm ID, 0.5-cm wall thickness) wound around a vertical structure with an inclination of 2^{0} to the horizontal in such a manner as to form a 1.65-m-high, 1.1-m-OD coil. The higher ends of the tubes were connected to a 20-L degasser. From the degasser, the culture flowed down through a single pipe connected to the lower ends of each tube. In this system, as in the NHTR devised by the same group, air was injected at the bottom of each tube and the bubbles flowing up the tubes provided mixing and gas exchange and prevented biofouling. Cooling was achieved by water spraying. With *A. platensis*, a mean volumetric productivity of 0.9 g L^{-1} d^{-1} and a photosynthetic efficiency of 6.6% (PAR) were achieved. The system operated with remarkable stability.

The cost of a unit of this type was estimated at about $150. A coiled, internal gas-exchange photobioreactor of about 3,000 L total volume was built in 1996 by Inalco S.p.A near Vicchio. The system consists of straight polycarbonate tubes, 5 cm in diameter, secured to an upright hexagonal frame with an inclination to the horizontal of 1.4°. The reactor is about 7 m high and occupies an area of about 40 m^2. Culture circulation is obtained by air injection at the base of the tubes. Temperature is controlled by a water-jacketed section. The reactor has been successfully experimented with in the cultivation of *Arthrospira platensis*.

Photobioreactors with flat culture chamber

Flat photobioreactors and flat culture chambers have often been used to grow phototrophic microorganisms in the laboratory because they provide a simple geometry and greatly facilitate the measurement of irradiance at the culture surface. Despite their apparent simplicity, few such systems have been used for mass cultivation of algae. The first flat culture unit devised for mass production of algae was the "rocking tray" used by Milner in 1953 to grow dense *Chlorella* cultures in a thin turbulent layer. This system consisted of a 2.65-m^2, 9-cm-deep tray covered with transparent wire-reinforced plastic sheeting and supported on a shaker. At a mean culture depth of 1.7 cm, an average productivity of 8.2 g m^{-2} d^{-1} was obtained in a 20-day run. Following this experience, flat reactors were practically neglected until the 1980s. Anderson and Eakin cultivated *Porphyridium cruentum* outdoors for polysaccharide production using 3-m^2 flat plates 5–20 cm in depth (made of stainless steel and covered with glass or plastic) that were inclined to the horizontal to maximize interception of solar irradiance. About 20 g of polysaccharide m^{-2} d^{-1} were produced during the summer, when temperature control was provided. In the mid-1980s, two groups working independently in France and in Italy introduced flat alveolar panels in algae cultivation.

These systems are constructed from transparent PVC, polycarbonate, or polymethyl methacrylate sheets that are internally partitioned to form rectangular channels called alveoli. These sheets are commercially available in standard thicknesses from 4.5 to 40 mm. Alveolar plates as culture systems for algae were first used by Ramos de Ortega and Roux at the Centre d'Etudes Nucl□aires de Grenoble. These authors experimented with 6-m-long, 0.25-m-wide, 4-cm-thick, double-layer panels made of transparent PVC for growing *Chlorella* in a greenhouse and outdoors. The plates were laid horizontally on the ground; the upper layer of channels was used for algal growth and the lower for thermoregulation. The culture suspension was circulated through a pump. A productivity of 24 g m^{-2} d^{-1} was achieved in the summer using units of 1.5-m^2 surface area. The authors stressed the fact that the double-layer plate permitted efficient thermoregulation of the culture and that significantly higher performances were obtained with the panels than with flexible sheets and semirigid tubes. The first panel reactors built and experimented with in Florence by Tredici and coworkers were placed vertically, with the channels (and hence culture flow) running parallel to the ground, a design later adopted in Germany by Pulz. A pump was used to circulate the suspension. In 1988, the Consiglio Nazionale delle Ricerche (CNR) of Italy presented this reactor at the ITALIA 2000 exhibition, held in Moscow.

The same year saw the awarding of a patent to a different design in which the plate was placed vertically, but with the channels perpendicular to the ground; mixing was achieved by bubbling air at the bottom of the reactor. This latter version has been used extensively in Florence for outdoor cultivation of cyanobacteria. The plates used in Florence are constructed from 16-mm-thick Plexiglas alveolar sheets commonly available on the market at a cost of \$30-40 m^{-2}. The main features that characterize these reactors are the high *S*/*V* ratio (80 m^{-1}), the vertical or tilted inclination from the horizontal of the channels, and the fact that nomechanical device is used to mix the cell suspension, circulation and degassing of the culture being attained through bubbling air at the base of each channel.

Elevated "bubble column" plates of 2.2-m^2 surface area have been used to study the influence of some of the main cultural parameters, such as areal density, turbulence, oxygen tension, and orientation of the photobioreactor, on productivity and biomass composition of *Arthrospira platensis*. Alveolar plates, due to their high *S*/*V* ratio, attain high volumetric productivities (>2 g L^{-1} d^{-1}) and permit operation at high cell concentrations (4–6 g L^{-1}); these systems, however, have typically attained lower areal yields (about 24 g m^{-2} d^{-1}) compared to tubular reactors (about 28 g m^{-2} d^{-1}). The lower performance achieved by cultures in plates has been attributed to the fact that these systems, unlike tubular reactors, do not achieve light dilution (unless they are placed at a high inclination with the horizontal) and thus cultures in plates suffer more strongly because of the light saturation effect and from photoinhibition. Alveolar plates, like other small flat systems, may be used profitably for research or in small production plants, however, due to the fact that the surface area of the cultivation unit is limited to a few square meters, they are not suitable for use in large commercial plants. More recently, at the IGV Institut fur Getreideverarbeitung, cultivation units based on 32-mm-thick alveolar panels were developed and operated by Pulz and coworkers. Typically, the plates are placed vertically with the channels arranged horizontally, just 20 cm apart, to form a compact structure called PBR.

The culture, circulated by a piston pump, must reach a velocity adequate to maintain turbulent flow inside the channels. The distance to the degasser is critical in avoiding toxic concentrations of oxygen. Units from 100 to 10,000 L, made of closely packed plates, have been developed and tested both indoors and outdoors with different microalgae and cyanobacteria. Productivity levels of 1.3 g l^{-1} d^{-1}, corresponding to 28 g m^{-2} of illuminated surface d^{-1}, have been obtained with *Chlorella* under natural illumination. Among the problems that have emerged are buildup of oxygen (up to 30–40 mg L^{-1} at the outlet of the circuit) and fouling. The main characteristic of the PBR is the closely packing of the panels: 42 plates, for a total volume of 6 m^3, can be packed on 100 m^2 of land. Areal productivities of up to 130 g m^{-2} d^{-1} have been reported for such compact units, but these figures seem excessive and indeed confirm the incorrectness of referring the yield of vertical reactors to unit of occupied land. The main merit of the PBR, besides its high degree of technical development, is the achievement of significant light dilution. PBR reactors are sold commercially by B. Braun Biotech Int. in sizes from 10 to 6,000 L.

Richmond and coworkers developed an inclined modular photobioreactor consisting of a series of flat glass chambers, 0.7 m high, 0.9 m long, and usually 2.6 cm in internal width, connected in cascade and tilted at the proper angle to maximize solar radiation capture. The culture volume varied from 6 to 50 L according to reactor width. Two mixing methods were tested: an airlift system consisting of five transparent PVC columns inserted vertically into the reactors with compressed air injected at the base of the columns, and air-bubble mixing by means of two perforated tubes running horizontally along the entire length of the reactor, one at the bottom and the other at mid-height. The latter method yielded better performance. By adopting a reactor width of 2.6 cm and vigorous mixing (2.5 L air L^{-1} culture suspension min^{-1}), a record areal productivity of 49.4 g m^{-2} illuminated surface d^{-1} was achieved. When the back and side surfaces of the reactor were covered to avoid entry of diffuse light, productivity was reduced by about 20%. Nevertheless, this figure is the highest ever reported for a cyanobacterium cultivated outdoors. Using this flat modular system, the Israeli group has addressed and contributed to clarifying such fundamental aspects of photobioreactor design as optimal cell density with reference to the light path, the effect of reactor inclination, and the importance of the mode and intensity of mixing.

Industrial Photobioreactors

Only a few large commercial-scale photobioreactors have been built and operated: the PBL photobioreactors in southern Spain in the late 1980s and the Hidrobiologica SA manifold photobioreactor set up in the mid-1990s in northern Argentina. Despite the very different designs, both the plants suffered from similar mistakes with respect to some fundamental technical aspects and the scaling-up phase.

Approved Photo Bioreactors

Although Pirt's measurements and extrapolations were not flawless, the quality of the basic work done at Queen Elizabeth College and the high projected productivities attracted investments and led to the creation of a company, Photo Bioreactors Ltd. (PBL), in 1986. A pilot plant was built and experimented with at the Reading University Research Farm. Three years later, PBL Spain was founded with investments from private industries and public Spanish sources. A second pilot plant was set up in Santomera, and finally a commercial plant for *Dunaliella* production was established in Santa Ana near Cartagena. Full-scale activity started in May 1990, with a vertical system made of 50 rows, each of which contained 50 1.2-cm-diameter, 50-m-long polyethylene tubes (for a total tubing length of 125,000 m). Circulation was provided by an airlift, and temperature was controlled by shading the reactors with nets or by water spraying. A near-horizontal system was built later according to the full-scale design patented by Pirt, using about 200,000 m of polyethylene tubing.

Several major technical errors are apparent in these two systems: The tube diameter was too small for effective mixing considering the design, the tube material degraded quickly under sunlight, oxygen degassing was inadequate, wall growth was almost unavoidable, and control of temperature was problematic.

The very high *S*/*V* ratio in relation to the length of the tubes, together with the insufficient circulation and the improper management of the culture, led to poor growth of the alga, biofouling, and heavy contamination. But, as far as we know, the main reason for failure was inadequate piloting: although the pilot plants had been carefully planned and constructed with an appropriate 10-fold increase factor in the different stages, the problems encountered during the pilot stage were not taken into due consideration and had not been solved by the time commercial activity started. In September 1991, after an almost complete crash of the culture, the workers were dismissed and PBL closed without ever entering into production.

Hidrobiologica Reactor

In 1996, Hidrobiologica SA, with an initial investment of about \$2 million, built the largest known photobioreactor about 15 km south of La Rioja (northern Argentina). The system consisted of 96 0.4-mm-thick polyethylene tubes, each 120 m long and 25.5 cm in diameter, laid parallel and horizontally on the ground and arranged like a manifold with feeding, connecting, and collecting channels made of concrete. The tubes, only partially filled, were maintained at very low positive pressure so that they assumed a roughly elliptical shape (about 35 cm wide and 9 cm high). The plant was divided into four modules, each of which comprised four six-tube submodules. An interspace of 0.5 m was allowed between one submodule and the next, while about 1 m separated the modules. The four modules were connected to common feeding and collecting channels. The surface occupied by the whole plant, including the gaps between tubes, was about 5,000 m^2, and the total culture volume was 600 m^3.

The culture (*Arthrospira platensis*) was circulated at a flow speed between 6 and 10 cm s^{-1} by a single axial flow pump with a maximum capacity of 900 m^{-3} h^{-1}. The pump lifted the suspension from the collecting channel, into which it was fed by the culture tubes, up to a second raised, sloped channel. The suspension flowed by gravity through this second channel, which ran parallel to the culture tubes for 120 m, to the feeding channel at the opposite end of the reactor. The feeding channel, about 42 m long, distributed the culture suspension to the inlets of the single culture tubes. During the initial period of activity, the cell concentration was typically maintained at about 1 g L^{-1} and the productivity was calculated to be around 0.2 g L^{-1} d^{-1}; the biomass produced was of excellent quality. In the long run, however, several problems became evident; the most significant were poor mixing and biofouling deriving from the low flow speed and the unequal distribution of the culture among the tubes.

The limited capacity to control temperature was also a major drawback in summer. It is likely that oxygen built up to dangerous levels despite the gas outlet provided halfway along the tubes, but no data are available. The plant is at present undergoing complete restructuring; the main objectives of this effort are to subdivide it into separate units, reduce the tube diameter, and improve mixing and culture circulation. The major errors, besides poor mixing and limited temperature control, were probably the absence of accurate piloting and the fact that only one large single unit was built and operated.

ECOLOGICAL APPROACH

The basic function of a photobioreactor is to provide a controlled environment in which to achieve optimal growth and maximum, sustainable biomass and/or product synthesis with a specific phototroph. This means that a photobioreactor cannot be properly designed without adequate knowledge of the growth physiology of the organism to be cultivated. Besides, since phototrophic microorganisms are highly diverse as concerns their morphology, biology, nutritional and light requirements, and resistance to stresses, a single photobioreactor design cannot adapt to all organisms and all conditions. Design criteria for photobioreactors should aim at achieving high efficiency of light conversion and at providing the necessary reliability and stability to the cultivation process by solving the main problems encountered in photobioreactor operation, such as overheating, oxygen buildup, biofouling, contamination, and photostage material deterioration. The fundamental design criteria for photobioreactors include reactor configuration, *S/V* ratio, mixing and degassing devices, temperature regulation, orientation, and the material used to build the photostage. These aspects are briefly discussed in the following sections. Reliability, ease of operation and scale-up, and construction and operating costs also take on particular relevance in relation to commercial application of photobioreactors.

Ratio between Illuminated Surface and Its Volume

The surface-to-volume ratio of the photobioreactor (i.e., the ratio between the illuminated surface of the reactor and its volume) determines the amount of light that enters the system per unit volume and the light regimen to which the cell population is subjected and is consequently one of the most important issues in photobioreactor design. The higher the *S/V* ratio, the higher the cell concentration at which the reactor is operated and the volumetric productivity of the culture. The cell concentration, in turn, greatly influences medium handling and harvesting costs and, to some extent, contamination level. The temperature profile and the hydrodynamic behavior of the culture suspension are also affected by this parameter. For example, in small-diameter tubular systems operated outdoors, the time required to reach the optimal temperature in the morning is reduced, and in the evening the relatively low culture volume cools down more quickly to ambient night temperatures, resulting in lower respiratory losses. For these reasons, in recent years we have witnessed a general trend toward a reduction of the diameter of tubular reactors and of the thickness of flat panels.

Yet, it must be pointed out that although high *S/V* ratio reactors may work well in the laboratory, they may become extremely inefficient when scaled up. In high *S/V* ratio photobioreactors, all the volumetric activities that depend on the input of light energy per unit volume (such as O_2 evolution, CO_2 absorption, nutrient depletion, metabolite excretion, and heat production) change at a high rate, and this could have long-term negative effects on the stability of the process. These effects are rarely observed because experiments do not last long enough for them to become evident. The industrial systems set up by PBL provide clear examples of the kinds of problems that may be absent at laboratory scale and yet arise when high *S/V* ratio photobioreactors are adopted in large-scale plants. It is not possible to fix an optimal *S/V* ratio for large-scale photobioreactors, since this parameter depends on the organism to be grown, on the type of mixing, and on other specific conditions. However, my laboratory experience suggests that a light path between 3 and 6 cm may represent a reasonable compromise.

Accumulation of Generated Oxygen

Accumulation of photosynthetically generated oxygen is one of the main factors that limit scale-up of photobioreactors. Dissolved oxygen concentrations over 35 mg L^{-1} (equivalent to about four times air saturation at 25°C) may be easily reached in outdoor dense algal cultures at midday. Such high oxygen levels are toxic to many phototrophs and, if coupled with prolonged exposure to full sunlight, may lead to the photooxidative death of the culture. Since volumetric oxygen production is directly related to volumetric biomass productivity (about 1 g of oxygen is evolved per gram of biomass synthesized), oxygen buildup becomes an especially serious problem in high *S/V* ratio photobioreactors operated outdoors. At maximal rates of photosynthesis, a 1-cm-diameter tubular reactor would accumulate about 8–10 mg of oxygen L^{-1} min^{-1}. In such systems, the maintenance of dissolved oxygen below the toxic concentration requires adoption of very short loops and this fact makes the serpentine design difficult to scale up. The situation is less dramatic in tubes of greater diameter, since the volumetric rate of oxygen evolution falls in proportion to the decrease of the *S/V* ratio, but it has been shown that even in 14-cm-diameter tubular reactors dissolved oxygen may reach levels that significantly reduce productivity.

In serpentine reactors, the time cycle (time required for the culture to reach the degassing point), the speed of circulation, and the length of the loop must be very carefully chosen in relation to the expected volumetric productivity to prevent buildup of toxic oxygen concentrations. For example, in a tubular reactor with a 2.6-cm ID and a culture flow speed of 0.3 m s^{-1}, the time cycle should not exceed 6 min, hence, the length of the reactor must be less than 100 m. To increase the length of the reactor it is necessary to increase the culture circulation speed or reduce the *S/V* ratio. Internal gas-exchange systems (bubble columns) offer some advantage in this respect. Unlike serpentine reactors, in which oxygen accumulates during the loop cycle and is removed only at the degasser, bubble columns present the advantage that the whole culture is subjected simultaneously to the degassing action of air. Mathematical models can be employed to predict the effect on oxygen concentration of various design variables, such as the length of the loop, the tube diameter, and the liquid circulation speed. A useful model for prediction of dissolved oxygen and carbon dioxide axial concentration profiles in serpentine tubular reactors has been developed and verified experimentally with *Porphyridium cruentum* in outdoor continuous culture by Camacho Rubio et al.

Mixing—The Culture Suspension

The type of device used to mix and circulate the culture suspension is essential in the design of a successful photobioreactor. Both the productivity of a photobioreactor and the cost of its construction and operation are influenced to a great extent by the type of device used for mixing. The beneficial effects that mixing has on growth and productivity of mass cultures of phototrophs are well known. Mixing is necessary to prevent cells from settling, to avoid thermal stratification, to distribute nutrients and break down diffusion gradients at the cell surface, to remove photosynthetically generated oxygen, and, above all, to ensure that all the cells experience alternating periods of light and darkness. It has been postulated that cells quickly brought into and out of the lighted portion of the culture could take advantage of the "flashing light effect" to increase photosynthetic efficiency. Conflicting results have been reported, however, for the light–dark fluctuations of medium frequency (i.e., in the order of seconds to minutes) that prevail in mixed-mass cultures as regards their stimulatory effect on productivity, and the true role of the mixing rate in determining algal productivity still remains to be clarified.

With a few exceptions, significant positive effects of intense stirring on productivity of algal mass cultures have been observed only in paddle-wheel-mixed raceway ponds. Carlozzi and Torzillo observed a 30–40% increase of the productivity of *A. platensis* grown in tubular reactors when the flow speed was increased from 0.18 to 0.75 m s^{-1}. Tredici and Chini Zittelli studied the influence of the bubbling rate on productivity of *A. platensis* in inclined flat plates. At an air supply rate of 1.21 L L^{-1} min^{-1}, productivity increased by 35% over that at a rate of 0.36 L L^{-1} min^{-1}. The authors suggested that the higher bubbling rate enhanced productivity by providing a more adequate light regimen to the cells, as well as by reducing the oxygen level in the culture.

Hu and Richmond have recently attained astonishingly high productivities with *A. platensis* cultivated in flat plates by adopting ultrahigh cell densities (higher than 15 g L^{-1}) and providing mixing at the maximal rate permissible without causing cell damage. The choice of the intensity of mixing and of the mixing system must be dictated by the characteristics of the organism to be cultivated. Unlike *Chlorella* or the marine *Nannochloropsis,* which are endowed with strong cell walls, cultures of filamentous cyanobacteria and fragile dinoflagellates cannot be mixed by pumps that subject the cells to high shear stress and so lead to fragmentation of thricomes and to cell lysis.

There is general agreement that bubble columns and airlift systems are low-shear-stress mixing systems and therefore are recommended for shear-sensitive organisms. Gudin and Chaumont have reported that productivity may increase up to 75% when pumps are replaced by airlift systems. However, it is known that even air bubbling produces some stress on the cells at the level of bubble formation and breakup, bubble coalescence, and bubble-bursting at the culture surface. Other noticeable advantages of bubble columns and airlifts, from the point of view of industrial application, are their simplicity of operation and low maintenance costs due to the absence of moving parts. Mixing devices and power requirements for mixing have been dealt with by Weissman et al. and by Torzillo. Torzillo has also analyzed the influence of the rheological properties of the culture on mixing and mass transfer. Gas–liquid mass transfer in a serpentine airlift photobioreactor has been described in detail by Ca- macho Rubio et al.

Thermal Regulation

Although photobioreactors are said to provide a better process control than do ponds, the maintenance of optimal temperatures in closed reactors is not an easy task, especially when the systems are operated outdoors. Photobioreactors behave as solar collectors: Any light absorbed by the reactor walls or by the culture and not used in photosynthesis is converted into thermal energy, and may lead to an increase of the culture temperature of up to 30°C or more above ambient temperature, causing the collapse of even rather tolerant species. With a few exceptions (e.g., cultivation of thermophilic species) photobioreactors operated outdoors require cooling. Among the solutions experimented with to control overheating in outdoor photobioreactors are shading, water-bath immersion, and water spraying.

Shading, to be effective, requires that a large proportion of the reactor be covered for several hours a day, and this greatly reduces the irradiance at the culture surface and, consequently, the productivity. Cooling by water spraying has been shown to be reliable and cost- effective when used in dry climates. Immersion of the reactor in a cooling water basin is very efficient, but its cost-effectiveness is rather doubtful. It may, however, be the sole available option in humid climates, where water spraying does not provide evaporative cooling. Although it varies with the reactor design, the local climate, and the specific organism used, the need for cooling can significantly increase operating costs. In cold climates and in temperate regions during winter, heating may instead be required to maintain the optimal growth temperature.

Solar Radiation

One of the main impediments to the efficient use of solar energy is the continuous variation of the available solar radiation throughout the day. Solar irradiance is very low in the early morning and in the evening hours and strong enough to inhibit photosynthesis at noon. Although the diurnal pattern of the solar irradiance incident on a horizontal reactor cannot be easily modified, some elevated systems can be oriented at various angles to the sun and thus offer the possibility to vary, continuously or discontinuously, the distribution of the light that strikes the reactor surface. The effect of the photobioreactor orientation on productivity of outdoor algal cultures was investigated by Lee and Low, who did not find the angle of inclination of the bioreactor to have any significant influence on the productivity of *Chlorella pyrenoidosa*. Because of the particular reactor design (two tubular panels were connected by hinges, erected at an angle to the ground, and placed in an east–west direction) and, above all, due to the location of the experimental facilities (about 1° latitude north), the findings of these researchers cannot be generalized. The influence of the orientation of flat photobioreactors in higher latitude regions has been studied by Tredici and Chini Zittelli and by Hu et al.

These studies have shown that the reactor tilt angle exerts a significant effect on both productivity and photosynthetic efficiency. Systems that are sun oriented so as to intercept the maximum solar radiation on a daily basis (e.g., south-facing plates inclined at 10–30° in summer and at about 60° in winter) typically achieve the highest productivities, even if a culture placed perpendicularly to the rays of the sun at midday may suffer due to excessively high irradiance. Vertical plates or systems oriented so as to dilute the impinging radiation achieve lower productivities but higher photosynthetic efficiencies. A major objective of photobioreactor design should be to reduce the so-called light saturation effect of photosynthesis.

One approach has been to install systems vertically (like Pulz's PBR series) or at a high inclination to the horizontal so that the reactor is at a large angle to the sun rays and solar irradiance is attenuated (diluted) at its surface. Another approach involves transmitting sunlight into the photobioreactor through light-diffusing prisms or optical fibers, but this option appears too complex to be effectively scaled up and applied commercially in the near future. By operating a near-horizontal flat panel and a near-horizontal tubular reactor in parallel, Tredici and Chini Zittelli have shown that tubular reactors can achieve, per se, higher photosynthetic efficiencies and productivities than flat reactors because their curved surfaces achieve a certain degree of sunlight dilution and thus reduce the light saturation effect.

Requirements

Materials for photobioreactors must have several properties: high transparency, high mechanical strength, lack of toxicity, resistance to weathering (durability), chemical stability, antifouling surface, and low cost. G. Chini Zittelli and M.R. Tredici have carried out a preliminary investigation of the durability of some plastic materials. The loss of transparency of six different tubing materials exposed to natural weathering in central Italy is shown. Polyethylene and polypropylene tubes are inexpensive, but both lose transparency very quickly and present serious limitations as regards biofouling and mechanical strength. Polycarbonate tubes are seriously weakened by alkaline solutions and become brittle and break easily after 2–3 years of exposure outdoors; they lose transparency mainly in the 300- to 500-nm range. Tubes of polymethyl methacrylate or Teflon have been shown to possess high resistance to weathering and good overall properties. PVC tubes lose transparency significantly after hydration.

Some of the flexible plastic materials tested have shown limited mechanical strength that has led to trivial problems (such as leakage due to punctures inflicted by cats and wild animals) that could become rather serious in large-scale systems. Glass might be an excellent material for photobioreactors, due to its high transparency, chemical stability, and durability. Glass tubular systems, however, would require mostly field assembly and numerous connections, and this would greatly increase installation costs; furthermore, fragility may constitute a serious limitation to the use of glass tubes in large-scale installations. The tubes required for building the photostage represent a major cost component of the photobioreactor but not as high as commonly thought. Suitable long-lasting tubular material is available for about $1 per meter or about $15 per square meter of photobioreactor.

CONCLUDING REMARKS

Photobioreactors are considered to have several major advantages over open ponds: They can (i) prevent or minimize contamination, (ii) offer better control over cultural conditions (pH, pCO_2, pO_2, temperature, etc.), (iii) prevent evaporation and reduce water use, (iv) lower CO_2 losses due to outgassing, (v) permit higher cell concentrations that result in reduced operating costs, and (vi) attain higher productivity. However, the ultimate and most important advantage is that photobioreactors permit cultivating algal species that cannot be grown in open ponds. On the other hand, certain requirements of photobioreactors, including the need for cooling, the need for strict control of oxygen accumulation and biofouling, and the need for frequent replacement of the material of the photostage, make these systems more expensive to build and operate than ponds. Thus, despite their advantages, the use of photobioreactors must be limited to production of very-high value compounds from phototrophs that cannot be cultivated in open ponds.

Photobioreactors and open ponds should not be viewed as competing technologies, since it is very unlikely that photobioreactors will have a significant impact on any product or process that can be attained in large outdoor open ponds. In recent years, significantly higher photosynthetic efficiencies and a higher degree of system reliability have been achieved due to better understanding of the growth requirements of phototrophic microorganisms in mass culture. Despite these advances, no photobioreactor has yet proven to be commercially successful, even for very-high-value products. This certainly does not depend on a lack of photobioreactor designs: There exist several valid basic concepts and myriad variations. The principle obstacle to commercial application of this biotechnology seems instead to lie at the level of scaling-up. Transferring those processes developed at the laboratory scale to the industrial scale is never a simple proposition, and it is particularly difficult in the case of photobioreactors because of the peculiar problems of these systems. One of the major causes of the failures of photobioreactor technology at commercial scale is a lack of adequate piloting.

The old rule of thumb used in the chemical industry, and still valid in fermentation technology, that each individual increase of scale should not surpass a factor of 10 has been ignored in practice. This was especially true in the case of Hidrobiologica SA, where the full-scale plant was set up after only limited experimentation on a very small scale. At the small scale, the photobioreactor performance is usually evaluated in terms of productivity, whereas issues such as contamination, material degradation, and the reliability and sustainability of the production process, which more than the productivity determine the failure or success of the industrial-scale activity, are ignored.

Only accurate piloting can shed light on the key problems and provide adequate solutions. It is not clear how much the failure of the PBL photobioreactors in Spain and of the 600-m^3 plant in Argentina depended on system design flaws or on management errors. However, in neither case had the basic technical and biological problems (mixing, oxygen buildup, temperature control, biofouling, contamination) that must be solved for successful full-scale operation been subjected to rigorous investigation at an adequately scaled pilot level.

Another mistake, common to both plants, is that to keep operating costs low, the units built and operated were very large. This deprived the systems of necessary flexibility and negated the possibility of making improvements by trial and error. Under these conditions, any problem that might arise (e.g., contamination or toxic oxygen tensions) affects the whole culture and jeopardizes all of production. It is worth pointing out that both industrial plants failed in the cultivation of two phototrophs (*Dunaliella* and *Arthrospira*) that are commonly grown in the much less rigorously controlled open ponds. The establishment of a commercial activity based on the cultivation of phototrophic microorganisms requires realizing that a photobioreactor design that is efficient at the laboratory level is not necessarily so at the level of industrial production—hence, the importance of accurate piloting. A few simple guidelines should be followed during piloting:

1. Experiments must last long enough to reveal the "aging" problems, including contamination, biofouling, and material deterioration, that become evident only after several months of continuous operation.
2. The units experimented with should be similar in size and type to the cultivation units envisaged for the full-scale plant.
3. The productivity of the system must be correctly evaluated and extrapolated.
4. Experiments must be carried out under the climatic and operating conditions that will prevail in the final commercial process.
5. Unless true axenic conditions are adopted, the possibility of contamination must not be underestimated.
6. Appropriate solutions to the main problems commonly encountered in photobioreactors (overheating, biofouling, oxygen accumulation, etc.) must be provided before beginning operation of the commercial plant.

7. The data obtained during the scaling-up stage must be subjected to rigorous analysis and used to reevaluate the economics of the process.

Today the efforts of many researchers, from both academia and industry, are directed toward development of new photobioreactor designs. In the majority of cases, such designs are anything but new and remain an academic exercise with little or no practical value. What we really need is a serious evaluation of the performance of photobioreactors at a relatively large scale. Through such analysis, and together with likely further improvements in engineering design, the development of new materials with suitable characteristics, and the improvement of cultivation techniques derived from a better knowledge of the physiology of phototrophs in mass culture, we can reasonably expect to achieve a degree of reliability in large-scale cultivation of phototrophic microorganisms such as to bridge the gap between laboratory experimentation and commercial application.

INDEX